Reinhard Heil, Andreas Kaminski, Marcus Stippak, Alexander Unger, Marc Ziegler (eds.)

Tensions and Convergences

Technological and Aesthetic Transformations of Society

[transcript]

Print kindly supported by the German Research Foundation.

Bibliographic information published by Die Deutsche Bibliothek
Die Deutsche Bibliothek lists this publication in the Deutsche Nationalbibliografie; detailed bibliographic data are available on the Internet at http://dnb.ddb.de

Layout by: Kordula Röckenhaus, Bielefeld
Edited by: Rebeccah Dean and Rebecca Schopfer
Typeset by: Reinhard Heil
Printed by: Majuskel Medienproduktion GmbH, Wetzlar
ISBN 978-3-89942-518-5

Distributed in North America by:

Transaction Publishers
New Brunswick (U.S.A.) and London (U.K.)

Transaction Publishers	Tel.: (732) 445-2280
Rutgers University	Fax: (732) 445-3138
35 Berrue Circle	for orders (U.S. only):
Piscataway, NJ 08854	toll free 888-999-6778

us Stippak,

Tensions and Convergences

Reinhard Heil (M.A.), **Andreas Kaminski** (M.A.), **Marcus Stippak** (M.A.), **Alexander Unger** (M.A.) and **Marc Ziegler** (M.A.) are fellows of the postgraduate college »Technisierung und Gesellschaft«, Technische Universität Darmstadt.

TABLE OF CONTENTS

Aesthetics of Technoscience

Working on Visibility

Technological Transformation of Space and Social Life

Cultural Assessment and Appropriation of Technology

The Aesthetic Dimension of Warfare

Preface

REINHARD HEIL/ANDREAS KAMINSKI/MARCUS STIPPAK/
ALEXANDER UNGER/ MARC ZIEGLER

This book collects and presents the results of the international conference "Technisierung/Ästhetisierung – Technological and Aesthetic (Trans)Formation of Society" that took place from 12th to 14th of October 2005 at the Technische Universität Darmstadt in Germany[1]. This congress was organised by the members of the interdisciplinary post-graduate college "Technisierung und Gesellschaft"[2] (Technology and Society) which was financially supported from 1998 to 2006 by the "Deutsche Forschungsgemeinschaft"[3] (German Research Foundation). Because of the manifold disciplinary composition that included philosophers, sociologists, philologists, historians, educators, civil engineers and media and communications scholars, it was a somewhat complicated, but nonetheless fruitful task when the post-graduate college initiated a call for papers on the occasion of its completion. Addressing the interaction of aesthetical and technological dimensions within the formation of contemporary society, this call for papers also widened the perspective of the college itself, in terms of content. More than 70 scholars from Germany and abroad – both the renowned and the young and promising – participated and presented their contributions at this symposium.

Centring on, for example, technologies of visualisation or the design of technological lifeworlds, the convention's contributions examine topics like the production of time and space, self and nature and individual and society with regard to technology. They focus on the productive tensions and convergences between aesthetical and technological concepts

1 See http://www.tu-darmstadt.de

2 See http://www.ifs.tu-darmstadt.de/fileadmin/gradkoll//index.html

3 See http://www.dfg.de

when implemented in everyday life and discuss the consequences of a large number of convergences of what could be regarded as the technological and aesthetical sphere within social order. Considering the conference's theoretical and methodological approaches, as well as the openness in form and content and the participants' different academic backgrounds, the lectures, unsurprisingly, did not create a homogeneous overall picture. The contributions collected within this volume mirror this irreducible heterogeneity. Because technology and aesthetics are far from being a one-dimensional phenomenon, it is our conviction that this heterogeneity is not to be equated with weakness or arbitrariness.

To sharpen and to bundle the perspectives, we suggest six general groupings for the object of investigation: modern alliances, technoscientific aesthetics, visibility, space, cultural assessment or appropriation and war. They represent six topical blocks that cover a broad spectrum of current and controversial topics in contemporary Philosophy, Sociology, History, Literary Studies, Art and Cultural Studies as well as Science and Technology Studies.

Deciding after the conference to publish a selection of the conference's talks, supplemented by a few additional contributions within a conference transcript, the editors believe this volume to be an invitation for a wider public to study complex phenomena like technology and aesthetics using various questions, methods and perspectives.

We thank the congress's participants, who contributed to this volume by converting their conference lectures into written form. We are indebted to all those who prepared, planned and conducted the conference. We also want to thank Rebeccah Dean and Rebecca Schopfer for proofreading the manuscript. Finally, we would like to thank the German Research Foundation and the post-graduate college "Technology and Society" for their generous financial support, which made this publication possible.

The following paragraphs give a short introduction to the single sections of the book and describe in brief the individual accounts.

Modern Alliances Between Aesthetics and Technology

This section collects analyses that focus on the relation between technology and aesthetics either from a macroscopic perspective, by outlining main strands on modern processes of social transformation, or by retracing specific aspects of typical modern phenomena or modern movements. Despite the thematic heterogeneity of the different accounts, all share the insight that the aesthetic dimension of technology has reached new qualities in the modern era. Strongly connected to the process of industrialisation and rationalisation, the fusion of aesthetics and technology nowadays creates new challenges and tensions on the level of social order.

Gernot Böhme starts his article *Technical Gadgetry. Technological Development in the Aesthetic Economy* by questioning the adequateness of the pragmatic approach to technology. Instead of following the Marxist theory of technology primarily as a useful social enterprise that belongs to the realm of necessity, he mainly refers to the joyful character of technology. Böhme analyses a strong expression of merely joyfulness-orientated technology in the profligate courtly technology of the 16th and 17th century. This joyful and lavish technology was not dominated by economic sensibility, but rather arranged to cause astonishment as well as to display power and prestige. Böhme outlines the annihilation of the difference between primarily useful and primarily joyful technology in the modern era. In modern societies, basic needs are fulfilled as far as possible, so lavishness becomes the main method of economic growth and, therefore, useful in the context of an "aesthetic economy".

Kathryn M. Olesko emphasises that a union of the technical and the aesthetic is also taking place in a field that is proceeding contrary to a lavishly aesthetic economy: the role of *Aesthetic Precision* in modern measuring practices. Following Pierre Bourdieu's account that seeing practices are primarily socially overdetermined, Olesko works out the reciprocal influences between industrial and technological practices of geometrical drawings as the most influencing aesthetic practice on precision measurement as well as on social, ethical and aesthetical judgements in Germany from the romantic era to the Third Reich. In precision measurement, therefore, modern forms of objective rationalisation amalgamate as well as new modes of subjective visual perception.

Peter Timmerman's article *Architecture in the Mirror of Technology* analyses the role of technology as a rhetorical means in the architectonic discourse in the first half of the 20th century, by contrasting statements from Le Corbusier and architects of the Futurist Movement. Timmerman's interest is to explicate the implicit, pre-modern, philosophical and metaphysical aspects of the modern and vanguard architects' reference to technology and architecture. He shows that Le Corbusier constructed a platonic image of technology as a role model for architecture, aiming for perfection and continuation that also allows racing cars to be regarded alongside the Parthenon as perfect machines. Unlike Le Corbusier, the Futurists emphasise the dynamical dimension of technology which leads to a rejection of history and a post-historical dynamical architecture of constant renewal that adheres to the Heraclitean philosophy of perpetual motion rather than to Platonic eternal continuity.

In *Readerless Media,* Heather Fielding suggests linking modernism with the commodity form by means of information theory. Referring to Henry James' theory of fiction, Fielding shows that the mediality of modern novels becomes deconstructed into a series of equivalent and quantitative units, a process that can best be analysed in terms of exchange value: He who analyses the aesthetics of modern novels at the same time analyses the commodity form of capitalism. The reader of a modern novel is separated from its use value just as the commodity form separates the exchange value from the use value by negating the irreducible heterogeneity of the prior natural form of the commodity. Both modern novel and commodity can therefore be seen as non-communicative media that fail to transmit the information of singularity. Before the jointed background of this exclusion, the discourses of information theory, modernism and the commodity form refer reciprocally to each other.

In *Mass Aesthetics, Technicist Expansionism and the Ambivalence of Posthumanist Cultural Thought,* Thomas Borgard also focuses on the modernity of aesthetics by outlining some ethical paradoxes that are closely connected to modernity. In a poly-dimensional inquiry, he reconstructs the correlations between the morality of modern and postmodern societies and fundamental transformations in the theoretical framework of Literary and Cultural Studies. The analysis thereby focuses on the relation of power and the often concealed setting of norms. Borgard works out an instrumental structure that equally links technology with ethics. As a conclusion, he claims to draw more attention to the paradoxical situation of a postmetaphysical morality as a task for current interdisciplinary reflections on the state of social order.

Aesthetics of Technoscience

Leading a life, designing contemporary lifestyles and the whole conception of the lifeworld ("Lebenswelt") throughout the last decades has become more and more dominated by values of primarily technology-orientated aesthetic dimensions. Current social modes of producing knowledge and a capitalistic application of science-based technology that can be subsumed under the concept of technoscience have reflexive effects on different aesthetical discourses. The phenomena of transhuman as well as transgenic art demonstrate very clearly the blending of artistic production, scientific research and social sensitivity. The technologically upgraded bodies of the transhuman artists and the genetically modified transgenic objects cast a new light on the relationship between art, technoscience and society. These spotlights on the aesthetic structure of posthuman subjectivity are flanked by an insight in new forms of human-computer interaction as a form of a new emotional and intuitive as well as reflexive communication between men and machines. The artistic suspension of ethics thereby calls for further reflection on the acceded influence of technology in the hermeneutical relation towards ourselves and the world. Furthermore, the aesthetics of a technoscientific society suggests rethinking the role of imagination in terms of a post-hermeneutical perspective.

Attempting to take evolution to a new, technology-based level, transhumanist artists focus on the human body and mind as the favoured objects of an aesthetical anticipation of a posthuman world. This makes the transhumanist artists into visionaries of a primarily technology-based transformation of society: They provoke necessary discussions about possible scenarios of the future. In *Transhumanist Arts. Aesthetics of the Future?*, Melanie Grundmann analyses parallels and differences in the vanguard role of the contemporary transhumanist artist and the 19th century dandy. Both the dandy and the transhumanist artist are mediators of the future who seek new perceptions of body, mind and world. Both emphasise in a reflexive manner the value of individuality and both are taboo-breakers. But while dandyism is an aesthetic reflection on the ugly, evil, dirty and sinful sensuality of the body, transhumanist artists tend to ignore the negativity of the ugly elements of human life in favour of a technologically upgraded, beautiful body.

While many transhumanist artworks still remain on the status of mere models for a prospective technical enhancement of bodies or a technical upgrade of minds, dramatic changes are becoming apparent in the practices of artists who create new and artificial laboratory-made life forms with the aid of biotechnology. Although it is a daily scientific practice to transform natural organisms into "epistemic objects" within a laboratory and to create machines on a macromolecular level to practice tissue engineering or to create transgenic objects, these processes become questionable in an ethical sense where they are moved from a laboratory into an art gallery. Ingeborg Reichle's article *Art in the Age of Biotechnology* points out how transgenic artists conceptualise the precarious category "nature" by using chromosomes, recombinant DNA or tissue as working materials for contemporary art. Reichle shows how far artistic approaches differ from scientific approaches to the phenomenon of life and DNA and outlines the ethical difficulties that are connected to the confrontation between transgenic art and society.

Communication, or more precisely: human-computer interaction, makes up the central aspect of the artistic work of Laurent Mignonneau and Christa Sommerer. The artists point out some key concepts for interactive designs and describe certain interactive technologies in detail. They challenge the *Cultural Aspects of Interface Design* with the aim to construct emotional, metaphoric, natural and intuitive interfaces by analysing different forms of intuitive knowledge that develop in daily interaction with objects and beings. By constructing non-linear, multi-layered and multi-modal interaction between humans and computers, the human user should experience his interaction as an artistic, educational and entertaining journey that facilitates the integration of several senses in the interactive process and a growing complexity of interaction the longer the interaction continues.

The growing technoscientific impact on human perception and experience demands a new evaluation of human imagination. In *Imaging Reality. On the Performativity of Imagination*, Marc Ziegler outlines imagination as a force enfolding in the indeterminate centre between technology and society. Therefore, technology is epistemologically regarded as a medium of social interaction that enables new hermeneutical and creative practices. Technoscience is the most productive socio-scientific force within this medial performing of technology. From a hermeneutical point of view, the technoscientific world is manifoldly deconstructed and disintegrated into senseless particles. This calls for a post-hermeneutical theory of imagination that is able to outline the necessary synthetic functions but also the forms of negativity that characterise a contemporary concept of human imagination.

Working on Visibility

For more than two decades now the theory of science has turned away from discussing the foundations of the validity of scientific research. Instead, questions of the practical dimension of science have arisen: What do scientists do in the laboratory? How do they prepare their epistemic objects in an experimental setting? In particular, how do they work on the visibility of their objects or events? The latter question appears in connection with Galileo's look through his telescope, Robert Hooke's look through his microscope, or Guericke's spectacles which made the vacuum visible. Along with these perspectives, however, new restrictions of focus came about in the relationship between science and technology. Working on visibility became an important scientific aim; technology seemed to be merely a means to this end. This section tries to view working on visibility from a broader perspective. For example, it enquires about the subjective techniques of scientists to make events visible, describes how metaphors and models produce invisibility, traces how spectacularly visualised scientific objects appear on television, specifies how lay people visualise the future as opposed to medical professionals and shows the role that engineering analogies play in explaining new phenomena.

The principle of causality is thought to be a cold, calculating concept of modern natural science and technology. Andreas Kaminski suggests the existence of an aesthetic dimension of causality. He reconstructs a triangle between *Causality, Custom and the Marvellous* in experiments and mechanical models. Causality and custom stand in a certain tension with each other: causality is imperceptible (Descartes, Hume); it is derived from custom (Hume, Pascal); and last but not least, custom obscures causality (Montaigne, Bacon). Therefore, methodologies suggest aesthetic techniques to reveal causes and effects and experiments stage them (Boyle, Hooke). The tension of custom and causality constitutes an economy of the marvellous (Pockels) which recurs in the admirable machines.

It seems that the clear-cut distinction between nature and technology has been worn away by the new Life Sciences. Amongst natural entities and technological artefacts, "biofacts" appear, as Nicole C. Karafyllis points out. A central criterion in distinguishing what belongs to which sphere was the question of growth. This raises the question: *Growth of Biofacts: The Real Thing or Metaphor?* Karafyllis argues that the difference is still true today but is covered by, firstly, technical metaphors, secondly, by epistemic models of growth, and thirdly, a lack of anthropological discussion. She agrees that although growth can be (and is) modelled with technical means, real and concrete growth nevertheless "belongs to an uncontrollable sphere of nature."

Claudia Wassmann reconstructs stages of a tense parallel history: scientific brain images and their integration in television. Her database consists of three documentary series on the brain which were broadcast on American television (1984, 1988, 2002). The analysis of the documentary series mainly leads to two results. Firstly, Wassmann points out how brain images are made evident as a normative instance; e.g. how the brain is referred to as the cause of mental disease, then later as the generator of the whole person and finally as a plastic entity with the possibility of self-shaping and development. Secondly, she reflects on the trends and ascertains an increasing gap: while the natural sciences gain concepts and complexity by means of brain images, the trend of documentary series is just the opposite. *The Brain as Icon* with an "inscrutable surface" enters the media stage.

Metaphor research led to the result that the conceptual metaphor for cancer is war and for gene it is language (information, book and so on). Andrea zur Nieden evaluates conversations in genetic counselling centres and interviews several affected women. Her contribution refers to these results of metaphor research and offers a specification: the visual representations and metaphors that are used to make the concept of genetic predispositions evident differ. The metaphors and visualisations of the affected women are heterogeneous, unsystematic, contradictory and prosaic. They do not correspondent to the conceptual and systematic metaphors of physicians. Andrea zur Nieden interprets this variety as a counterpart to the contingency of genetic predisposition. *Images of 'Genetic Risks' as Anticipation of the Future* provides answers to this contingency.

Up to now science studies have precisely described the metaphors, analogies and models science used, while researchers in technology studies have turned their attention only to engineering models. This means that such structures have been inadequately examined in the theory and history of technology. In his article *From the Historical Continuity of the Engineering Imaginary to an Anti-Essentialist Conception of the Mechanical-Electrical Relationship*, Aristotle Tympas contributes to remedying this situation. Therefore, he traces engineering analogies for the relationship of mechanical and electrical phenomena between the 1880s and the 1940s, when an inversion from mechanical-electrical to electrical-mechanical analogies was executed. The novel phenomenon is always examined using knowledge of the familiar one, as he points out.

Technological Transformation of Space and Social Life

Criticism on the disregard of the spatial dimension of the human existence in the discourse of social science led to a more intensive research on the topic 'space' that became even more urgent with the emergence of the information society and the growing influence of ICT (information and communication technology). Among other things, the 'spatial turn' called attention to the fact that spaces organise social life and are able to influence peoples' behaviour and development through their structure but that on the other hand, space is not an entity beyond the social world (like the Euclidic space), but shaped and 'symbolically charged' by human beings in a social process. This ambivalent perspective on space has not only affected the social sciences, but also architecture. It has also raised the question of how changes in society can be understood and how they affect the organisation and structure of specific social spaces. The articles in the following section deal with the tension of determination through and annexation of spaces in post-industrial society and their connection to development in the field of technology and architecture.

In her article *Architectural Structuralism and a New Mode of Knowledge Production* Martina Heßler discusses the connections between technology, science and architecture. With her research on the 'Forschungsstadt' (science-city), she focuses on the question of how the workspace for scientific research was restructured by the emergence of the information society and its new demands on knowledge production. She argues that these demands could not be met by functionalism and its static architecture, but rather required a more dynamic approach like that offered by architectural structuralism. Instead of determining spaces and rooms to perform specific functions, the Forschungsstadt was designed as a flexible, never-finished open space. Heßler points out that this design corresponds to the necessity of flexibility and innovation which are characteristic in an information society.

In *Some Notes on Aesthetics in a 'Städtebau' on a Regional Scale*, Wolfgang Sieverts focuses on a different 'space' of our post-modern life world; the neglected area of the 'Zwischenstadt'. He argues that while half of the population lives and works there, it is only regarded in economic terms, but not as a cultural space that has to be aesthetically shaped. Following the pre-industrial tradition of shaping landscapes as a whole, Sieverts appeals for an extended notion of architecture that is able to develop the cultural potentials of the Zwischenstadt. According to Sieverts, such an interdisciplinary and multidimensional 'socio-aesthetic' approach does not only have to strive for a deeper understanding of the environmental narratives, but also needs a new understanding of space in the context of socio-economic change, which allows for mediating, designing and planning.

In *Between Determination and Freedom: The Technological Transformation of Education and Learning*, Alexander Unger is concerned with the future design of another important space: the institutionalised space of learning and education. While learning is commonly understood as a 'real' social interaction, lately these processes have become more and more framed by ICT. By comparing Programmed Instruction with the ongoing process of establishing e-learning environments, Unger points out that Virtualisation can be realised in the spirit of two different 'cultures': on one hand, the culture of perfect regulation that focuses on the substitution of real spaces and persons and, on the other hand, a culture of open systems that focuses on participation and co-operation. The author argues that even in virtual environments, uncertainty, which is produced in spaces which are either nonregulated or infrequently regulated, is irreplaceable in imparting the competence necessary for success in the information age.

In *(Self)Normalisation of Private Spaces – Biography, Space and Photographic Representation*, Katja Stoetzer researches the connection of biographical experience and the shaping of spaces along with the amount of adjustment or resistance to the norm. She starts from a relational understanding that conceptualises space as something that emerges in the simultaneous process of arranging and synthesis as the perception of these arrangements. In this context she focuses on student residences which have been photographed by the inhabitants and serve as a visual stimulus for an interview about the specific form of spacing. In this, Stoetzer is not only able to encompass biographical references and visually communicated agendas of normality in the spatial arrangement, but also to reveal the reproduction of 'public' standards in private spaces as their creative variation.

Cultural Assessment and Appropriation of Technology

The paragraph *Cultural Assessment and Appropriation of Technology* is dedicated to two relatively young, related and partly merging approaches. Both concern the societal handling of technology and emphasise that it is important to be aware of the complexity of both present and future technologies. They appeal for a complex assessment and consideration which implicate the participation of experts, politicians and citizens. One approach concentrates on alternatives to traditional methods of technological assessment while referring to the fact that every effort to deal with technology, especially in everyday life, always involves a form of technological assessment. The question is also posed what it means to include aesthetic aspects in the judging, normalising and shaping of technological futures. The second approach (Hård 2003; Hård, Stippak 2005) stands for the cultural history of technology. Unlike the traditional history of technology, which mainly deals with the invention and production of material goods, the cultural history of technology devotes itself in particular to the practical and mental appropriation of artefacts, technology and technological systems. Thus, it also stresses associations, perceptions and discourses which are connected to the abovementioned phenomena. These discourses are comprehended as active and creative processes of appropriating, and integrating modern artefacts, techniques and technology into society.

In *The Aestheticisation of Futurity – from Facts to Values – Authority to Authenticity,* Nik Brown illuminates the role of expectations for the acceptance of techno- and bioscience. With two examples he clarifies that expectations concerning new technologies can be affected. Here, aestheticisation is a major strategy to do so. The first example concerns the report for a consortium of biotechnology firms that emphasises the necessity of changing the branding of biotechnology by avoiding "logic" debates about truth and risks, instead steering them towards intangible and aesthetic considerations of futures values connected with hope and promise. The second example shows the power of this strategy. Brown argues that the latent potential of technology can be more important for its acceptance than its present value, and aesthetically mediated futures can be more powerful than logical reasons.

In *Technological Transformation of Society in Children's Books,* Dominique Gillebeert focuses on the strategies which authors of so-called realistic children's literature in particular use to pass on their perspective and assess phenomena linked to a modern society that is going through a time of change. These include the experience of family life, i.e. its changes and diversity, gender casting and its destabilisation, education, media consumption, the influences of economic issues on childhood, and technological childhood. Considering children's books as an important element of a multi-layered cultural system, Gillebeert raises the question which values, norms, ideas and opinions these books imply and how they convey those to their readership.

Wolfgang Krohn drafts a concept of technology in which many tendencies are integrated and transformed. His basic approach is described in *Aesthetics of Technologies as Forms of Life.* These technological life-forms lay the foundation for the identity of individuals and society. They do not use them to achieve something; they exist in technology (as in a medium). The crucial point is: Technologies are not artefacts. However, they *are* practices which involve artefacts (and causal rules), conventions (and social order), competences (and self-controlling embodiment). The aesthetics of technology refers to the relation of these levels. Technology has turned out well if these levels fit together (one of many interesting references to Aristotle made by Krohn).

In the first part of *On the Symbolic Dimension of Infographics and its Impact on Normalization,* Jürgen Link addresses the concept of "normalism" as the entirety of all discursive as well as practically intervening procedures, dispositives and institutions, through which "normalities" are produced and also reproduced in modern societies. In the second part, Link deals with the role infographics played in the context of normalism and of normalisation processes. The last part of the text comments on topical problems of demographic trends, both on the global level and the level of the Western metropolitan areas, with reference to the dichotomy between "protonormalism" and "flexible normalism".

In *The Mental and Practical Impact of Pre-Bacteriological Quality Criteria for Water in the 1870s,* Marcus Stippak examines to what extent and in which ways efforts to reach a consensus on generally accepted scientific quality criteria for water facilitated the implementation of centralised and mechanical water supply systems. With reference to two case studies Stippak shows and investigates the decisive role pre-bacteriological "standards" played despite their already then obvious limits. Appropriated by municipal authorities as a representatively legimitating argument, these limits initiated a new and positive perception of technology. By creating an idea of hygiene transformed into technology they also caused a fundamental technological transformation, especially within urban areas.

The Aesthetic Dimension of Warfare

The section *The Aesthetic Dimension of Warfare* chooses the aesthetic qualities of military technology as a central theme. Like technology in general, weapons are also culturally constructed and endowed with symbolical meaning. Placed in exhibitions, art, literature and the media for instance, military technology announces, among other things, superiority, power and clean wars on the one hand, and inferiority, defeat and death on the other. Armament manufacturers make use of aesthetic qualities to advertise and sell their products. Likewise, the military command takes the style and appearance of a certain weapon technology into account when it has to decide which technology should be financially supported and bought. Soldiers, in turn, are oftentimes fascinated by modern technology and develop an emotional attachment to their armament. Furthermore, technology is in itself a medium that offers the possibility of experiencing war aesthetically. The authors concentrate on clarifying how weapons are illustrated and culturally constructed. They also analyse in which context the aesthetic qualities of warfare matter, and which aesthetic aspects qualify a certain type of warfare as modern or not. Furthermore, they work out how soldiers perceive aesthetic dimensions. Finally, they ask whether the identified aesthetic dimensions allow for conclusions about modern warfare's nature to be made.

As the title *War and the Beautiful: On the Aestheticizing of the First World War in Film – Yesterday and Today* already indicates, Silke Fengler and Stefan Krebs discuss the aspects of how World War One was, and still is, medially constructed and aestheticised. In the first section they show that the medial newcomer film, especially the propaganda film, tried to create the officially desired picture of a "clean war". In this context they point to the films' language, their content and sometimes more, sometimes less, obvious messages. While referring to a young World War One documentary in the second section, Fengler and Krebs examine the historical documentaries' use of this prejudiced film material.

In *Soldiers on the Screen. Visual Orders and Modern Battlefields,* Stefan Kaufmann argues that the process of shaping a soldier has also been an aesthetic project concerning its physical appearance. Referring to the visible soldier of the pre-industrial and its counterpart, the invisible "steel figure" of the industrial war, he analyses and compares two opposed manifestations of soldierly self-determination. These represent not only two antithetic military concepts of visual order but also two opposed notions of how soldiers should be physically and mentally trained, disciplined and prepared for battle. During the "Cold War", a new aesthetic representation was already created: the networked soldier of the information age. It seizes stimuli from Hollywood movies and computer games.

Based upon interviews with female and male soldiers, Cordula Dittmer analyses two topics in *Military Bodies, Weapon Use and Gender in the German Armed Forces*. Examining military socialisation as a general disciplinary technique, she works out how soldiers of both sexes and different branches become familiar with weapons. She underlines the many-facetted incorporation process for which the soldiers gain identification with their weapons. Secondly, Dittmer adopts a gendered perspective. In this connection, she concentrates on the question whether weapons still symbolise traditional male connotations like "a rifle is a soldier's bride" and how female soldiers are dealing with these symbolic aspects.

Evaluating and interpreting published and unpublished war diaries from the First and Second World War, Christian Kehrt explores in *A Horribly Beautiful Picture* the *Aesthetic Dimensions of Aerial Warfare*. After analysing the military pilots' represented image as "modern warriors", he works out the symbolic meaning of military airplanes as a technological materialisation for power and modernity. Drawing on Immanuel Kant's notion of the sublime, he then focuses on the moments in which military pilots perceived war aesthetically. In this connection,

Kehrt also discusses to what extent technology as a medium affected these aesthetic perceptions and how far the expressed sublime experience reflected real or idealistic circumstances.

In *Postmodern Aesthetics: Manipulating War Images,* Raphael Sassower underlines the ambivalent attributes postmodern aesthetics implies. Unlike its modern pedant, postmodern aesthetics enables one to dissolve or to soften traditional assessment criteria; by accepting multiple sets of criteria, it also allows for manifold interpretations. With reference to theoretical thoughts on art theory as well as on Just War Theory and referring to war images of World War I and II, the Vietnam War, the first and second Iraq Conflicts, the conflict in Afghanistan, and the Palestinian-Israeli Conflict, he discusses this interpretive openness. Facing an enormous potential for manipulation, Sassower questions whether allegedly out-dated modern aesthetics should be given up in favour of their postmodern counterparts.

List of References

Hård, Mikael. 2003. "Zur Kulturgeschichte der Naturwissenschaft, Technik und Medizin – eine internationale Literaturübersicht." *Technikgeschichte*, vol. 70: pp. 23-45.

Hård, Mikael, Stippak Marcus. 2005. "Die Aneignung der modernen Stadt durch Künstler und Schriftsteller um 1900. " In Matthias Luserke-Jaqui, ed. *"Alle Welt ist medial geworden." Literatur, Technik, Naturwissenschaft in der Klassischen Moderne*. Tübingen, Francke Verlag, pp. 35-52.

Technical Gadgetry. Technological Development in the Aesthetic Economy

GERNOT BÖHME

Technology and Usefulness

We should begin by reminding ourselves of a prejudice so ingrained as to seem self-evident: technology is useful. We tend to believe that technology cannot be understood in any other way than in terms of the particular goal it serves. Whether it be a tool, a machine or a whole system of installations: to infer its function is to understand what it is. More recent conceptions of technology as medium, as infrastructure, or as material dispositive have done little to change this. No matter how many attempts are made to overcome this perception of technology as a means to an end, it creeps in time and again. We have to turn to art – to Tinguely's machines, for example – to hit upon different ideas.

This intrinsic relationship of technology and usefulness is a product of the theory of technology, not of its actual history. We can put it down to the lasting influence of Karl Marx that our understanding of technology is so closely allied to the concept of usefulness. There were certainly books on technology before Marx, but he was presumably the first to develop a *theory* of technology, to have a conception of technology as a social enterprise. This theory falls in the period of the Industrial Revolution, historically the first epoch in which technological development was coupled with far-reaching social change. It is therefore quite understandable that Marx should define technology as a force of production. He understands technology in the context of labour, as the appropriation of nature. Technology is a means of labour that is subservient to productive efficiency. The process of technological development set in train during his lifetime radically altered the form of labour, and with it the relations of production. Whereas technology, in the form of manual tools, had for centuries been a medium for mediating human labour force and skills in the appropriation of nature, the Industrial Revolution gave rise to two new developments, which Marx elucidated with reference to characteristic inventions: the spinning machine, which signifies the mechanization of manual labour, and the steam engine, which signifies the substitution of humankind as a force of nature. These developments revolutionised the relations of production. While workers, namely those employed under the conditions of prefinanced utilities and in manufacturing, might previously have been dependent on capitalists for their livelihood, as a rule they were still in possession of their tools of trade, and surplus value accrued through their labour, through their crafting of the material. With the Industrial Revolution, the relations of production come to be transformed in so thorough-going a fashion that the worker no longer owns the means of production, with surplus value being largely created through the operation of machinery. That can still be called an intensification of labour productivity, but because it is brought about through the deployment of capital in the form of machinery, the capitalist claims the right to appropriate surplus value for himself. Productive technology thereby produces and reproduces the class relationship that inheres between labour and capital.[1]

This way of understanding technology has determined the mainstream theories of technology ever since. Technology belongs to the realm of necessity, not to the realm of freedom. It is a means for the human appropriation of nature and exhausts its purpose in maintaining life or improving living conditions. It is subordinate to certain goals, and these are ultimately the reproduction of the human race.

This conception of technology has also had institutional ramifications, particularly in Germany, insofar as the history of technology is generally regarded as a part of economic rather than art history. Even those theories of technology which depart from the Marxist model, at least as far as their political orientation is concerned, share with Marx an understanding of technology as an institution oriented toward usefulness. This is true, for example, of the first modern philosophy of technology, that of Ernst Kapps (1877). In his view, technology is to be understood in terms of organ relief, organ augmentation and organ replacement. This anthropologically oriented theory of technology consequently takes the functions of the human organism as its point of departure. Technology subserves these functions by increasing their effective-

1 Marx's remarks on technology are to be found scattered throughout his major works. For an excellent overview, albeit one still undertaken in the spirit of socialist hopes, see Kusin (1970).

ness and range, or even replacing with them entirely. Here, too, technology serves the survival and, ultimately, the reproduction of the species.

Arnold Gehlen's (1957, 1962) explicitly anthropological theory of technology operates in a similar fashion. Gehlen takes up the view of man as a privative being first put forward by the Sophist Protagoras. Because man is insufficiently fitted out by nature, he has no option but to acquire auxiliary means of survival – the gifts of Prometheus, according to the Greek myth. Here, the useful character of technology, and its subordination to the realm of necessity, are even more obvious than in Marx: deprived of technology, the human race would have no chance of survival. In Gehlen, not just the means of production, shelter and defence are incorporated into this perspective, but even social institutions, which Protagoras still views as gifts of the gods.

Useful and Joyful Technology

As mentioned above, the prevalent conception of technology as a useful institution derives from the theory of technology, not its real history. For while it is perfectly legitimate to cite the techniques of mining, metallurgy, construction and war as very early examples of useful technology, since classical antiquity these have existed alongside playful technologies aimed at instruction and entertainment. We find the distinction clearly articulated in an early modern book on technology, Salomon de Caus's *Of forcefull motions. Description of divers machines both useful and joyful*, which appeared in French and German in 1615. Salomon de Caus was engaged at the time as an engineer and garden architect at the Heidelberg court. He is the creator of the famous *hortus palatinus*, the garden perched midway up the mountain to one side of Heidelberg Castle.

De Caus makes clear in the title of his book that it is a question here of technology in the ancient sense of *mechanike–techne*, i.e. devices which cunningly wrest from nature motions which nature would not evince of its own accord. Technology is *para phusin*, concerned with forceful, i.e. enforced motions. De Caus divides this art of getting nature to bring about something which runs counter to its basic tendency into two categories: the useful and the joyful. These are attributed in the frontispiece of his book to the ancient engineers Archimedes and Heron of Alexandria, respectively, although this is an overly neat distinction, given that Heron also developed theodolites for canal and tunnel construction, built a crane for lifting heavy loads and wrote a treatise on the design of fortifications. He is nonetheless better known for the Heron fountain, which uses heated air to drive up water, as well as for his automata and stage constructions. Archimedes, on the other hand, is renowned for pioneering useful technologies. Legend has it that he helped defend Syracuse by using mirrors to set fire to enemy ships. By formulating the law of lever he placed the technology of scales on a rational footing, and his discovery of the law of water displacement made it possible to determine specific weights. Archimedes is accordingly portrayed with scales and a bowl of water in which a crown is lying, with a depiction of the defence of Syracuse taking up the background. Heron, on the other hand, receives as attributes the Heron fountain, the syphlon and flutes. Both engineers are subordinated to Hermes/Mercury as the god of communication with Hermes' staff and panpipes and Hephaistos/Vulcan as god of the smithy and metallurgy with a bellows.

This distinction of different technology types is rigorously followed through in the course of the book. For example, de Caus discusses the use of pumps in mining in a manner similar to Agricola's chapter on the instruments of mining (1556). For the most part, however, de Caus deals with *joyful* technologies: here it is a question of automata for garden design – mechanical gods, birds which start to twitter at daybreak and water-driven organs.

There is an important reason for this predominance of joyful technologies in de Caus. The useful arts were largely controlled by guilds. Here tradition and, to an extent, explicit rules determined how work was to proceed, which materials were to be used, how a product was to look, and the like. If they wanted to ply their trade without such interference, engineers – literally, inventors (*inventores*) – had no choice but to turn to the court, where they were under no compulsion to join a guild. In his foreword, de Caus consequently praises Charles the V of France and Henry the VIII of England for "once again restoring […] the free arts" (2). Together with the technology of war, the free artists of the courts were now principally occupied with the technology of pleasure.

With the identification of a special, non-useful technology type we have thus discovered a specific source for the modern development of technology. While the mainstream history of technology, sticking to useful technology, adopts the ascendant bourgeoisie as the main agent of development, and industry as its economic framework, we can now discern a second source of technological development in the princely courts and the profligate economy. It is chiefly responsible for the technology of entertainment, but the inventions made in the guild-free climate of the courts naturally exerted an influence far beyond this sphere – and freedom from guild interference means that they are occasionally superior in creativity to their useful counterparts.[2]

As a rule, courtly technology in the modern age veered towards fine mechanics and intricate machinery. It betrays something of the mannerism which characterized courtly handicraft in the 16th and 17th centuries; thence, too, the fetish for automata that continued well into the 18th century (see Sutter 1988). The pinnacle was attained with Vaucanson's duck, which could not only run and cackle, but also feed and – to all appearances, at least – digest. On the other hand, however, the greatest achievements in the construction of massive technical installations also derive from the courtly sphere. The machine of Marly was, at the time, the greatest hydraulic plant in the world. Driven by 14 waterwheels, each 11 meters high, it pumped water from the Seine over a difference of level of some 163 meters. And all this just to supply the Castle of Marly and the fountains of Versailles with water! It should be emphasized that this feat of engineering had no economic benefit and was only possible in a profligate economy such as the court of Louis XIV.

Courtly technology was directed to the 'curious' and improbable; it was meant to astonish, to show something that the everyday experience of nature would not lead one to expect. As such, it belonged in the context of courtly art collections, curiosity chambers and bestiaries, which were supposed to exhibit the strange and exotic rather than the natural (Klemm 1973; Bredekamp 1993). A deformation or monster was worth more than a superb example of a known species. It should be remembered that this was the milieu that gave rise to modern natural science, concerning itself not with natural nature, but with the unexpected in nature. Nature was therefore investigated under conditions which wrested from it appearances that, according to the classical understanding, ran counter to nature – *para phusin*, i.e. under technical conditions. That was Galilei's revolutionary achievement. And as is well known, Isaac Newton had to subject nature, for example the falling of an apple, to a process of artificial estrangement in order for it to require explaining in the first place (see Böhme 1986).

The spirit of *curiositas* did not just have a seat at courts, although the means at a court's disposal and the absence of compulsory guild membership meant that it could be indulged there with unrivalled lavishness. The compulsion to representation also played a part. The art collections and curiosity chambers, the bestiaries, the automata in gardens and grottoes, the aquatic games and the magic of the theatre all subserved the

2 A rare reference to this, specifically in relation to antiquity, may be found in Wolgang König's entry on 'Automaton technology in Alexandria' (1999: Vol. I, pp. 202-204).

glory of the court. This relation between what Blumenberg calls 'theoretical curiosity'[3] and display determined modern science as a whole in the period of its nascency. From demonstrations with the telescope and the microscope via anatomical theaters to the public exposition of experiments, the new science was characterized by a marked tendency to the theatrical. Instruction went hand in hand with entertainment. This science, which put a premium on novelty and curiosity, was as much at home in courtly cabinets as at the annual fair. This holds particularly true for pneumatics, vacuum experiments and the first experiments with electricity. The most impressive was probably the experiment of Otto von Guericke, who demonstrated that not even 16 horses could pull apart two metal hemispheres between which – approximately, we would say – a vacuum had been created. That Guericke carried out similar experiments at the *Reichstag* in Regensburg, where he appeared in his official capacity as a diplomat of the City of Magdeburg, shows that here, too, representation played a role (see Krafft 1978). Certainly, scientific questions were also at stake – the fluid qualities of air, the existence of the vacuum, atmospheric pressure – yet what Guericke demonstrated was above all a technology, and not a useful one at that, at least at the time. The experiment was conducted in 1656 or 1657 and first published in the voluminous work of the Jesuit father Caspar Schott entitled *Technica Curiosa* (1664). Both title and contents of the book demonstrate that while it is a question of serious empirical research here – research that draws its self-understanding, moreover, from its opposition to scholastic science –, the main focus is on unusual and amazing, curious and admirable facts that have been brought forth through technology and human ingenuity. By emphasizing this in his foreword to the reader, Schott reproduces the division of technologies into useful and *joyful* that we have already encountered in de Caus: "[…] nothing extols the excellence, the inventive gifts, the diligence of a man more highly than the astonishing technical achievements of art, which are produced by him for the use or entertainment of his kind…" (Schott, 1). Guericke's airpump experiments, which we are inclined to regard as scientific – perhaps because at that time through the foundation of the Royal Society a professional expertise in Experimental Philosophy was developed – appear in Schott's tome alongside plans for a *perpetuum mobile*, the outline of a universal language, and fountains whose intricate construction permits water to rise above the level of its source. All this not for the use, but for the amazement and diversion of the public.

3 In his endeavour to legitimate the modern age, however, Hans Blumenberg (1966) disregards the flowers sprouted by courtly *curiositas* both in its technical products and in its dealings with nature.

Vaucanson's duck, the great machine of Marly, the hemispheres of Magdeburg – these all belong in the domain of *joyful* technology.

Aesthetic Economy

It is telling that historians of technology, grown accustomed to regarding technology from the viewpoint of usefulness, seem unable to broach the topic of *joyful* technology, whose traces we are following here, without harping on about the exorbitant and (for them) squandered costs involved in its production. Thus it is written with regard to the automata constructions of Heron: "In fact the automata belonged in the context of the Alexandrine court and had the task of contributing, through their startling effects, to the entertainment of courtly society. [...] For the Alexandrine mechanics there existed the possibility of constructing and testing complicated machines without heed to economic concerns" (König 1999: 205). Apropos the Great Machine of Marly, it is said: "This marvellous feat astounded all who saw it not just through its dimensions, gigantic by the standards of the day, but ran up construction costs that, at almost 4 million livres, bordered on the astronomical [...]. Considerations of cost played only a secondary role, since those who commissioned such projects were quite prepared to pay enormous sums for the satisfaction of their wishes" (Troitzsch 1999: 39, 41). Fritz Krafft (1978), in his report on the Magdeburg experiment, likewise cannot refrain from mentioning that the total expenses incurred by Otto von Guericke, some 20,000 gulden, far outstripped his income.

We thus have cause to ask whether this type of technology, namely *joyful* technology, the technology of entertainment, is best viewed within the parameters of a different kind of economic thinking. Now there are certainly other types of economy which are opposed to our conception of use, such as the economy of the gift and the potlatch. Of more significance, however, is that for over a hundred years there has been another view of capitalist economics, a view different to that which defined the mainstream from Adam Smith via Karl Marx up to Max Weber. In their perspective, capitalism is a product of the bourgeois class; it is an economics of scarcity, of materially or morally restricted consumption. The accumulation of capital necessary for this economic form, and its expenditure for the increase of production and productivity, depends upon profits being reinvested rather than consumed. According to Max Weber, the Protestant, and more specifically the Puritan ethic constitutes the background for corresponding behaviour. The counter-concept to this perspective comes from authors such as Thorstein Veblen, Werner Sombart and Georges Bataille. In his book *The Theory of the Leisure Class* (1899), Veblen described the dissolution of the Puritan ethic in the leading stratum in capitalism, insofar as demonstrative expenditure was an integral part of its habitus. Werner Sombart surpassed him in his book *Love, Luxury and Capitalism* (1913) by identifying the origin of capitalism precisely in courtly luxury and excessive consumption. Georges Bataille brings this line of heretical theories of capitalism to a close. If his more general ideas on a post-Victorian ethics predisposed him to excess and transgression, in his general economics he finally comes to characterize capitalism in terms of a periodic oscillation between accumulation and profligacy (see Wex 1999).

Now it must be said that all these theories do not exactly contradict the Marxist analysis, if for no other reason that that the side of consumption likewise belongs to capitalism. Sombart is quite right to point out that the demand arising from courtly luxury was a source first of manufactured and then of industrial production – albeit one that did not lead to capital accumulation on the part of consumers. On the other hand, Marxist authors also recognised that an over-accumulation of capital can come about, followed by an obliteration of capital, for example through warfare. What ought to be retained from this debate, however, is the importance of consumption – and excessive consumption at that – for the functioning of capitalist economies. This fact could obviously only come to light in a particular developmental phase, one that Walter Benjamin already discerns in late nineteenth century Paris, that of high capitalism. We are dealing here with a phase in which the basic needs of the general populace already have been, or easily could be satisfied, imperiling the growth necessary for capitalism. In this situation, the economy must place its bets on those needs which are intensified rather than stilled by their satisfaction, that is to say, on desires. I have designated the form of capitalism which principally stakes on desires the *aesthetic economy* (Böhme 2003). The term is justified by the fact that in this phase, a new value type emerges which is a hybrid of exchange- and use-value: the staging-value of a commodity. Commodities are scarcely goods to be used and used up any more, they are rather display objects, status symbols, gadgets.[4] In this phase, consumption becomes largely that of luxury, and *inside* capitalist society to boot, not outside it, as Sombart supposed.

Remarkably, all of the above-mentioned authors neglect technology, and hence fail to perceive its role in a profligate economy. Thus Som-

4 These terms are already used by Jean Baudrillard in his book *La societé de consummation, ses mythes, ses structures* (1970), an important step in the post-Marxist analysis of capitalism.

bart, for example, speaks of silks and brocades, silverware and gifts, he speaks, too, of the costly building projects financed by Louis XIV, yet there is no mention of expenditure on automata, waterworks and especially the great Machine of Marly. Equally scant attention is paid to the sums lavished on pyrotechnic displays. None of these authors takes into consideration that there existed non-productive technologies which were quite simply a 'waste of money', or served nothing but the self-transcendence of the maker, or were merely consumed as such. In order to gauge the significance of this field one must additionally take into account that a great many technologies were originally developed for purposes of display, popular entertainment, prestige or sport before any potential for a useful, i.e. economically profitable application could be detected in them. Air travel, electronics and the automobile all spring to mind.

This state of affairs has probably only become clear today because we in the rich industrial nations of the West now live in the aesthetic economy, a phase of capitalism which suffers from a constant lack of growth, and in which consumption must be increased by all means available. In this phase, the consumption of technology has become a central economic factor, and that means that the second category of technology mentioned above, *joyful* or entertainment technology, which historically played a fairly marginal role, has moved to the centre of the economy.

Technological Consumption

In the aesthetic economy, the developmental phase in which the industrialized nations of the West currently find themselves, the difference between useful and *joyful* technology seems to disappear. The aesthetic economy is characterized by the appearance of a new value type, the staging-value of the commodity. Yet this is in effect an amalgam of use- and exchange-value: it transforms those characteristics of the commodity which serve its marketing into use-values. A similar conflation is to be observed in the case of recent technological products: they become useful by virtue of what makes them pleasurable. It goes without saying that this use lies principally on the side of the producers and marketers: the technological objects offered by the entertainment industry have become an important sector of the economy. In this respect, one thinks above all of the accoutrements bound up with acoustic and visual consumption, which are to be seen as late derivatives of the courtly apparatus of merry-making – with the difference that such consumption has long since been democratized. Yet quite different devices, such as computers and mobile phones, are also to be reckoned to this category. Originally conceived for economic and scientific use, they have managed to mutate into mass commodities only because they could also be used in a non-useful manner, as toys. Other technologies explicitly conceived as toys developed alongside them, such as gameboys, play stations and public gaming machines.

Gaming, television, DVD and music consumption have become a necessary sector of needs satisfaction for the broad public because – to borrow from Veblen – our society is shifting from a working to a leisure society. Herbert Marcuse (1955) showed that the coming (or already arrived) leisure society is not the realm of freedom beyond labour which, according to Marx's prognosis, would see to the satisfaction of basic needs. The capitalist economy ensures that even the domain of free time is regulated by the principle of *performance*. In concrete terms, this means that our needs to communicate with each other, to listen to music, to participate in events through images, etc. must be transformed into insatiable desires. That is effected by means of an ever-ascending spiral of technical innovation. Image and sound quality, the multifunctionality of appliances, their reciprocal connectivity, their communicative capacity – all these are subject to constant improvement, with the result that appliances become unusable even if they continue to function normally: they suffer, as Marx says, an *ideal obsolescence*; they become out-dated because they fail to meet current performance standards, prove incompatible with the latest equipment, or are no longer able to process information streams with the requisite speed. The consumption of technology is thereby driven forward in leaps and bounds. In a great many areas, technology today is no longer simply a means of production or entertainment; rather, the development of the entertainment industry in the aesthetic economy has led to technology as such becoming a consumer good.

This characteristic of modern technology probably appears in its purest form in our present-day automata. We are currently experiencing the third great wave of enthusiasm for automata, after the Alexandrine and the Baroque. Products of artificial life technology have displaced automatic and remote-controlled toys as the leaders in the field today. The Tamagotchi, which enjoyed enormous worldwide commercial success and generated well-nigh classic symptoms of commodity fetishism, is a case in point. An egg-shaped minicomputer that simulated a living being on its display, it thrust the user into the role of care-giver, demanding that she devote continuous attention over a period of weeks to the development and welfare of this virtual being. Depending on the skill and quality of this care, which was also meant to entail pedagogic measures, the Tamagotchi evolved into a more or less good-natured character with a longer or shorter lifespan. The virtual being's programmed finitude was also one of the reasons the underlying technology was consumed – partly due to the fact that technically, an arbitrary resetting operation was impossible, partly due to the piety of the owner.[5]

The Tamagotchi, too, has since fallen prey to ideal obsolescence. There are now far more sophisticated products of robot technology to keep pace with our delight in artificial life. The dog *Aibo*, for instance, demonstrates an adaptive intelligence which rewards the particularly solicitous owner with a personal relationship. In addition, industry has ensured that this product of entertainment technology has useful functions in the classical sense, such as that of an alarm clock or calendar of appointments.

It is interesting that, even in the present phase of capitalism, the marketing strategies used to sell technical gadgets emphasize as much as possible the utility of these devices. That means that our economy has only half-heartedly become a profligate economy, just as our society persists in the fiction of being a working society. Innovations such as smart clothing and smart houses are praised for their usefulness, even though their primary purpose is to induce pleasure in technology. A piece of clothing which independently monitors the bodily functions of the wearer, adapting itself to them or keeping the wearer up-to-date with necessary information, ought by rights to be useful only in situations of war. A house which allows the bathwater to be set running from inside the car could likewise be said to be equipped with superfluous technology. But having them is fun. In the aesthetic economy, technical objects and installations, no less than aesthetic gadgets, become display pieces, status symbols, means to intensify life, and props for self-staging. They no longer take care of basic needs, corresponding instead to desires which are fuelled rather than satisfied through consumption.

Conclusion

It might be objected that the distinction between useful and *joyful* technology fails to take military technology into account. To be sure, military technology is one of the principal dimensions of technical innovation. It owes this position to a parameter condition which it shares with

5 See the Darmstadt Masters Thesis of Thorsten Dubberke, "Ich bin ein Tamagotchi. Ein neuer Typ der Beziehung zu technischen Gegenständen" (1998); see also www.virtualpet.com

entertainment technology: it was neither subject to regulation by guilds nor bound to the usual standards of fiscal responsibility. On the other hand, it always counted as a useful technology, however perverse that may seem. In light of both these characteristics, one might be tempted to allocate it an intermediary position between useful and joyful technology. Although that is prohibited by the destructive potential of military technology, the idea nonetheless helps illuminate one facet of military technology which has been there all along: just like entertainment technology, it is used for representative purposes; its development – regardless of wartime application – stands in the service of national prestige; and the demonstration of its achievements – from military parades to air-force fly-events to atom bomb tests – has the character of a popular festival. These displays place military technology, which already belongs in the context of profligate economics, in disturbing proximity to entertainment technology.

Yet in this advanced stage of capitalist development, the borders between technology types tend to blur anyway, or better put: almost every technological development contains features of entertainment technology, precisely because it no longer serves the satisfaction of basic needs. At any rate, in important areas such as communications technology, air and space travel, technological development today can no longer be adequately understood from the perspective of usefulness.

Translated by David Roberts, Monash University and Robert Savage, Monash University.

List of References

Agricola, Georg. 1978 [1556]. *Zwölf Bücher vom Berg- und Hüttenwesen.* Düsseldorf: VDI-Verlag.

Baudrillard, Jean. 1970. *La société de consummation, ses mythes, ses structures*. Paris: Denoel.

Blumenberg, Hans. 1966. *Die Legitimität der Neuzeit.* Frankfurt/M.: Suhrkamp. The Legitimacy of Modern Age, MIT Press 1985.

Böhme, Gernot. 1986. 'Kant's Epistemology as a Theory of Alienated Knowledge', in R.E. Butts, ed. *Kant's Philosophy of Physical Science.* Dordrecht: Reidel. 335-350.

Böhme, Gernot. 2003. Contribution to the Critique of Aesthetic Economy, in: Thesis Eleven 73, May 2003, 71-82.

Bredekamp, Horst. 1993. *Die Geschichte der Kunstkammer und die Zukunft der Kunstgeschichte.* Berlin: Wagenbach.

De Caus, Salomon. 1977 [1615]. *Von gewaltsamen Bewegungen. Beschreibung etlicher, so wol nützlichen als lustigen Machiner*. Hannover: Curt R. Vincentz Verlag.

Dubberke, Thorsten. 1998. *'Ich bin ein Tamagotchi.' Ein neuer Typ der Beziehung zu technischen Gegenständen*. Darmstadt Technical College: Master's Thesis.

Gehlen, Arnold. 1962. *Der Mensch. Seine Natur und seine Stellung in der Welt*. Frankfurt/M.: Athenäum.

Gehlen, Arnold. 1957. *Die Seele im technischen Zeitalter*. Reinbek: Rowohlt.

Kapp, Ernst. 1877. *Grundlinien einer Philosophie der Technik*. Braunschweig: Westermann.

Klemm, Friedrich. 1973. *Geschichte der naturwissenschaftlichen und technischen Museen*. Munich: R. Oldenbourg.

König, Wolfgang. 1999. 'Die Automatentechnik in Alexandria', in *Propyläen Technikgeschichte*. Berlin: Propyläen-Verlag.

Krafft, Fritz. 1978. *Otto von Guericke*. Darmstadt: WBG.

Kusin, A.A. 1970. *Karl Marx und Probleme der Technik*. Leipzig: VEB Fachbuchverlag.

Marcuse, Herbert. 1955. *Eros and Civilization*. A philosophical Inquiry into Freud, Beacon Press 1974.

Schott, Caspar. 1664 [1977]. *Technica Curiosa, sive Mirabilia Artis*. Hildesheim: Olms.

Sombart, Werner. 1913. *Liebe, Luxus und Kapitalismus. Über die Entstehung der modernen Welt aus dem Geist der Verschwendung*. Berlin: Wagenbach.

Sutter, Alex. 1988. *Göttliche Machinen. Die Automaten für Lebendiges*. Frankfurt/M.: Athenäum.

Troitzsch, Ulrich. 1999. 'Technischer Wandel in Staat und Gesellschaft zwischen 1600 und 1750', in *Propyläen Technikgeschichte*. Berlin: Propyläen-Verlag.

Veblen, Thorstein. 1969. *Theorie der feinen Leute*. Cologne: Kiepenhauer und Witsch.

Veblen, Thorstein. 1994. *The Theory of the Leisure Class*. Dover Publications.

Wex, Thomas. 1999. 'Ökonomik der Verschwendung. *Allgemeine Ökonomie* und die Wirtschaftswissenschaft', in Andreas Hetzel, Peter Wickens, eds. *Georges Bataille, Vorreden zur Überschreitung*. Würzburg: Königshausen und Neumann.

Aesthetic Precision

KATHRYN M. OLESKO

Measuring practices embody social customs, power relations, economic standards, moral precepts and religious beliefs, as well as scientific techniques. Measurement is thus a social activity uniting a variety of institutionalized practices and beliefs. Marcel Mauss has dubbed such complex social activities "total social phenomena" that also possess aesthetic dimensions with the power to engage the senses and elicit emotions (Mauss 1967). The union of the technical and the aesthetic is especially evident in one form of measurement, precision measurement. This essay examines three moments in the history of precision measurement in the German states: the Romantic Era, Imperial Germany and the Third Reich. It argues that the development of precision measurement was linked to aesthetics, especially aestheticized forms of visual perception.[1]

Precision and Perspectival Drawing in the Romantic Era

Of the cultural fields within which the definition and protocols of precision took shape in the German states, none was more important than the field of visual practices associated with measuring. Spanning architecture and urban planning, surveying and military surveillance, boundary determinations and property assessments, fine arts and theater, crafts and manufacture, civil engineering and machine construction, public works and safety, as well as the observation of nature, the reform of customary visual practices of the late-eighteenth and early-nineteenth centuries converged upon the common objective to decode what were then viewed as optical illusions, real and contrived, in nature and in everyday life. Li-

1 This essay is based on material in my book, *Precision in German Society, 1648-1989* (forthcoming), where additional references may be found.

terature, especially the stories of E. T. A. Hoffmann, as well as visual attractions like the diorama, panorama and theater, sensationalized and accentuated the prevalence of visual ambiguity.

Artificial schemes to combat visual illusions and achieve transparency in representation were many, but the most successful for training the eye and controlling visual ambiguity was perspectival drawing. Extraordinarily popular in the German states from about 1700 to 1850, it was at first no more than an academic exercise cursorily deployed for visually pleasing effects. Later, and especially after the Seven Year's War, perspectival drawing depicted the metric of spatial reality for surveyors, civil engineers, gardeners, landscape architects, military ballistics experts, masons, gold- and silversmiths, jewelers and other craftsmen. By the early nineteenth century, the architects and artists were virtuosos in perspectival representation, but so were precision instrument-makers like Theodor Baumann and civil engineers like Johann Albert Eytelwein.

In the German states, perspectival drawing linked refined measurement to visual perception in two ways. It required, first, an understanding of the quantitative relationship between reality and representation through scales of measurement, a concept not always well understood. Stories of a *Landbaumeister* under Friedrich the Great, who nearly brought to completion a building 1/27 the size ordered because he had read the scales on Friedrich's perspectival drawing wrong, reminded craftsmen and engineers of the disastrous consequences of errors that could be made in translating measurements from concept to reality (Anon. 1792). Second, because it trained the eye as a visual instrument, perspective facilitated a shift in metrical standards from visual estimation (*Augenmaß*) to accuracy (*Akkuratesse*).[2] Practical men, especially in the military, regarded visual estimation as sufficient for most purposes, but when errors had to be kept to a minimum only the measurement-based geometry of perspective sufficed. A technical term, *Akkuratesse*, as used in perspectival drawing treatises after 1770, meant geometrical, metrical and visual detail, all varieties of quantitative precisions.

All forms of perspective played with the metric of visual reality, but none linked precision and drawing more strongly than the adaptation of perspectival architectonic drawing to technical areas, creating what be-

2 One of the features of precision measurement that make its history inviting but also vexing is that the terms "precision" [*Präzision*] and accuracy [*Akkuratesse* or *Genauigkeit*] were not strictly distinguished until sometime in the nineteenth century. For most of the century, *Genauigkeit* could mean either. By around 1900, precision meant the convergence of measured values, whereas accuracy meant the degree of closeness to the true number.

came known in the Germans states as geometrical drawing. Applied to machinery and weapons construction after 1820, geometrical drawing – especially in Berlin where artists, engineers and scientists actively explored the multiple dimensions of visual culture – became the foundation of measuring practices within a visual culture obsessed with the precise representation and reproduction of material reality. An especially important member of this group was Meno Burg, who linked drawing and precision in a series of textbooks for architects and artillery personnel published between 1822 and 1845. Customary perspective, he argued, could create "the most striking resemblance" to nature, but this only depicted objects "as they appear to the senses." In geometrical drawing, by contrast, "one is, so to say forced" to portray the "true [numerical] relation of parts to the whole" by presenting the object from an imaginary infinite distance away (Burg 1822: 60, 61, 63).

Pierre Bourdieu points out that "the 'eye' is a product of history reproduced by education" and that ways of seeing things are bound to a social space (Bourdieu 1984: 3, 5). The new aesthetic in which perspectival drawing played such a prominent role after about 1800 inaugurated an ocularcentrism predisposed to recognize and appreciate visually minute details amenable to precise quantitative description. In an age characterized by the central role of reading in the creation of the public sphere and civil society – new foundations of the political order – those responsible for representing the material world argued for the superiority of images over words.

Arguments for the primacy of the visual over the aural were echoed in the cultural field of precision measurement. Burg, in the first and second editions of his handbook on geometrical drawing based on precision measurements, underlined the nagging inadequacies of verbal descriptions of objects which could never provide "all measures and relations." The advantage of drawing was, quite simply, the attainment of "correct measure" and the "true relationship of the parts to the whole" which speech itself could never provide (Burg 1822: 60, 61, xxxii). Burg identified another organ of sight, the mind's eye, which, when equipped with the principles of architectonic drawing, could "see" each part of an object in order to dismantle, reassemble, and manufacture it. In fact, the point of his instruction was to *imagine new things* that could be manufactured. Perspectival drawing was used to order a reality that already existed; geometrical drawing, to create a manufactured one that might be.

Yet aesthetic precision was not only a matter of compulsions of the eye. Proponents of perspectival drawing aligned themselves with the Romantic concern for the aesthetic shaping of the self; with the belief that effective teaching enhanced the student's ability to make moral distinctions by learning how to deal with detailed nuances; with the overriding importance of having the proper "taste"; and with the need to observe sharply, to be attentive and to know how to focus on a task. The ability to make "good, reasonable and useful judgments about anything" was one benefit of the sharpening of intelligence fostered by perspectival drawing, but so too was the ability to make aesthetic judgments, to have a "purified taste" that identified beauty with accurately drawn lines and angles. A quantitative perspective might have been primarily for technicians, but these sensibilities were for "every person, the scholar and the uneducated, the artist and craftsman and even women" (Werner 1763: 72-73).

One did not, however, have to read these texts or even learn perspectival drawing to acquire some of the visual acuity and sensitivity to precision measurement that it promoted. Public architectural and civil engineering projects that remade rural and urban landscapes as well as scenography in theater, along with shop windows and public dioramas and panoramas introduced several different social groups to a form of visual acuity sensitive to the detailed, angular construction of three-dimensional reality. With its emphasis on proportion and the "proper" balance of objects, this form of aesthetic precision projected a sense of social order: everything in its place, in proportion to its perceived importance.

Precision in the Imperial Era: Commerce & Manufacture

The 1851 World Exhibition at the Crystal Palace highlighted the poor standing of Germany's manufactured items in the world market. Neither clandestine infiltrations of foreign manufactories nor state-sponsored trade journeys abroad had done much to improve manufacturing practices in the German states. After the Paris Exhibition of 1867, where French and Italian goods met with acclaim but German goods one again proved uncompetitive, the architect Martin Gropius (Walther Gropius's great-uncle) pressed the Prussian government to remedy immediately the poor standing of German craft and industrial manufactures. His proposal: make drawing an even stronger foundation of manufacturing innovation. Instructional methods used hitherto to "improve the precision of the eye and the hand in the product" were useful, but "such mechanical exercises never develop the natural feeling and intellectual capacity" necessary for the production of goods that could be competitive globally. Deeper intellectual qualities had to be cultivated, he pressed, including attentiveness, judgment, the ability to compare objects and sustained reflection. Moreover, instructing workers was simply not enough. "The prosperity of all branches of industry whose products appear in artistic form," he concluded, "is dependent upon the instruction in taste provided to the *entire population*" (Gropius 1868: 1).[3]

Recommendations similar to Gropius's had a long history, but it was not until the late eighteenth century that state academies and Sunday drawing schools began to cultivate "pure taste and feeling," an aesthetic sensibility to and appreciation of good living, and (significantly) public safety. Drawings of all types – perspectival, architectonic, orthogonal, and even free hand drawing – supporters maintained, could overcome the shortcomings of "weak minds" unable to hold at once all plans for the construction of an object. After all, shouldn't citizens be protected from faulty products, errors in building construction, and the expensive uncertainty of handwork through the prior execution of a drawing that put "every part in order" and achieved the "best proportion, the most accurate connection of the parts to the whole"? Neighbors, one author assured, would actually "stop and stare" at a building or object constructed according to precise measurements (Seiz 1796: 16, 19, 27). Precision protected, but it was also pretty.

By the late 1850s, geometrical drawing like Meno Burg's had become the standard of instruction in crafts and trades, while now perspectival and freehand drawing was largely, although not exclusively, the reserve of artists and school children. The growth of an industrial, capitalist economy lay behind the shift in emphasis. Art instructor Christoph Paulus remarked that geometrical drawing was not an exercise in logical thinking for the mathematician, but a way of constructing precision for "the market of industrial life" (Paulus 1866: iv). Whereas earlier in the century perspective had spread accuracy uniformly across the landscape, now after 1851 geometrical drawing made possible manufactured items precisely constructed, and these objects now punctuated the visual field. Of note were those designed to stabilize vision: binoculars, weapons sights, and telescopes.

A widespread belief – that perspectival vision and drawing were subjective, psychologically and physiologically grounded and even after the First World War, psychical (Wedepohl 1919) – had more to do with changes brought to the visual landscape by commercial culture and mass production than it did with any of the built-in illusions of perspective itself. By the turn of the century a superfluity of picture books, illustrated

3 The suggestions of Gropius and others eventually became the foundation of the nationalist *Deutscher Werkbund* of 1907, and of *Bauhaus* in the 1920s. Both movements preserved the connections between precise measurements, stylized vision, and industrial design.

catalogs, items of mass production and more generally the mass circulation of images made visual attentiveness not only nearly impossible, but inappropriate in a world illuminated both day and night. The fleeting glance threatened to replace sharp observation and sustained attentiveness even in the classroom, complained Kiel's art history professor, Adelbert Matthaei. The images that greeted children when they left school and entered the street were nothing like the simple visual relations of earlier times. Shop windows with their connotations of unrest, dispersion and haste only accentuated the "imprecise vision" which had replaced the "carefully assembled power of observation" that drawing instruction tried to sustain. He even noted that leading professors of medicine and the natural sciences, as well as those involved in trade, had noticed that in the present generation, the ability to observe no longer corresponded to the actual needs of the present. Medical students, for instance, were "not able at first to see fine nuances in curvature, size, and especially color." The "best defense" against visual illusions for both science and daily life, he argued, was "a concentrated intensive method of observation" that could best be achieved by learning "*precision geometrical drawing*." In an age of manufactured items whose form first took shape on pages filled with geometrical drawings, Matthaei chose precision as an antidote to everyday visual disarray (Matthaei 1895: 14, 16).

Aesthetic precision in the Romantic era was about sense and social order. In the Imperial era, it was about sensibility and the order of the global market, about the material preferences and behavioral compulsions of members of an advanced capitalist culture, the success of which depended as much on tasteful consumption as it did on manufacture itself. For the producer, the engineer, and the craftsman, adapting to the marketplace was not only a matter indeterminate "feeling" or "individual taste" but also of fostering "intuitive faculties and the powers of imagination through geometrically precise drawing." "The nation that will make it to the top in the global struggle of international competition," one art teacher wrote, "will be the one that has at its disposal the greatest number of seeing, trained, eyes" (Grothman 1911:14). Although Jonathan Crary (1999) has recently identified a chronic visual inattentiveness marking the late nineteenth century , bearers of the economic order at the time used aestheticized precision, in geometrical drawing, to stabilize vision, restore attentiveness and cultivate order and accuracy in work by training the eye for measure and order (e.g., Burg 1845).

By the First World War manufacturing firms themselves had taken the initiative to train employees in the production of precision goods competitive on the world market where they became the finest examples of objects "Made in Germany." The firm of Carl Zeiss in Jena, manufacturer of precision optical instruments founded in 1846, is one example. Fifty young men were chosen each year to become precision mechanics through a grueling seven hour test. The eye and the hand worked in tandem to demonstrate manual dexterity, technical understanding, and manual skill in following instructions. Of special importance were exercises in wire bending and paper folding; both extended aesthetic precision to the sense of touch. In the end a score sheet listed the evaluations of the eye (sensitivity to color, sharpness of vision, and ability to perform visual estimations), manual dexterity and work habits in wire bending, paper folding, drawing and calculating (Anon. 1930). The examination was a testament to the types of behavioral compulsions that made aesthetic precision a persistent social fact, now bound to an economic order.

Aesthetic Precision in the Third Reich: Salvation and Standardization

For the young men able to pass through the sieve of Zeiss's seven hour test to become a precision mechanic, a new life began, one marked by respect, achievement and moral fiber, all the offspring of precise labor. So highly was this position regarded that parents fretted over failure and its consequences; for their son would then become *merely* a lithographer, typesetter, or bookseller (Anon. 1930: 10). Now imagine the same test, or a variant, years later in a Nazi concentration camp. The same visual and manual skills used to construct precision manufactures in the Imperial period were used in the Third Reich to achieve ends both more bizarre and of greater consequence.

The behaviors and attitudes associated with precision measurement in concentration camps beg for examination. Take, for instance the example of Primo Levi who recalled the "absurd precision to which we . . . had to accustom ourselves" in daily life: the record keeping, the punctual and regular trains, the roll calls, the tattooed numbers, the camp rules, the precisely ordered life, the executions efficiently carried out. Yet he also noted that while one form of precision was absurd, another could be emancipatory. At the work camps especially, but also in the forced labor programs of Volkswagen and other firms, to be able to use a vernier caliper, precision balance, or other precise measuring instrument, to have the skills, for instance, in which Zeiss tested its candidates for precision mechanics, such as precise wire bending , meant nothing short of salvation. Levi's bed companion Chijim, a watchmaker by profession, was selected to be a precision mechanic in the Buna rubber works linked to Auschwitz, and his friend the Galician Schepschel knew "how to make braces with interlaced wires," a skill cultivated for other purposes at Zeiss. The rational pragmatic efficiency of Germans became for these prisoners nothing short of an existential resurrection from hell. When the Nazis asked Levi of his knowledge of measuring procedures, Levi's sensory self came alive. He felt himself "growing in stature." And then the decisive moment: "Something protects me. My poor old 'measurements of dielectrical constant' are of particular interest to this blond Aryan who lives so safely." While staring "at the fair skin of [the Nazi's] hand, writing down my fate on the white page in incomprehensible symbols," Levi learns he is assigned to the laboratory, joining "the few who are able to preserve their dignity and self-assurance through the practice of a profession in which they are skilled." Death nonetheless could come as easily as a misplaced decimal point: "at the first error in measurement," he reveals, "I will go back to waste away in the snow." (Levi 1996: 16, 47, 93, 106, 107, 140) What precision practices offered prisoners in camp was nothing less than an opportunity to actualize a variant of the Romantic aesthetic shaping of the self through the moral reaffirmation of human life.

During the Third Reich an aesthetic precision in the form of extreme standardization – of behavior, but also of appearance – became a defining feature of social reality. One sees this not only in the bizarre application of precision measurement to racial features – the function of which was to eliminate idiosyncrasy (or variations from the norm) in the visible social realm – but also in the entire process of *Gleichschaltung*, the coordination of the population to totalitarian state objectives, which could be viewed as the political apotheosis of the standardizing impulses of the nineteenth century. When we look at the propaganda material of the Third Reich , we find aesthetic elements linking precision as a social trait to artistic forms of expression, much in the same way that perspectival and geometrical drawing were used to develop and spread the sense and sensibility of precision in the nineteenth century. Films like Leni Riefenstahl's *Triumph of the Will* projected aesthetic images of society constructed on the basis of precise standardizations that were the foundation of an artificially constructed totalitarian political order.

Conclusion: Subjectivity, the Senses and Aesthetic Precision

During each period, forms of precision measurement took shape within different visual fields. In each one, practitioners used an aesthetic precision to spread accuracy across different sectors of reality and to create images of order. They did so by uniting precision measurement with visual practices, thus shaping the sensory self as well as standards of taste, morals, judgment and even behavior. Yet this aesthetic precision was not unique to Germany. Even after the Second World War the Swiss publication *Microtechnic* (1947) reaffirmed the aesthetic qualities of visual practices associated with precision measurement:

> Paramount achievements in precision mechanics may be called for good reasons 'works of art.' Only a great master can create them. To the expert [...] they convey a feeling of admiration and satisfaction, similar to that which all of us experience when contemplating the finest masterpieces of nature, such as, for instance, a snow crystal, a butterfly's wing, or an eye (Tank 1947: 12).

Late into the twentieth century the same associations and emotional responses were found in German commercial culture in advertisements for Braun shavers, Bosch dishwashers, Rotring pens and other consumer goods. Aesthetic precision – created from the union of rational procedures of precision measurement with aesthetic practices of drawing and visualization – thus continued to guide judgments of taste by shaping the sense of vision.

The history of precision measurement is most often embedded in the narrative of Weberian rationalization. Yet the historical linkage between precision measurement and aesthetic practices outlined here suggests that precision is not merely about rationalization and objectivity, but also – and perhaps more importantly – about new modes of subjectivity, especially forms of visual perception that viewed the technical in aesthetic terms.[4] The tandem development of objectivity and subjectivity in precision measurement suggests that its history depends as much on the subjective history of the senses as it does on rationalization.

List of References

Anon. 1792. "Fehler, der in dem Bauwesen darus entstehen kann, daß man vergisst, auf die Verscheidenheit des Bau-Maaßes zu achte." *Allgemeines Magazin für die bürgerliche Baukunst* 2(1), 1792: 254-255.

4 Early objections to precision during the Romantic era may thus have very well been against its aesthetic dimensions as well as its purported claims to truth; the former was an alternative to Romantic aesthetics, with its strong moral foundations.

Anon. 1930. *Lehrlings-Auslese und Ausbildung bei der Firma Carl Zeiss Jena.* Jena: n.p.

Bourdieu, P. 1984. *Distinction: A Social Critique of the Judgment of Taste.* Cambridge, Mass.: Harvard Univ. Press.

Burg, M. 1822. *Die geometrische Zeichnenkunst, oder vollständige Anweisung zu Linearzeichnen, zum Tuschen und zur Construktion der Schatten.* Berlin: Duncker und Humblot.

Burg, M. 1830. *Das architektonische Zeichnen, oder vollständiger Unterricht in den beim Zeichnen der Architekturgegenstände und der Maschinen vorkommenden Constructionen.* Berlin: Dunker und Humblot.

Burg, M. 1845. *Die geometrische Zeichnenkunst oder vollständige Anweisung zum Linearzeichnen, zur Construction der Schatten und zum Tuschen für Künstler und Technologen, und Selbstunterricht.* Berlin: Duncker und Humblot.

Crary, J. 1999. *Suspensions of Perception: Attention, Spectacle, and Modern Culture.* Cambridge, London: The MIT Press.

Gropius, M. 1868. Geheimes Staatsarchiv Preußischer Kulturbesitz Rep. 120E, Abt. I, Gen. Nr. 11 Bd. 1: Die Pflege des Zeichen und Modellunterrichts zur Hebung der Gewerbe- und Kunstindustrie. Von 25/6 1868 bis 18/2 1869, 1-8r (22 April 1868).

Grothman, H. 1911. *Die Erziehung des Auges: Der Unterricht im Zeichnen an den höheren Schulen Preußens in seiner gegenwärtigen Gestalt.* Berlin: Verlag des Albrecht Dürer-Hauses.

Levi, P. 1996. *Survival in Auschwitz.* New York: Simon & Schuster.

Matthaei, A. 1895. *Didaktik und Methodik des Zeichen-Unterrichts und die künstlerische Erziehung in höheren Schulen.* München: C. H. Beck.

Mauss, M. 1967. *The Gift: Forms and Functions of Exchange in Archaic Societies.* New York: Norton.

Paulus, C. 1866. *Zeichende Geometrie zum Schulunterricht und zum Privatstudium.* Stuttgart: J. B. Metzler.

Seiz, [no first name]. 1792. "Ueber den Einfluß der Ausbildung der Handwerker auf Baukunst und Staat." *Allgemeines Magazin für die bürgerliche Baukunst* 2(1): 14-27.

Tank, F. 1947. "Zum Geleit." *Microtechnic* 1: 12.

Wedepohl, T. 1919. *Ästhetik der Perspektive.* Berlin: Verlag Ernst Wasmuth.

Werner, G. H. 1763. *Zur Erlernung der Zeichen-Kunst durch die Geometrie und Perspectiv.* Erfurt: n.p.

Architecture in the Mirror of Technology. The Rhetoric of Le Corbusier and the Futurist Movement

PETER TIMMERMAN

Introduction

The avant-garde architect Le Corbusier and the architects of the Futurist Movement were fascinated by modern technology. At the dawn of the 20th century they enthusiastically glorified cars, airplanes and ocean liners. Architecture should respond to the drastic modernization of the world – so they argued. For that aim they compared architecture with modern technology. The Futurist architect Antonio Sant'Elia claimed that the "Futurist House will be like a gigantic machine" (Sant'Elia 1914). In 1923, Le Corbusier described the house as a "machine for living in" (Corbusier 1986: 4).These metaphorical statements were clearly meant to influence public opinion about architecture.

This paper investigates how Le Corbusier and the architects of the Futurist Movement used technology as a rhetorical means to persuade their audience about the importance of modern architectural forms. I will use Perelman's *The New Rhetoric* as a conceptual framework to analyze their statements about technology and architecture. The rhetorical techniques they applied produced favorable *images* of technology. Interestingly enough, these images were very different. Le Corbusier constructed an image of technology in which perfection and universal principles dominated. The Futurists, on the other hand, created an image in which change and dynamism were important. It will become clear that these constructed images of technology reveal a much broader worldview. To put it philosophically, they also express metaphysical assumptions about the essence of reality.

This paper starts with an outline of the conceptual framework based on Chaïm Perelman's *The New Rhetoric.* Within this framework, the initial question – how did these architects convince their audience of modern architecture by referring to technology? – can be reformulated into more detailed sub themes. These sub themes deliver the guidelines for a detailed analysis of the statements of Le Corbusier and the Futurists about technology. The paper ends with conclusions about their implicit world-views.

The New Rhetoric

According to Perelman and Aristotle, there exist three types of argumentation (or demonstration); 1. Logical (or analytical) demonstration, which is based on logical deduction from axioms; this is the realm of full proof and certainty. 2. Dialectical argumentation, which consists of a dialogue between two experts (like scientists); the result is plausible, but lacks logical rigor. 3. Rhetorical argumentation, which is about things which can neither be logically proved nor scientifically established; rhetoric deals with the realm of *conviction* and *persuasion.*[1] In the seventeenth century rhetoric received a bad name as consisting only of hollow phrases[2]. Perelman shows that this is a regrettable misunderstanding. He reestablishes rhetoric in its original – Aristotelian – meaning and stresses that rhetorical discourse deals with the most important things in life: love, beauty, and the distribution of power and justice. With this in mind, an interesting position was taken by Friedrich Nietzsche: "For Nietzsche, rhetoric is […] part of a greater artistic act – the act of ordering the chaos of life." (Whitson 1993: 136). This broad dimension of rhetoric can certainly be recognized in the writings of Le Corbusier and in the Futurist manifestos. By constructing their favorable images of technology, they not only glorified technology, but also made very fundamental decisions about what really matters in life. To put it in Nietzschian terms, Le Corbusier and the Futurists constructed their own order out of chaos.

Perelman states that an orator has three means to convince his audience of his view of the subject at stake: good arguments (referred to by

1 Perelman 1969: 185-410. See part three of *The New Rhetoric* – "Techniques of Argumentation" – which deals with the three types of argumentation.

2 This very poor interpretation of rhetoric dates back to Petrus Ramus who, in the seventeenth century, claimed that both logic and dialectic argumentation is 'objective', whereas rhetoric is 'subjective'.

Aristotle as 'logos'), the beliefs and feelings of the audience ('pathos'), and his own presence ('ethos').

To start with *logos*: this especially applies to the arguments that are used. Perelman distinguishes three types of arguments:

1. *Quasi-logical arguments.* These arguments reduce reality to a logical universe in which all ambiguity has vanished. The conclusion is transferred to daily reality and *seems* to have the same logical rigor. Ordinary language, however, is much more complex and ambiguous. The famous statement of Heraclitus "we step and do not step twice into the same river" (Perelman 1982: 54) is strictly spoken as a logical contradiction. In ordinary language this is only *apparent* since the phrase "the same river" can be interpreted in two different ways, solving the contradiction.

2. *Enthymemes: arguments based on the structure of reality.* These arguments are based on so-called 'objective' existing relationships between elements of reality. These principles are not subjects of discussion and can be assumed 'evident' so that they form a solid basis for further arguments. This is the realm of the *enthymemes*, a rhetorical syllogism in which one of the premises is assumed true. Good examples can be found in Aristotle's *On Rhetoric*: "If not even the gods know everything, human beings can hardly do so." (Aristotle 1991: 192). This statement presumes that all listeners agree that "If something is not the fact in a case where it would be more expected, it is clear that it is not a fact where it would be less." (Aristotle 1991: 192.).

3. *Paradigms: arguments which establish the structure of reality.* In this type of argument concrete facts are used as a model or basis for a more general rule or paradigm. This is also the realm of analogies and metaphors, which orient thought by structuring some unknown reality. Or, to put it another way, they reinterpret a familiar thing (architecture) in terms of the new (technology). "Every analogy highlights certain relationships and leaves others in the shadows." (Perelman 1982: 119) Metaphor is very important for our analysis, since "truly creative […] thought, whether in the arts, the sciences, religion, or metaphysics, must be invariably and irreducibly metaphorical." (Berggren 1962: 472).

Pathos refers to the emotions aroused in the public by the speaker or writer. A very important condition to arousing these emotions is that the orator or writer must envision his audience. This means that he has to know the prejudices and opinions held by the audience. "The emotions are those things through which, by undergoing change, people come to differ in their judgments and which are accompanied by pain and pleasure […]" (Aristotle 1991: 121).

Finally, *ethos* refers to the character of the speaker. How does he present himself? Is he reliable? Does he meet the expectations of the audience? Does his behavior accord with the point he wants to make? A good orator realizes that his own appearance has great influence on how his message is perceived by the audience.

With rhetoric as a conceptual framework, we can now refine our initial question – how do architects use technology to legitimize their architecture? – by paying attention to the three rhetorical domains. So, in the analysis of Le Corbusier and the architects of the Futurist movement, we focus on the use of arguments (logos), the use of emotions and beliefs held by the audience (pathos) and on the architect's own presence (ethos).

Le Corbusier and the 'machine for living in'

One of Le Corbusier's most legendary statements is that a house is a "machine for living in" (Corbusier 1986: 4). This intriguing statement contains two aspects, which need further clarification. First, the statement suggests an analogy between technology and architecture, resulting in a new metaphor for the house: the "living-machine". As we learned from Perelman, this argumentative strategy establishes a certain structure of reality. Second, in order to find out what new structure of reality is opened up here – what characteristics of machines are transferred to architecture (by means of the metaphor) – it is necessary to find out exactly what Le Corbusier means by 'machine'. A detailed analysis of *Towards a New Architecture* and *Aircraft*, his two books about technology and architecture, will shed light on these issues.

Praising technology

Towards a New Architecture and *Aircraft* can be interpreted as *eulogies* on technology. These books demonstrate the excellence of modern technology. This is done, in the first place, by showing splendid photographs of all kinds of technology: sports cars, ocean liners, grain elevators, an electricity turbine and cranes. *Aircraft* focuses on airplanes: water planes, sail planes, bombers, biplanes, helicopters. Le Corbusier emphasizes the perfection and elegance of these machines by showing close-up photographs, from wheel suspensions to fuselage constructions. These vivid photographs make clear, in a very direct way, what his argument is about. It is remarkable that the photographs are all sharp and clear; even flying airplanes are frozen in mid-air by short shutter times. The same clarity can be recognized in the bird's-eye views Le Corbusier shows. From images to words: what does he *say* about the airplane? "The airplane flies straight from one point to another." (Corbusier 1988: 65). This observation might seem obvious, but this obviousness serves a rhetorical purpose. As Perelman shows, good orators always start with 'agreed upon facts'. Le Corbusier needs this evident fact as a solid basis for his conclusion that the airplane is a superb example of "freedom produced by technical knowledge" (Corbusier 1988: 65). He tries to point out that the airplane is a powerful machine, which makes man independent of capricious nature. But the statement also has another meaning. The perfect trajectory of the airplane should be interpreted as a metaphor for a good design approach. No unnecessary detours (like ornamentation), straight towards the end goal, with great accuracy. With such statements, Le Corbusier tries to evoke a new approach towards architecture. He even wants to make way for a new machine-civilization: "The world's miseries are due to the fact that functions are nowhere defined or respected. A stock of old functions, anachronistic or confused – residue of a civilization in the throes – clogs the wheels and slows down the joyous and productive impulse of the new machine-civilization." (Corbusier 1988: 2) This statement is surrounded by photographs of proud airplanes, which underline the promises of the new machine age. What strikes the most is Le Corbusier colorful language; he describes present society as a society in great agony. Note that he nowhere gives an argument as to *why* this is the case. The image itself convinces, because it uses the feelings of the audience. Who, after all, likes a society in great pain? A detailed argument would be tedious and would lose the attention of his readers. Besides these feelings, Le Corbusier also assumes the implicit beliefs shared by his audience, namely that progress is a virtue.

The Parthenon as machine

How does an *avant-garde* architect like Le Corbusier, who clearly adheres to progress, deal with the historical past? It shall become clear that the Futurists have a very radical answer to the question of history. They simply delete it in order to start with a fresh beginning. Le Corbusier, on the other hand, manages to present history as a powerful argument for his machine eulogy. How does he manage to reconcile progress and history? The answer is found in an intriguing passage in *Towards a New Architecture*. Here Corbusier presents two Greek temples side-by-side.

Below them, two sport cars are shown. (Corbusier 1986: 134-135). On the left page, we see the Paestum (600-550 BC) and a 'Humber' (1907); on the right page, the Parthenon (447-434 BC) and a Delage 'Grand-sport' (1921). Le Corbusier comments on these combinations as follows: "When once a standard is established, competition comes at once and violently into play. It is a fight; in order to win you must do better than your rival *in every minute point* […] Thus we get the study of minute points pushed to its limits. Progress." (Corbusier 1986: 134-135). By comparing Greek temples with modern cars Le Corbusier manages to show that progress is a thing of all ages. Or to put it more succinctly, progress is *universal* and *timeless*. Greek temples advance, as cars develop to higher standards. It is worth noting that this comparison affects Le Corbusier's *ethos* in a positive way, since he manages to reconcile two positions at the same time: being a modern man and being sensitive to history. In the same move, he also addresses a large audience, from people impressed with modern technology – in the 1930s a lot of people! – to lovers of ancient history.

The comparison between temples and cars continues. Le Corbusier starts to see the Parthenon as a *machine* and describes it in *mechanical* terms. "All this plastic machinery is realized in marble with the rigour that we have learned to apply in the machine. The impression is of naked and polished steel." (Corbusier 1986: 217). With this comparison, Le Corbusier manages to put technology on the same level as the greatest endeavors of human kind. He concludes that in the Parthenon and the Delage, the same 'spirit' is at work.

In order to see how Le Corbusier characterizes this universal 'spirit', it is necessary to examine Le Corbusier's opinion of engineers. Le Corbusier's discussion of the engineer is meant to further glorify technology. He uses an enthymeme which assumes one of its premises – outstanding qualities of the creator are transferred to his creation – to be obvious. The 'only' thing that Le Corbusier has to demonstrate is that the methods of the engineer are outstanding.

Constructing the perfect machine

Engineers, according to Le Corbusier, always start with 'clearly stated problems'. By means of logical analysis and by applying universal natural laws, these problems are solved systematically. To illustrate his argument, he refers to Ader who tried to fly like a bird. "To wish to fly like a bird is to state the problem badly […]. To invent a flying machine having in mind nothing alien to pure mechanics […] was to put the problem properly". (Corbusier 1986: 113) When a problem is stated properly and solved by means of universal laws, the resulting machine is also universal. "Simply guided by the results of calculation (derived from the principles which govern our universe) [...] the ENGINEERS of to-day [...] make the work of man ring in unison with universal order." (Corbusier 1986: 31). Le Corbusier stresses the use of universal laws and the transparency and purity of the design process, while keeping silent about other facts.[3] In doing this Le Corbusier constructs a favorable image of technology, in which machines appear as 'objective' artifacts with universal features. Although this might seem a bit exaggerated, why is this image so convincing and attractive? Le Corbusier uses the fact that a lot of people prefer universal principles to individual or coincidental cases. This Platonic preoccupation, which claims universals ('Ideas') to be more real than individuals ('empirical reality'), is deeply rooted in western civilization.

With the above analysis in mind, Le Corbusier's point about the house as 'a machine for living in' can now be explained. This metaphor transfers the qualities of modern technology to architecture, which implies that in good architecture all redundancy is scrapped by fierce selection. Architecture must be pure, 'to the point' (no decoration), and must follow universal principles of the cosmos (instead of individual artistic whim).[4] Perelman stated that the argumentative strategy of using metaphors establishes a certain structure of reality. In the case of Le Corbusier, it means that the reality of architecture is structured in a new way by comparing it to technology.

The adoration of the machine in Futurism

The machine is also a model for Futurist architecture. For the Futurists, however, 'machine' meant something completely different than for Le Corbusier. Instead of a 'static' universal image, they constructed a rather dynamic image of technology. In the Futurist manifestos, fast and noisy technologies like trains, racing cars and airplanes perform a leading role. Their speed and dynamism is emphasized by 'speedy' language: "A racing car whose hood is adorned with great pipes, like serpents of explosive breath – a roaring car that seems to ride on grapeshot is more beau-

3 In practice, engineers seldom use pure universal laws to solve problems. In most cases they use specific empirical laws which apply only to a very limited number of cases.

4 For a more elaborate discussion of the features of Corbusian architecture, see: Timmerman 2006.

tiful than the *Victory of Samothrace.*" (Marinetti 1909). Static constructions are described in metaphors, in order to make them dynamic: "bridges that stride the rivers like giant gymnasts, flashing in the sun with a glitter of knives." (Marinetti 1909).

The Futurists use the rhetorical technique of contrast much more strongly than Le Corbusier. If you want to glorify something (in their case, modern technology) you can do this directly, but you can also despise the situation in which it is absent. In the case of Futurism, this comes down to a squib against *all* history. They depict history as retarded in order to let the future, and of course the promises of technology, shine gloriously. They especially attack institutions meant to cultivate the past, like museums and libraries. "Admiring an old picture is the same as pouring our sensibility into a funerary urn instead of hurtling it far off, in violent spasms of action and creation." (Marinetti 1909). He summons to destroy the past: "Come on! Set fire to the library shelves! Turn aside the canals to flood the museums!.. Oh, the joy of seeing the glorious old canvases bobbing adrift on those waters, discolored and shredded!.." (Marinetti 1909).

In the Futurist manifestos, it is very difficult to find solid argumentation *why* exactly history is a bad thing. The Futurist merely persuades on the basis of the existing emotions of their audience. With their exuberant style they manage to make history ridiculous, in such a way that no one would seriously adhere to it. We must not forget that around 1900, when Italy was lagging behind the rest of Europe, this fresh Futurist sound created a lot of resonance.

It is obvious how the *avant-garde* Futurists solved the tension between historical continuity and progress: they skipped the historical dimension. The unexpected promises of the Future have their primary interests. Progress is perhaps not the most appropriate concept to describe Futurism since progress suggests a development towards a better situation. Le Corbusier is a man of progress, since for him machines and architecture had developed towards higher standards. The Futurists, on the other hand, are merely interested in change for the sake of change. They seem to adhere to some sort of 'absolute' or 'universal' dynamism, in which everything changes continuously. This desire for constant change and renewal can be clearly recognized in the *Manifesto of Futurist Architecture* published by Antonio Sant'Elia in 1914. He described the Futurist City as "an immense and tumultuous shipyard, agile, mobile, and dynamic in every detail" (Sant'Elia 1914). This city is characterised by a "constant renewal of the architectonic environment" and by "beneficial demolitions". Sant'Elia concludes: "Every generation must build its own city." (Sant'Elia 1914). In Futurism there is no place for utopian blue prints of ideal cities or societies.[5] On the contrary, they reject fixed projects and instead long for the unknown and unexpected.

It is worth noting that the Futurists made strong use of their ethos in order to convince their hearing. They not only glorified war – "the world's only hygiene" (Marinetti 1909) – but actually went to the trenches during the First World War. Sant'Elia was even killed during an assault in 1916 at 28 years old. Ironically their ethos has become historical because it is linked with an important historical fact. The ethos of Le Corbusier's, 'the man of universal continuity', is far more timeless. This is one of the reasons why his texts are still appealing. Another reason why Futurism paled next to Le Corbusier has to do with the size of the audience they addressed. As a result of their radical anti-historic position, the Futurists lost all admirers of the great achievements of the past. For Marinetti, a racing car surpassed the Greek *Victory of Samothrace* in beauty; for Le Corbusier, a modern car exhibited the same beauty as the Parthenon.

What characteristics of machines are transferred to architecture in the case of Futurism?

It became clear that the Futurists constructed a dynamic image of technology by applying several rhetorical techniques. Architecture (and city design) should exhibit the same features. In the case of architecture, this is of course a problem since buildings are essentially static and heavy. The Futurists resolved the paradox of 'dynamical architecture' in several ways: 1. by *suggesting* that buildings move (using steep perspectives, elliptical and oblique lines); 2. by creating buildings which generate a lot of movement (like railway stations, busy cities and the Fiat-garage with a racetrack on top of it); 3. by experimenting (in the late 1930s) with light-weight structures which could be easily deconstructed and transported.[6] Though solving the paradox of dynamical architecture generated several interesting experiments, it did not result in a very consistent architectural style.

Creating order out of chaos: Plato and Heraclitus

With Perelman's rhetorical framework in mind, we can conclude that the texts from Le Corbusier and the Futurists involve much more than modern technology legitimizing modern architecture. Obviously they do this,

5 Le Corbusier's city designs – *The Contemporary City, The Radiant City* and his capital for Chandigarh – clearly exhibit utopian features.

6 For a detailed analysis of the dynamical aesthetics of Futurist architecture see: Timmerman 2006.

but if one reads carefully between the lines a rich world of implicit philosophical assumptions can be discovered. Perelman stressed that each rhetorical argumentation implies a selection of facts and values and a specific way of presenting the theme. On the basis of the same factual machines (cars, trains, and airplanes), the Futurists and Le Corbusier constructed different images of technology. In order to construct these images, they had to make fundamental choices about what really matters in life. Le Corbusier believed that perfection, purity and unchanging universal principles were the most basic features of modern times. Architecture should reflect these essentials. The Futurists, on the other hand, held the belief that dynamism and change are the essence of the modern world, which architecture should express. Their choices reveal *metaphysical* assumptions since they claim knowledge about the essence of reality. Their metaphysics can be interpreted in terms of great philosophical systems. The position of Le Corbusier resembles the philosophy of Plato and the Futurists' metaphysics resembles that of Heraclitus. The mirror of technology reflects ancient philosophies as well as modern architecture.

Acknowledgements

I would like to thank Henk Procee for introducing me to the fascinating world of rhetoric, and for carefully reading the manuscript. Furthermore, I thank Caroline van Eck for her useful comments.

List of references

Aristotle. 1991. *On Rhetoric. A theory of civic discourse.* Oxford, New York: Oxford University Press.

Berggren, Douglas. 1962. "The Use and Abuse of Metaphor." *Review of Metaphysics* 16: 450-472.

Corbusier, Le. 1986. *Towards a New Architecture.* New York: Dover Publishers. (Originally published as: Corbusier, Le. 1923. *Vers une Architecture.* Paris: Crès.)

Corbusier, Le. 1988. *Aircraft.* New York: Universe Books. (Originally published as: Corbusier, Le. 1935. *Aircraft.* Great Britain: The Studio Publications)

Perelman, Ch. & Olbrechts-Tyteca, L. 1969. *The New Rhetoric. A Treatise on Argumentation.* Indiana: University of Notre Dame Press.

Perelman, Ch.1982. *The Realm of Rhetoric.* Indiana: University of Notre Dame Press.

Sant'Elia, Antonio. 1914. "Manifesto of Futurist Architecture." Available from http://futurism.org.uk. Originally published as: Sant'Elia, "L'Architettura Futurista. Manifesto." *Lacerba* (Florence): 1 August.

Marinetti, F.T. 1909. "The Founding and Manifesto of Futurism". Available from http://futurism.org.uk. Originally published as: Marinetti, 1909. "Fondazione e Manifesto del Futurismo." *Figaro* (Paris): 20 February.

Timmerman, Peter. 2006. "Architecture, Technology and Metaphysics." *History and Technology*, forthcoming.

Whitson, Steve & Poulakos, John. 1993. "Nietzsche and the Aesthetics of Rhetoric." *Quarterly Journal of Speech* 79: 131-145.

Readerless Media: Modernism, Information Theory, and the Commodity

HEATHER FIELDING

From the very beginning of the foundational text of information theory, Claude Shannon's 1948 essay "A Mathematical Theory of Communication," it becomes clear that the "information" in information theory is a catachresis. According to the Oxford English Dictionary, prior to its information theoretic use, one primary definition of "information" emphasized "the action of telling or fact of being told something." Shannon's definition of information turns radically away from this focus on information as the communication of "something": "[f]requently the messages have *meaning*; that is they refer to or are correlated according to some system with certain physical or conceptual entities. These semantic aspects of communication are irrelevant to the engineering problem" (Shannon 1948: 379). Rather than linking a message's information content to its effectiveness in communicating meaning to a receiving subject, Shannon argues that information is a mathematical quality of the message itself, determined by the statistical relationship between the symbols of the message. Shannon draws on a body of theoretical work on telegraph transmission from the 1920s, which attempted to eliminate consideration of what the symbols in a message mean and how a receiver might interpret them from theories of information transmission; toward this end, these engineers often posited mechanical rather than human receivers.[1] Shannon's theory takes these authors' arguments to

1 The OED dates the first information theoretic use of the term to 1925. This usage reversed earlier definitions of information that put emphasis on a receiver's interpretation of a message; information is now "separated from, or without the implication of, reference to a person informed." For other early attempts to elide the problem of the human receiver in communication theory, see Hartley and Nyquist, whose work was influential on Shannon.

another level: any reader or interpreter is irrelevant to information transmission. According to his definition, the more information a text contains, the less able a reader would be to interpret it anyway: a high information text is composed of improbable symbols that do not follow the rules of the code, a "q" followed by an "x" rather than a "u." An easily comprehensible message carries little or no information; a high information text is, by definition, incomprehensible. Shannon uses the thermodynamic vocabulary of entropy, or disorder, to describe this statistical improbability. In linking information to the random disarray of entropy, he makes his most paradoxical and famous move: he defines information as structurally identical to pure noise. The message containing the greatest amount of information is structured like the random noise generated by the transmitting apparatus, and not like a message that would successfully convey information in the vernacular sense.

In "A Mathematical Theory of Communication," Shannon turns to literary modernism as a model for his theory of uninterpretable information: *Finnegan's Wake* is, he argues, the exemplary text composed of symbols that are highly improbable, the opposite of a message composed in Basic English. Its information content does not reside in what it communicates to an interpreter, but entirely in the statistical improbability of the symbols of the text. In this brief mention of Joyce's novel, Shannon offers an implicit theory of the modernist novel as a noncommunicative text, a medium that should be analyzed without reference to the problem of reading, a literature that structurally occludes its reader. Media theory has followed Shannon's reading in its account of modernism; the leading media theoretic analyses of modernism locate it in or near a shift in paradigms of mediation away from both hermeneutics and the subject as the destination of mediation. In Bernhard Siegert's terms, for example, Kafka's modernism marks the limit point of a media structure governed by information transmission, which links information to a subject that receives it; what replaces media-as-transmission is information processing, in which a message's ultimate reception becomes irrelevant. Friedrich Kittler has influentially developed Shannon's thesis about the noncommunicative nature of the modern media into an argument that in the discourse network of 1900, media are necessarily full of noise. Media transmit information and noise equally and interchangeably, resulting in messages that cannot be assimilated to a hermeneutic model of literary communication: "nonsense syllables […] constitute media in the modern sense: material produced by random generation" (Kittler 1985: 211). Importantly, Kittler links modernist noise to materiality and the psychoanalytic category of the real. As information theory defines noise as the transmitting technology writing itself into symbolic communication, a modernist text considered as a noisy form of mediation must be read as a material object rather than a symbolic vehicle: "indelible and indigestible existence [of words] on the page is all that the page conveys" (Kittler 1985: 212).

Information theory's quantification of language and its turn away from the categories of the reader and meaning have led other thinkers to associate it not with material objects but with the abstraction of the commodity and exchange value: for Baudrillard, information theory is an exemplary component of the bourgeois elision of the signified that he calls the "political economy of the sign." However, literary analyses of modernism and information theory have largely followed Kittler's materialist account of modern media in deploying information theory in the interest of a reading of modernism that turns away from the problems that have interested theorists of modernism and reification or the commodity, such as abstraction.[2] This paper will suggest that information theory can open up a new way to read the relation between modernism and the commodity form, beginning from the idea that the modernist novel and the commodity are both media that fail to transmit information to some kind of reader. Looking at Henry James's theoretical account of his own late, recognizably modernist work, I will suggest that James begins to analyze the novel in information theoretic terms when he works out a theory of the novel as a medium which does not give a reader access to its content. James explicitly draws a connection between this kind of mediation and the commodity: restructuring the novel so that it does not mediate between its content and the reader has the effect of making the novel work like exchange value. In his account of this literary structure, James is simultaneously performing an economic analysis of the commodity form; his theory of the novel as a readerless medium gives him a vocabulary to talk about economic value. I will then look briefly at Marx's account of the commodity as a medium,

2 The critical richness of Shannon's information theory lies in the fact that while on the one hand his equation of information and noise can be read as a materialist account of information, at the same time information is an abstraction. Not only is information quantifiable, but it must be represented by a logarithm, which describes the symbols on the page in terms of all other possible symbols: "[t]he significant aspect is that the actual message is one *selected from a set* of possible messages" (Shannon 1948: 379). Information is the probabilistic relation of the actual message to the larger set, and not an inherent property of the actual message itself. Conceived of as information, the message's physical body is itself not important. Disorder – the quality that information and noise share – derives from this probability measurement, so that an improbable symbol is disorderly; consequently, Shannon dematerializes noise as much as information.

which James's theory of economic value resembles: both writers set up scenarios in which exchange value fails to mediate between use value and an economic subject or "reader" of the commodity, ultimately preventing this "reader" from having access to the commodity's use value.

I will focus my analysis on two sites in the prefaces to the New York edition of James's work. In the prefaces, James theorizes his work through the category of the "center of consciousness," his radical version of point of view. In his accounts of point of view, James works through a theory of how the novel mediates between its literary content (its plot and characters) and the extratextual reader, how the novel makes that content visible to or hides it from a reader.[3] I will juxtapose his account of his center of consciousness technique in the preface to *The Portrait of a Lady*, a novel published in 1881 and often read as part of his quasi-realist phase, and his very different account of the same technique in the preface to *The Ambassadors*, a novel published in 1903 and often read as his headlong jump into modernism. I will argue that while James argues that his realist novel deploys point of view as a device of transmission, providing a linking mechanism between the novel's author and reader and its content, when he rethinks his later work the metaphor of transmission strains until point of view prevents rather than enables a connection between reader and content.

In the preface to *The Portrait of a Lady*, James figures the center of consciousness character as "a pair of eyes, or at least [...] a field-glass, [which] forms again and again, for observation, a unique instrument, insuring to the person making use of it an impression distinct from every other. [...] The house of fiction has, in short, not one window, but a million" (James 1986: 290). The center of consciousness acts as a literally optical point of view onto the world of the novel, providing "the person making use of it" – specifically, not a character, but a person, the novel's extratextual reader – with access to the novel's content. The reader, that is, looks at the events of the novel through the eyes of the central character.[4] James's vocabulary of instrumentality, which presents the eyes as a

3 It is important to keep in mind that in James's theory point of view mediates between the novel's content, what James often terms the novel's "subject," and the reader, *not* between the novel and the real world. In using information theory to read James's theory of how the novel mediates between its content and the reader, I also depart from the strategy of the best critical account of James and communication theory. Patricia Thurschwell uses information theory to account for the relations between James's characters, not between the novel and the reader.

4 Such a reading of point of view as an optical structure – as the reader's eyes, so to speak – has been useful to Christina Britzolakis, who builds on the claims of Bersani and Cameron that James's center of consciousness

field-glass that can be "[made] use of" by someone else, detaches the eyes from a natural relationship to one body, that of the center of consciousness. It is significant that James pairs an organic image, eyes, with a technological one, the field-glass: while these eyes can be transferred like the field-glass from viewer to viewer, they still are human body parts, differentiated from the dead technology of the window or the field-glass. Eyes necessarily belong to a body, in James's figure: at the end of the sentence James reattaches the eyes to another body, that of the reader.

While the interpretation of center of consciousness as a literal point of view seems to have explanatory potential in relation to James's account of *The Portrait of a Lady*, it is not so effective at explaining the central metaphor that describes the center of consciousness in the preface to *The Ambassadors*. To theorize Lambert Strether's function as center of consciousness in *The Ambassadors*, James turns away from the eye/field-glass metaphor to a figure of projection, which he specifically associates with the magic lantern earlier in the essay:

> The thing was to be so much this worthy's intimate adventure that even the projection of his consciousness upon it from beginning to end without intermission or deviation would probably still leave a part of its value for him, and a fortiori, for ourselves, unexpressed. (James 1986: 368-9)

In this filmic figure, the center of consciousness is now projected onto the novel's content. The character's consciousness is no longer a component of the viewing apparatus; instead, it becomes part of the image that would, purportedly, be viewed, if it were not for the fact that the metaphor also eliminates any potential viewer from this narrative situation. An essential component of James's scenario of point of view in the preface to the earlier novel is the person who can look through the eyes of the center of consciousness and see the world of the novel; but this preface leaves out the character's eyes, and there is no one to look through them anyway. Because this figure eliminates a viewing situation, this kind of point of view cannot provide a reader with access to the novel's content in the way that Isabel Archer does when she becomes a pair of eyes.

The difference in the kind of technology James is calling upon in the two metaphors – the difference between a low-tech field-glass or a

---

technique is not built upon on a theory of a psychological, interior, centered subject. The protagonist can be a novel's optical device even if consciousness becomes less a quality centered in a character and, as Cameron argues, a transsubjective system of power relations.

quasi-organic set of eyes and a high-tech projector – supports this claim that projection displaces the viewer onto the viewable image. While all of these components of these metaphors are technological in some respect, the window, the eye and the field-glass rely on a subject looking through them in order to work, according to the logic of James's metaphor. The window and the field-glass produce images only if there is a viewer present to look through them; the viewer forms a necessary component of this image-producing apparatus. However, the activity of projection mediates between the parts of an image, between the frames of a film, the parts of a magic lantern projection, or a slide and its manifestation on a screen, regardless of a person watching. It throws out an image, as the very word "project" suggests; the window and the field glass are far more passive figures. In a certain sense, projection is more automatic than the glass or the window – it requires no subject to construct the image, and it works on its own without a reader, viewer, or author. Point of view as projection works even if it is not actively providing a reader with access to the novel's content.

But what exactly is left of point of view once no viewer is any longer a part of the scenario? In both metaphors, the visual device acts like a medium: it constructs a connection between two elements that are separated across a distance. The eye and window connect the subject safely ensconced in a domestic space, the many windowed house of fiction, to a scene that is necessarily exterior to that space. However, the projection metaphor compresses the neatly compartmentalized distance between the center of consciousness and the novel's content. The filmic figure requires some space between the character and the content – projection only makes sense if there is distance between what is being projected and the screen, a distance over which projection can take place – but it interiorizes this distance, and no longer requires that this mediation take place between an inside and an outside. Rather than connecting the novel to an outside, the extratextual author or reader, projection connects two intratextual elements, the point of view and the novel's content. A connection between two distant places, an inside and an outside, the novel's content and the reader, is replaced by an internal relationship between two parts of the novel, between Strether the character and the content of his novel.

In the second half of the passage I have been examining, James uses a specifically economic vocabulary of abstraction and efficiency to suggest that projecting Strether's consciousness onto the content of the novel will in fact prevent that content from becoming visible.

> [E]ven the projection of his consciousness upon it from beginning to end without intermission or deviation would probably still leave a part of its value for him, and a fortiori, for ourselves, unexpressed. I might, however, express every grain of it that there would be room for – on condition of contriving a splendid particular economy. (James 1986: 369)

While *The Portrait of a Lady* could provide full access to its fictional world, *The Ambassadors* cannot; in sharp contrast to the abundant room of the mansion of fiction with its million unique windows, the later novel is a confined space. But this passage does not merely suggest that projection allows the novel to become more efficient; more specifically, that efficiency takes the form of an economic abstraction. The projection of Strether's consciousness onto the novel's content "expresses" that content as a value. The expression of a value is a literary enterprise, but it is also an economic procedure, the abstracting activity of a commodity-driven monetary economy. When value is "expressed" efficiently, a heterogeneous, incalculable quality is transformed into a homogeneous, calculable quantity; a good produced by concrete labor is turned into a price. Strether's consciousness does not make the novel's content visible to a reader; instead, it "expresses" that content as an abstract value, reducing the content's qualitative heterogeneity to a manageable, countable size.[5] This process of economic abstraction is at work in the way this passage describes the novel's content as a number of grains to be lodged into a finite space: when Strether's consciousness is projected on top of the fictional world of the novel, it breaks that heterogeneous scene into a series of comparable and equivalent units. James's economic vocabulary of value expression suggests that these units are not unlike dollars and cents, but they are quite dissimilar to the unique and thus uncomparable windows of the house of fiction. Those windows endlessly accumulate qualitatively different versions of the novel's content; this projection thins that content out until it fits into a confined space and a neat abstraction.

When James theorizes modernist point of view as a device that blocks the novel's content rather than providing access to it, he is working out an aesthetic answer to a demand that he acknowledges is simultaneously economic and aesthetic; he is analyzing the commodity econ-

5 Tim Armstrong has also linked Jamesian efficiency to the commodity form, though he argues that James thinks of the commodity in terms of transparency, "what you see is what you get" (Armstrong 1998: 56). In this passage, James seems to be making the opposite argument: the novel or commodity is an efficient expression when what you see is a price tag, not the heterogeneous, unmanageable use value that is hidden behind it.

omy of capitalism at the same time that he is analyzing the form of the novel. That is to say, this noncommunicative form of mediation seems to James to be a link between how the novel and the money economy function. I will conclude by briefly looking at Marx's theory of the commodity to flesh out how James's theory of novelistic mediation might work as an economic analysis. 40 years prior to the prefaces of the New York edition, Marx theorized a similar sort of noncommunicative mediation as an integral process in the formation of exchange value. He described the abstraction associated with exchange value as a form of mediation that blocks a connection between heterogeneous terms. When the commodity form generates a relation between two different products of qualitatively different forms of concrete human labor, the qualitative body of one commodity becomes what Marx specifically identifies as a "mirror" for the value – the amount of abstract, not concrete, human labor contained in the object – of the second term.

"The natural form of commodity B becomes the value-form of commodity A, in other words the physical body of commodity B becomes a mirror for the value of commodity A." (Marx 1990: 144)

The corpus of B, its use value, is reflected onto A; through B we see A. But when we see A through the body of B, A has also magically been made comparable to B, and A's qualitative difference, the concrete specificity of the kind of human labor of which its value is an expression, disappears behind the abstracted similarity of the two commodities to each other. Moreover, this mirror blocks mediation by hiding visual data, because it elides information about what it seems to make visible as it translates A into an exchange value and obliterates the qualitative heterogeneity of A. To go even further, the scenario has no room for a spectator, someone to look into the mirror and see A's value: "B becomes a mirror" that cannot communicate information to a viewer but instead subtracts information from A. Marx's mirror metaphor suggests that the commodity form is structured like a mediating device that blocks rather than transmits information.

Shannon and James theorize the modernist novel according to parallel concepts of the novel's status as a medium: both think that the modernist novel does not mediate between its content and an extratextual reader. When James rethinks how point of view works in his late novels, he simultaneously offers a theory of exchange value as a form of noncommunicative mediation that draws on Marx's analysis of the commodity. For James, this media structure is necessarily both literary and economic at the same time; this literary point of view that fails to mediate between a reader and the novel's content cannot be understood outside of an economic context. This structure of thought that Shannon identified in modernism thus opens up a way to think about the relationship of modernism and the theory of the commodity, through the category of mediation. At the same time, this relation in James's work between modernist point of view and the commodity suggests that the connection between modernism and information theory works through the theory of the commodity, that these three discourses on mediation must be understood together.

List of References

Armstrong, Tim. 1998. *Modernism, Technology, and the Body: A Cultural Study*. Cambridge: Cambridge University Press.

Baudrillard, Jean. 1981. *For a Critique of the Political Economy of the Sign*. Trans Charles Levin. New York: Telos Press.

Bersani, Leo. 1989. *A Future for Astyanax: Character and Desire in Literature*. New York: Columbia University Press.

Brillouin, Leon. 1956. *Science and Information Theory*. New York: Academic Press.

Britzolakis, Christina. 2001. "Technologies of Vision in Henry James's *What Maisie Knew*." *Novel* 34: 369-390.

Cameron, Sharon. 1989. *Thinking in Henry James*. Chicago and London: University of Chicago Press.

Hartley, R.V.L. 1928. "Transmission of Information." *The Bell Systems Technical Journal* 7: 535-563.

James, Henry. 1986. *The Art of Criticism: Henry James on the Theory and the Practice of Fiction*. Ed. William Veeder and Susan M. Griffin. Chicago and London: University of Chicago Press.

Kittler, Friedrich. 1985. *Discourse Networks, 1800/1900*. Trans. Michael Metteer with Chris Cullens. Stanford: Stanford University Press.

Marx, Karl. 1990. *Capital*. Trans. Ben Fowkes. New York: Penguin Classics.

Nyquist, H. 1924. "Certain Factors Affecting Telegraph Speed." *The Bell Systems Technical Journal* 3: 324-347.

Shannon, Claude. 1948. "A Mathematical Theory of Communication." *The Bell System Technical Journal* 27: 379-423, 623-656.

Siegert, Bernhard. 1999. *Relays: Literature as an Epoch of the Postal System*. Trans. Kevin Repp. Stanford: Stanford University Press.

Thurschwell, Pamela. 2001. *Literature, Technology, and Magical Thinking, 1880-1920*. Cambridge: Cambridge University Press.

Mass Aesthetics, Technicist Expansionism and the Ambivalence of Posthumanist Cultural Thought

THOMAS BORGARD

Ending History? The Construction of the Present as Self-Legitimation

It has been pointed out that current literature studies suffer from an unfruitful gap between everyday philological practice such as retrieving archives respectively editing texts and their interpretation, often referring to systematic approaches towards a theory of literature or theory of culture (Proß 1999). Whereas the first field covers empirical or factual issues, in the second the problem of authority arises. It is deeply rooted in the hermeneutical tradition and the Hegelian as well as Marxist idea of overcoming an amiss reality by constructivist approaches. Their result is an integrative view as it, for example, has been developed by the Frankfurt School and from which society as a whole is supposed to benefit. Since the 1950s, in contrast to this idea of an ethical surplus of historiography (Rusch 1987), rapidly developing sociological functionalism has rejected the idea of societal order being governed by moral values. Accordingly the primacy of space over time is alleged, decomposing the normative corpus of any ontological, ethical and aesthetic hierarchy. Only after the classical liberal oligarchy, bourgeois culture and communism have been defeated, what is now called postmodernism accentuates against purely technical or instrumental functionalist rationality the capability to make playful use of the handed down material. But still, what appears to be a new approach towards historicity utterly prevails within its specific aesthetics and its terms of informal creativity the achieved conceptual structure of analysis, combination and sequence, ambivalently suiting the demands of technicism as well as of mass-cultural promiscuity.

The named occurrences pose the question of legitimation or of the legitimatory function of academic concepts. The reason for this are persisting patterns of normativity within the sociocultural formation of the West even after historical substantialism has been overcome. They are made available by technoepistemic transformation of the social sciences through cybernetics, information theory and system theory to meet the rising need of a pluralist society for social planning and control. At the same time they conceal their own irrational potential which is, first, the result of gradually replacing history by mathematics and, second, of the ambivalent contemporaneity of ascetic production and hedonistic consumption. Both affect the notion of the knowledgeable society, the economic concept of rational choice and some appropriate turns in Cultural and Media Studies.

Development towards functionalism is equally initiated by aesthetic formalism in architecture and the artistic avantgardes as well as by the emancipation of modern sociology from 19th century historicism. Its advocates no longer refer to given features of a single human being, but to specific phenomena of the group showing only momentarily stable structures of interrelationship. Sociological analysis now dedicates itself to the "interdependency of the parts" as Georg Simmel points out, dissolving the "substantial" into "function" or "movement" and reducing quality to the quantity of combinatory elements (Simmel 1989 [1890]: 130; Simmel 1989 [1900]: 366). At the same time the American context generates "Scientific Management" (Taylor 1911) and "pragmatic" or "objective psychology", decomposing the human gesture into its partial effects and into a sequence of reactions or parts of movements in order to ensure frictionless machine operation.

Theories of reducing complexity for the practical reasons of social engineering together with the logical consequences of the turn to philosophical conventionalism breed a considerable lack of interest in some peculiarities of the contemporary epistemic situation (Smith 1973). Responsible for this are the disadvantageous effects of change over from temporality to spaciality, respectively from "consecutiveness" to "coexistence" (von Wiese 1955 [1924-1928]: 27). Ironically just at the moment of the liquidation of the bourgeois strain, an old philosophical concern is still going strong. It anticipates the unity of institutional forms and moral norms as can be seen in Talcott Parsons' definition of society as a system:

> The core of a society, as a system, is the patterned normative order through which the life of a population is collectively organised. As an order, it contains values and differentiated and particularised norms and rules, all of which require cultural references in order to be meaningful and legitimate. (Parsons 1966: 10)

Parsons' concept only takes notice of those aspects of interaction that preserve the system. And it epistemologically favours communication over action, communication being understood as exchange of meaningful symbols. Expulsion of action theory from the theory of structure and function coincides with the commitment of Parsonian normativism to behaviorism:

> I believe that basic innovation in the evolution of living systems [...] does not occur automatically with increases in factors or resources at the lower (conditional) levels of the cybernetic hierarchies, but depends on analytically independent developments at the higher levels [...]. Similarly, I believe that, within the social system, the normative elements are more important for social change than the 'material interests' of constitutive units. The longer the time perspective, and the broader the system involved, the greater is the *relative* importance of higher rather than lower factors in the control hierarchy, regardless of whether it is pattern maintenance or pattern change that requires explanation. (Parsons 1966: 113)

Parsons' culturalist view of evolving social order discharging "material interest" and his understanding of adaptation to norm depend on the specific change of work practice since the 1940s which is widely described as the shift to the tertiary sector of industry and to service economy. What seems highly problematic here, first, is the treatment of the historically symptomatic as a universal concept and, second, the limitation or even exclusion of anthropological knowledge. This happens frequently in the so-called information age, following the build up of commercial "knowledge industries" (Jamison 1994: 2-7) and the invention of the electronic calculating machine (Bolter 1984), meeting the demands for self-legitimation of a mass society that claims consensual progress through an increasing connectedness and interdependence.

During the second half of the 20th century, the leading disciplines developed in two ways: towards rational planning (respectively behavioral control) and towards a precarious reconstruction of cultural communication. Their members more or less all agreed by rejecting the humanities ("Geisteswissenschaften"). After alignment to cybernetics, sociology lost some of its specific competences which it had acquired during the "value judgment dispute" in the works of Max Weber, in Karl

Mannheim's theory of "documentary interpretation" (Lichtblau 1996: 492-539) and in the "histoire totale" of the French Annales School (Honegger 1977). The main advancements here were the rejection of nationalism from a viewpoint that included non-European cultures and the reflection of the antagonism between history and nature in connection with the apparent crisis of the humanities. Moreover, the *relationalist* character of their cultural synthesis did not give way to the belief that the acceptation of value relativism would inevitably lead to epistemological scepticism concerning the possibility of scientific rational understanding. Responsible for the subsequent backlash following the socio-political forming of mass democracies after around 1940 was a massive import of theories into sociology which did not originate in this discipline itself but were raised in evolutionary biology and computational engineering (Shannon 1948; Wiener 1950; Ashby 1956; Easton 1964; David 1965). Unlike the concept of Parsons, these cybernetically modified theories, dreaming "the dream of a subjectless machine" (Frank 1983: 398), no longer resorted to "normative elements" as they explained order and progress by the means of "selection", "adaptation" and "feedback dynamics". The concept of "positive feedback" was well suited to bridging the gap between technocracy and hedonistic lifestyles. And this is the reason for its immense success starting from the postwar building of a "Consumerized Republic" (Cohen 2003) and culminating during the cultural and sexual revolution of the 1960s.

Today the semantics of systems, modules, networks and spaces has replaced the former entities of man, nature and history. Moulded by the devision of matter, signal and sense and initiated by several heterogenous driving forces, the process of deconstruction coincides with a change in awareness regarding the relationship between the horizontal (geographical) and vertical aspects of cultural dynamics. Every technological innovation casts a new light on the moment at which it emerges. As the cycles of innovation have speeded up since the 20th century, they produce and intensify a discontinous stance of generational self perception (Mannheim 1928), even favouring intellectual notions of "the end of history" (Fukuyama 1992). Now not only causal explanation in history is devaluated, but there is also a general shift to conceiving of science as *practice*. Through this process mainstream sociology and economics tend to suffer from severe tensions between the closeness of their abstract methodologies and the openness of preceedings (Hodgson 2001). Moreover, the overall confusion of the issues and different ontological levels of the formal and the material (or normative), for example, believing in the ethical effects of the number of media contacts, disguises the "phase lock that restricts the emergence of new knowledge, and which tends to confine intellectual innovation to established formats." (Simpson 1994: 109) Especially the idea of "differentiation" marks a neo-evolutionary postulate that derives from a specifically Western historical experience (Smith 1973: 32-36) and neglects the fact of widespread structural assimilation through workplace computerisation (Kondylis 1999: 40). But since the straight forward optimism of some socioeconomic metatheories can be challenged, interest in micrological descriptions of interactions arises. At this point the legitimate critique of authoritarian interpretation in an ethnomethodological perspective leads to "textualisation" which is understood as the production of a potentially infinitely variable "collage". Its assigned nonrepresentational and discoursive structure indicates that literary theory here provides itself with an artful image (Clifford 1988; Bachmann-Medick 1996).

Machine Efficiency, Enjoyment of the Unspecific and the Unspiritual Spirit

A crucial point regarding some specific features of postmodern Literary and Cultural Studies seems to be the difference between the question "does a correct or true interpretation of a literary text exist" and the question "how are we able to distinguish adequate from inadequate interpretations?" Whereas the first question can be neglected approving of the advancement towards ontological scepticism, the idea that interpretations are ultimately ficticious – as is claimed by "New Historicism" – simply cannot be accepted (Iggers 1996). The explosiveness of this situation can only be understood by analysing its fundamental structure.

In approaching this task it is to be reminded that the members of the avantgarde movement combined their cult of the machine with constructivist ideas and the stylisation of a powerful hedonistic and permanently changing subjectivity. With unfolding mass culture, the strong position of the artistic subject is gradually replaced by the aestheticisation of the everyday object supporting the impression of an existing World Culture. Now on a planetary level, contrasting Classical Liberalism's partition of the private and the public sphere, the identity of art and life is accentuated, potentially accessible by all and sundry. Instead of creating hierarchies, particles that formerly were integrated into the neohumanistic canon are unhinged in order to be selected and newly combined to form a transitional "network of representations" (Hartman 1997: 32f.). Ferdinand de Saussure's semiology, Roman Jakobson's and Nikolai Trubetzkoy's phonology and Claude Lévi-Strauss' mythopoetic elementarism give good examples of this process (Reckwitz 2000: 224). Structurally this changeover not only applies to concepts of the mind, but also to the categories of subjectivity and gender. The individual is now regarded as being able to hold multiple identities and concepts of self-awareness and meaning (Butler 1990; Fischer-Lichte 2000: 128).

Looking at consumption as a means of self-realisation the implementation of technicism and every day culture refers to the necessity of an ethnically heterogenous American society to develop its highly assimilatory capacity. The prominent features of this sociocultural "pantheism" (Tocqueville 1835-1940: vol. 2 Ch. VII) influence the modelling of today's time-fragment into a phenomenology of the life-world by which "communication" is an acclaimed social technique in a twofold sense. It not only forms an integral part of managerial efficiency control in the realm of computerised production, but also signifies life-style promiscuity within corresponding mass (media) consumption. Nevertheless, the *Gestalt* of the whole with its two sides forming a single coin is rarely ever taken into consideration. This becomes apparent in the process of legitimation through rationality linking with deontic concepts such as "Good Governance" or "New Public Management" and by resorting to the metaphysics of the "invisible hand". This leads to a chronic underestimation of the persisting ranges of conflict and of social dysfunctionality. One has to take into consideration that with rising (ethical) pluralism the problem of "how to make the right decision" amounts to the most urgent of all questions, stimulating high demand for consultancy services. They follow the idea that after the downfall of the spiritual monopoly of the Christian Church, social interaction had to be regulated on the basis of proceedings ("Verfahren"). Focussing upon the theoretical basis of this prime technology, linguistics, semiotics and computer science urge to unify their field of cognition. But as the resulting sociological, economic and managerial concepts show, they are not only insensitive about the constraints of rationalisation, rooted in the specific logic of mass entertainment and mass consumption (Schulze 1992), but also about some basic anthropological facts and persisting social patterns resulting both from the need to preserve the self and from the presence of fear.

Every human society endeavours to stabilise its particular status quo to safeguard it from potential risks. With the evolutionary development of the intellect, awareness of imminent dangers increases, the human mind becoming a permanent source of unrest and willingness to reduce fear by appropriate means. That "nature" or the "beastly instincts" are overcome by the "spirit" is therefore a common myth. On the contrary: as a consequence of the human capability to anticipate what might happen in the future the aggressive potential of the instinct is refined. Participants in the power game, however, far more often use their intellectual abilities during conflict than brute force. Under these fundamental conditions the struggle for power is projected onto the ideational level. It is now most important to understand the interconnection of power, the setting of norms and their often deliberate concealment.

The Problem of Unforseen Consequences: Unveiling the Culturalist Fallacy

Occidental ethics traditionally tends to delimit the concept of rationality and action in a specific way, for example, in sharp contrast to the Chinese lore of the "strategem" (Jullien 1999; von Senger 2000). The Western paradigm is set by Aristotle who draws a distinction between cunning (*métis*) and practical wisdom (*phronesis*) linking it with morality (Schröder 1985: 236). Because he delimits his comprehension of rational goal setting to ethics (Rohbeck 1993: 43f.), all purely instrumental aspects of self-control are neglected. This leads to a one-sided estimation of discipline and a problematic view upon the relationship between theory and practice which is guided by the model of "ideal form". Consequently, the intellect permanently sees itself exposed to an antagonism: between what it causally announces and what it wishes or intends to be the final goal so that the problem of unforseen consequences arises (Borgard 2005b: 97ff.). It is not only responsible for the dynamic character of history but also constricts psychological explanations and reveals the ontological priority of the fact of society. Nevertheless, there exists a continuous tradition in which this important insight is totally subdued to ethics prominently marked by the invisible hand and Hegel's idea of the cunning of reason ("List der Vernunft"). These concepts allege that history is guided throughout by the principle of progress no matter how evil the wills and deeds of individual subjects may be. By the end of the 19th century, this scheme came under heavy pressure. Due to the invention of the automotive machine the potential of social conflict was rising, or to say it more clearly: Without forseeing or wanting it the economic activities of the *Bürgertum* unleashed the social question which eventually would precipitate the ruin of their original social stratum. Consequently the idea of progress around 1900 became dubious, generating a specific type of tragic consciousness that was furnished with readings of Schopenhauer, Nietzsche and Freud. It was therefore not by chance that Simmel described "objective culture" as the result of depersonalisation and reification using Marx's term of alienation (Lenk 1972: 10-24). His assumptions nevertheless showed three problematic results: first, the tendency of transforming science into art (Lichtblau 1996: 22) and, second, from the viewpoint of Max Weber they underestimated the *gestaltistic* character of "objective" validity, transcending any subjectively intentioned validity (Weber 1988 [1913]: 427; Weber 1988 [1921]: 541; Lichtblau 1994: 538). A symptom for this was Simmel's massive use of illustrations and analogies which posed the third problem of identification, i.e. of the (scientific) object with the (scientific) method (Frisby 1981). It will later also be found in Niklas Luhmann's theory of self-description (Borgard 2001).

At this point it is interesting to see how Weber's critique of Gustav Schmoller's ethicised economics (Schluchter 2005: 91) opened up the imagination of the social fact to everything that man is capable of. In contrast to Parsons, who traced back to Émile Durkheims sociology of morality, Weber did not identify the idea of reciprocity with sociability. Morals and values, so he firmly underlined, can form the basis of friendly agreement as well as of extreme hostility (Weber 1980 [1922]: 13). With respect to the question how the tension between theory and history could be relieved, Weber was totally unsentimental about the elitist caste of the German cultural mandarins (Ringer 1969) who bemoaned de-substantialisation (Curtius 1929) and "loss of the center" (Sedlmayr 1948). Instead he reconstructed their precarious situation by keeping in mind "the transformation of intended *and* unintended outcomes of action into macrosocial phaenomena" which also form the border of understanding and observation (Schluchter 2005: 12). That this assumption was correct could be proved by looking at the fact that negotiating and quarreling over what is and should be done never ceased.

The Weimar Republic is a very good example of a state of civilisation close to Civil War generating intellectual consent to ethics, divinatory practice in literature and culturalist discourses being insufficient. For example, in his novel *Berge Meere und Giganten* (1924), Alfred Döblin develops a perspective that also contrasts with the recent assumptions of Samuel P. Huntington in his *Clash of Civilizations* (Huntington 1993/1996). Döblin rejects the idea of cultural invariants being the source of conflict. Instead, vital interests determine how cultural meaning is (variably) accentuated and not the other way round. With regard to orientation in nature and space, the text unfolds a grandiose panorama of contingency and its accomplishment. Cultural signs stabilise the scope of interaction of the human being characterised by deficiency, but they also form the instruments of power and deception. In his didactic play *Die Maßnahme*, Bertolt Brecht reflects within this context on the logic of the strategem that adapts to the contingency of the empirical world and characterises an advantage of superior knowledge in the society of the Chinese: "Klug ist nicht, der keine Fehler macht, sondern / Klug ist, der sie schnell zu verbessern versteht." (Spinner 1994: 175-239.)

Here the relation of technology to action is not established by an ethical maxim but by the concept of occasional reason that indicates a cleft between means and ends, respectively aims and consequences. In doing so Brecht stresses a paradoxical moment within the realm of ethics itself. It is the result of a subtle relationship between ethics and technology in a positive (a) and in a negative sense (b). In a positive sense (a) ethics, although many times rejecting pure instrumentality, has to use the scheme of means and ends in order to decide which procedure will best suit to achieve its targets. Accordingly, the question of means and the question of ends must be handled seperately (Kondylis 1999: 569-589). And here the negative sense (b) of the link between ethics and technology becomes evident. It is clear that an ethically inclined person wants to carry into effect a moral target, but the effectiveness of the means is morally neutral. Like technology including its artefacts, it bears a mere instrumental aspect. On this level there is no difference between the rational abilities of a peaceful saint, of a medical doctor or even say of a dictator, who deliberately resorts to violence to establish and stabilise his power. In other words: the absolute formal distinction between irrational power and rational ethics is completely invalid. This can be seen in the biographies of Brecht and Carl Schmitt who coincide over decisionism although charmed by two opposite tyrannies. Their cynicism, although sometimes hard to swallow, still draws a conclusion worth while reflecting as modernisation creates a situation in which every single value system can render itself absolute. So then only violence, as Hermann Broch asks in one of his darkest moments, may be capable of effecting real change. Regarding values, the "ethical picture" can easily "turn down" (Broch 1978 [1939]: 32, 38, 40-43), the intellectual and the dictator finding themselves agitating on the same ontological level. What is furnished in the name of culture now to a lesser extent appears to be more liable to poetry than the battlefields of vital interests, or in other words: the idea of modernising the aesthetic form against traditional straight forward narrative in order to transport an ethical message leads to an unsolveable paradox. This is the central theme of Broch's magistral work *Der Tod des Vergil* (Borgard 2003; 2005) and here, too, the equally instrumental character of technology and ethics comes into focus. It is Walter Benjamin's essay *Zur Kritik der Gewalt* (1921) that indicates most clearly the underlying problem:

> Ist nämlich Gewalt Mittel, so könnte ein Maßstab für ihre Kritik ohne weiteres gegeben erscheinen. Er drängt sich in der Frage auf, ob Gewalt jeweils in bestimmten Fällen Mittel zu gerechten oder ungerechten Zwecken sei. Ihre Kritik wäre demnach in einem System gerechter Zwecke implizit gegeben. Dem ist aber nicht so. Denn was ein solches System, angenommen es sei gegen alle Zweifel sichergestellt, enthielte, ist nicht ein Kriterium der Gewalt selbst als eines Prinzips, sondern eines für die Fälle ihrer Anwendung. Offen bliebe immer noch die Frage, ob Gewalt überhaupt, als Prinzip, selbst als Mittel zu gerechten Zwecken sittlich sei. Diese Frage bedarf zu ihrer Entscheidung denn doch eines näheren Kriteriums, einer Unterscheidung in der Sphäre der Mittel selbst, ohne Ansehung der Zwecke, denen sie dienen. (Benjamin 1965 [1921]: 29f.)

Set against this background, present-day's technoepistemic functionalism and economic panpragmatism seem to be conceptualised merely as self-fullfilling prophecies, underestimating the possibility that as former chaos may turn into contemporary order, contemporary order may just as easily turn into future chaos. Economistic concepts of social order after 1945 apparently do not depend on seemingly universalist references to human rights and tolerance, but only on momentarily stable common welfare. In future they may turn into their very opposite under the conditions of global population growth. So if traditional metaphysics was dead and truth a construct, then what was real either represented a narcissistic language game or – if the nominalist values were questioned – an expression of social power. Tom Wolfe's novel *The Bonfire of the Vanities* (1987) and Brett Easton Ellis' *Glamorama* (1999) recently fathomed the depths of this problem. Together with the twofold character of the problem of technology and ethics (respectively of unforseen consequences) it should as here is suggested be more considered within a new interdisciplinary effort to reshape cooperation between Literary Studies, Cultural Studies and the Social Sciences.

List of references

Ashby, Ross W. 1956. *An Introduction to Cybernetics*. London: Chapman & Hall.

Bachmann-Medick, Doris, ed. 1996. *Kultur als Text. Die anthropologische Wende in der Literaturwissenschaft*. Frankfurt/M.: Fischer.

Benjamin, Walter. 1965. Zur Kritik der Gewalt [1921]. In *Zur Kritik der Gewalt und andere Aufsätze*. Frankfurt/M.: Suhrkamp, pp. 29-66.

Bolter, David J. 1984. *Turing's Man. Western Culture in the Computer Age*. Chapel Hill: University of North Carolina Press.

Borgard, Thomas. 2001. "Das Problem der Selbstreferenz, die Subjekte des Handelns und das 'proton pseudos' der Systemtheorie: Herbart, Hegel, Kleist." In Andreas Hoeschen and Lothar Schneider, eds. *Herbarts Kultursystem. Perspektiven der Transdisziplinarität im 19. Jahrhundert*. Würzburg: Königshausen & Neumann, pp. 107-131.

Borgard, Thomas. 2003. "Hermann Brochs Roman 'Der Tod des Vergil' als Gegenstand einer analytischen und funktionalen Geschichtsschreibung." In Michael Kessler, ed. *Hermann Broch. Neue Studien. Festschrift für Paul Michael Lützeler*. Tübingen: Stauffenburg, pp. 117-167.

Borgard, Thomas. 2005a. "Planetarische Poetologie. Die symptomatische Bedeutung der Masse im amerikanischen Exilwerk Hermann Brochs." In Thomas Eicher, Paul Michael Lützeler and Hartmut Steinecke, eds. *Politik, Menschenrechte – und Literatur?* Oberhausen: Athena, pp. 205-229.

Borgard, Thomas. 2005b. "Sind die Folgen der Technik die Zwecke der Literatur? Ökologisches Denken zwischen historischer Forschung und metaphysischem Bedürfnis." In Catrin Gersdorf and Sylvia Mayer, eds. *Natur – Kultur – Text. Beiträge zu Ökologie und Literaturwissenschaft*. Heidelberg: Carl Winter, pp. 79-108.

Broch, Hermann. 1978. "Zur Diktatur der Humanität innerhalb einer totalen Demokratie [1939]." In Paul Micheal Lützeler, ed., *Hermann Broch Politische Schriften*. Frankfurt/M.: Suhrkamp, pp. 24-71.

Butler, Judith. 1990. *Gender Trouble. Feminism and the Subversion of Identity*. New York, London: Routledge.

Clifford, James. 1988. *Predicament of Culture. Twentieth Century Ethnography, Literature, and Art*. Cambridge (Mass.): Harvard University Press.

Cohen, Lizabeth. 2003. *A Consumer's Republic. The Politics of Mass Consumption in Postwar America*. New York: Vintage Books.

Curtius, Ernst Robert. 1929. „Soziologie – und ihre Grenzen". In *Neue Schweizer Rundschau* 36/37: 727-736.

David, Aurel. 1965. *La cybernétique et l'humain*. Paris: Gallimard.

Dyson, Esther; Gilder, George; Keyworth, George; Toffler, Alvin. 1994. *Cyberspace and the American Dream: A Magna Charta for the Knowledge Age*. http://www.hartford-hwp.com/archives/45/062.html.

Easton, David. 1964. *A Framework for Political Analysis*. Englewood Cliffs: Prentice Hall.

Fischer-Lichte, Erika. 2000. "Performance-Kunst und Ritual: Körper-Inszenierungen in Performances." In Gerhard Neumann, Sigrid Weigel, eds. *Lesbarkeit der Kultur. Literaturwissenschaften zwischen Kulturtechnik und Ethnographie*. München: Fink, pp. 113-129.

Frank, Manfred. 1983. *Was ist Neostrukturalismus?* Frankfurt/M.: Suhrkamp.

Frisby, David. 1981. *Sociological Impressionism. A reassessment of Georg Simmel's Social Theory*. London: Heinemann.

Fukuyama, Francis. 1992. *The End of History and the Last Man*. New York: Free Press.

Hartman, Geoffrey H. 1997. *The Fateful Question of Culture*. New York: Columbia University Press.

Hodgson, Geoffrey M. 2001. *How Economics forgot History. The Problem of Historical Specificity in Social Science*. London, New York: Routledge.

Honegger, Claudia, ed. 1977. *M. Bloch, F. Braudel, L. Febvre u.a. Schrift und Materie der Geschichte. Vorschläge zur systematischen Aneignung historischer Prozesse*. Frankfurt/M.: Suhrkamp.

Huntington, Samuel P. 1993/1996. *The Clash of Civilizations and the Remaking of World Order*. New York: Simon & Schuster.

Iggers, Georg G. 1996. "Die 'linguistische Wende'. Das Ende der Geschichte als Wissenschaft?" In *Geschichtswissenschaft im 20. Jahrhundert. Ein kritischer Überblick im internationalen Zusammenhang*. 2nd ed. Göttingen: Vandenhoeck und Ruprecht, pp. 87-96.

Jamison, Andrew; Eyerman, Ron. 1994: *Seeds of the Sixties*. Berkeley, Los Angeles, London: University of California Press.

Jullien, François. 1999. *Über die Wirksamkeit*. Berlin: Merve.

Kondylis, Panajotis. 1991. *Der Niedergang der bürgerlichen Denk- und Lebensform. Die liberale Moderne und die massendemokratische Postmoderne*. Weinheim: VCH Verlagsgesellschaft.

Kondylis, Panajotis. 1999. *Das Politische und der Mensch. Grundzüge der Sozialontologie vol. 1*. Berlin: Akademie Verlag.

Lenk, Kurt. 1972. *Marx in der Wissenssoziologie. Studien zur Rezeption der Marxschen Ideologiekritik*. Neuwied, Berlin: Luchterhand.

Lichtblau, Klaus. 1996. *Kulturkrise und Soziologie um die Jahrhundertwende. Zur Genealogie der Kultursoziologie in Deutschland*. Frankfurt/M.: Suhrkamp.

Lukes, Steven. 1973. *Emile Durkheim. His Life and Work. A Historical and Critical Study*. Stanford: Stanford University Press.

Mannheim, Karl. 1928. „Das Problem der Generationen". In *Kölner Vierteljahrshefte für Soziologie* 7: 157-185, 309-330.

Parsons, Talcott. 1966. *Societies, Evolutionary and Comparative Perspectives*. Englewood Cliffs: Prentice-Hall.

Proß, Wolfgang. 1999. "Kater Murr ordnet in Goethes Archiv die Fragmente von Michelangelos Karton von Pisa." In Henriette Herwig, Irmgard Wirtz and Stefan Bodo Würffel, eds. *Lese-Zeichen. Semiotik und Hermeneutik in Raum und Zeit. Festschrift für Peter Rusterholz*. Tübingen: Francke Verlag, pp. 178-220.

Reckwitz, Andreas. 2000. *Die Transformation der Kulturtheorien. Zur Entwicklung eines Theorieprogramms*. Weilerswist: Velbrück Wissenschaft.

Ringer, Fritz K. 1969. *The Decline of the German Mandarins. The German Academic Community, 1890-1933*. Cambridge (Mass.): Harvard University Press.

Rohbeck, Johannes. 1993. *Technologische Urteilskraft. Zu einer Ethik technischen Handelns*. Frankfurt/M.: Suhrkamp.

Rusch, Gebhard. 1987. *Erkenntnis, Wissenschaft, Geschichte. Von einem konstruktivistischen Standpunkt*. Frankfurt/M.: Suhrkamp.

Schluchter, Wolfgang. 2005. *Handlung, Ordnung und Kultur. Studien zu einem Forschungsprogramm im Anschluss an Max Weber*. Tübingen: Mohr Siebeck.

Schröder, Gerhart. 1985. *Logos und List. Zur Entwicklung der Ästhetik in der frühen Neuzeit*. Königstein: Athenäum.

Schulze, Gerhard. 1992. *Die Erlebnisgesellschaft. Kultursoziologie der Gegenwart*. Frankfurt/M.: Campus.

Sedlmayr, Hans. 1948. *Verlust der Mitte. Die bildende Kunst des 19. und 20. Jahrhunderts als Symptom und Symbol der Zeit*. Salzburg: Otto Müller.

Senger, Harro von. 2000. *Strategeme*. 2 vols. Bern, München, Wien: Scherz.

Shannon, Claude. 1948. "A Mathematical Theory of Communication." *Bell System Technical Journal* 27: 379-423, 623-656.

Simmel, Georg. 1989. "Über sociale Differenzierung [1890]." In Heinz-Jürgen Dahme, ed. *Georg Simmel. Aufsätze 1887 bis 1890. Gesamtausgabe vol. 2*. Frankfurt/M.: Suhrkamp, pp. 109-295.

Simmel, Georg. 1989. "Philosophie des Geldes [1900]". In David P. Frisby and Klaus Christian Köhnke, eds. *Gesamtausgabe vol. 6*. Frankfurt/M.: Suhrkamp.

Simpson, Christopher. 1994. *Science of Coercion. Communication Research and Psychological Warfare 1945-1960*. New York, Oxford: Oxford University Press.

Smith, Anthony D. 1973. *The Concept of Social Change. A Critique of the Functionalist Theory of Social Change*. London, Boston (Mass.).

Spinner, Helmut F. 1994. *Der ganze Rationalismus einer Welt von Gegensätzen. Fallstudien zur Doppelvernunft*. Frankfurt/M.: Suhrkamp.

Taylor, Frederick Winslow. 1911. *The Principles of Scientific Management*. New York: Norton.

Tocqueville, Alexis de. 1835-1840. *De la démocratie en Amérique*. 2 vols. Paris: Charles Gosselin.

Weber, Max. 1988. "Über einige Kategorien der verstehenden Soziologie [1913]." In Johannes Winckelmann, ed. *Gesammelte Aufsätze zur Wissenschaftslehre*. 7th ed. Tübingen: UTB, pp. 427-474.

Weber, Max. 1988. "Soziologische Grundbegriffe [1921]." In Johannes Winckelmann, ed. *Gesammelte Aufsätze zur Wissenschaftslehre*. 7th ed. Tübingen: UTB, pp. 541-581.

Weber, Max. 1980. *Wirtschaft und Gesellschaft. Grundriss der verstehenden Soziologie* [1922]. 5th ed. Tübingen: Mohr Siebeck.

Wiener, Norbert. 1950. *The Human Use of Human Beings. Cybernetics and Society*. Boston: Riverside Press.

Wiese, Leopold von. 1955. *System der Allgemeinen Soziologie als Lehre von den sozialen Prozessen und den sozialen Gebilden der Menschen (Beziehungslehre)* [1924-1928]. Berlin: Duncker & Humblot.

Transhumanist Arts. Aesthetics of the Future? Parallels to 19th Century Dandyism

MELANIE GRUNDMANN

Parallels between the Dandy and the Transhumanist Artist

Like the Transhumanist Artist, the Dandy is an aesthetic and artistic as well as a social phenomenon. Aesthetically, the Dandy is a piece of art. He stylizes himself and puts a high value on form. The Dandy is highly self-conscious, self-restricted and self-made. He is subject to many rules he puts on himself. Furthermore, the Dandy is highly individualistic, independent and claims to be an original. The historic context in which he first appeared is nineteenth century England and France. Society was abruptly changing: The aristocracy lost its elite power due to the uprising mass of people. Industrialization and modernization brought about changes in the economy, politics and culture. The upcoming press allowed the discussion of different opinions. Earlier, the French Revolution brought about the idea of personal liberty. Despite growing economic and political freedom, the moral development lagged behind. As John Stuart Mill discussed in 1859 in his essay *On Liberty,* the rules of society are applied to the great mass of people. The people form a homogeneous mass, which needs to be controlled. Individuality and originality are neither favoured nor wanted. Dandyism, now, was a highly intellectual phenomenon. The Dandy analysed the state of society; he knew the changes that were about to happen. He adopted full individuality and independence as they were proposed. He did not care for morals, standards and conventions. Therefore, he appeared anti-social. He did not accept the rules that were applied to society. The Dandy loved to surprise and to shock. Therefore, he started to 'misbehave', always on the edge of breaking taboos. The Dandy wanted to be different and original. In this claim of individuality Dandyism became the first subculture of modern society. Society itself mocked the Dandy, who deliberately positioned himself 'outside'. Since he was not only a social but also an artistic phenomenon (many Dandies were artists), his otherness ultimately influenced the cultural pattern of the nation. Furthermore, the Dandy's social concern for individuality became apparent in his art. Dandyism cannot be separated from nineteenth century aestheticism, which brought forward a radical new aesthetic. The realization that the perception of the world depends on the senses became known in the nineteenth century and it had a deep impact on the people. Many artists felt estranged from the world, alone, and thus, by the end of the century, a radical inwardness became apparent which led to an artistic silence.[1] The new conception of aesthetics that thus developed away from the beautiful and sublime to a mere and individual sensual experience paved the way for the beauty of the ugly and the evil. Art came to portray the evil in the world, the sinful fleshliness of men, the modern sufferings, the dirty urbanization. Everything that was in the world could be portrayed in art. Social critique needs art as an independent context. The demand of *l'art pour l'art* – art free from moral rules – guaranteed the spreading of the Dandy's message.

Transhumanist Art has an equivalent approach. After the industrialization and the European revolutions in the nineteenth century, we now face Globalization and the alteration of society through technology. The latter is the main concern of Transhumanism. Transhumanism is "a philosophical and cultural movement concerned with promoting responsible ways of using technology to enhance human capacities and to increase the scope of human flourishing." (Bostrom 2003: 45).

The parallels that I see with regard to the transition from the nineteenth to the twentieth century, and from the twentieth century to the twenty-first century is a change in the perception of the world. The philosophers of the seventeenth and eighteenth centuries experienced the world primarily through the mind. The turning point is the Romantic era in which the ingenious artist emerged. Dandyism and Aestheticism play a major role in positioning the body as a central theme in nineteenth century arts and literature. Today, both the body and the mind can be enhanced by technology. Artists, as mediators and visionaries, often are the first ones to accentuate these changes. Nineteenth century Dandy-

1 See the critique on language as a means of perception in the philosophy of Wittgenstein. This critique is earlier discussed by artists and writers such as Hugo von Hofmannsthal, Arthur Schnitzler and Stéphane Mallarmé to name just a few. Hofmannsthal's *Der Brief des Chandos* holds as the generally accepted document of this discourse. The focus on the senses was made public by the philosopher Ernst Mach in *Die Analyse der Empfindungen und das Verhältnis des Physischen zum Psychischem* (1886).

Aesthetes not only portrayed the new sensual experience of the world, they even suffered from it. The body and its senses were established as a central theme, which resulted in charges like censorship and public ostracism. The beginning of the twenty-first century, now, sees a comparable new perception of the world through technology.

Transhumanism

Transhumanists are willing to use technological means to overcome the human condition. They do not want to wait until evolution jumps to the next level. Since mankind continually evolves technologically, it is just a question of personal willingness to use technology on and inside one's own body in order to overcome human barriers. This obviously brings about ethical and moral questions but these cannot be discussed here. Nonetheless, Transhumanism is a proactive movement. Its members intend to shape the future, primarily through technology. Transhumanism values and aims to secure individual liberty, tolerance, prosperity, activity, the constant shaping of oneself, society and the future, and autonomy. Transhumanists approve of expanding cognitive, emotional and physical conditions through technological and medical means. Transhumanism, therefore, is connected with plastic surgery, sex change, neuropharmacology, prostheses, implants, organ transplants, birth control, abortion, in vitro fertilization, anti-aging, smart clothing, etc., which is all existent already. IT-Tools, such as wearable computers, smart agents, information filtering systems and visualization software are in active development worldwide. Further areas of expertise include Cryonics, which places people in low-temperature (-196°) storage in order to revive them once technology allows this and the cell damage can be repaired, Virtual Reality, human cloning, Biotechnology/Genetic Engineering, Artificial Intelligence (AI), Nanotechnology, and ultimately the uploading of a persons' brain onto a computer or a brain-computer interface.

Transhumanist Art

In my analysis of Transhumanist Arts I will focus on two manifestos:

The Transhumanist Arts Statement
The Extropic Art Manifesto of Transhumanist Arts

Extropianism is a sub-branch of Transhumanism. It was established by Max More and Natasha Vita-More, the latter being probably the most important mediator of Transhumanist Arts and author of the above mentioned manifestos. Extropianism is highly positive, individualistic and optimistic. It focuses on boundless expansion, self-transformation, dynamic optimism, intelligent technology and an open society. (See Bostrom 2005: 12) According to the Transhumanist Arts Statement,

> Transhumanist Arts represent the aesthetic and creative culture of transhumanity. […] Our aesthetics and expressions are merging with science and technology in designing increased sensory experiences. Transhumanist Artists want to extend life and overcome death. […] We are ardent activists in pursuing infinite transformation, overcoming death and exploring the universe.

Some conclusions can be drawn from this. Firstly, due to the intense connection with technology, Transhumanist Artists must have a certain knowledge about science. Indeed, the border between artist and scientist becomes blurry with Transhumanist Artists. Secondly, like the Dandy-Aesthete, the Transhumanist Artist actively seeks new perceptions of the world. Baudelaire experimented with drugs and Stelarc tries out a third hand. The Dandy-Aesthete and the Transhumanist Artist, as mediators of the future, both emphasize values such as individuality, independence, self-awareness and self-responsibility. The Dandy paved the way to bring the body not only into art, but into everyday perception. With the dawn of the twentieth century skirts got shorter, bathing suits turned into bikinis and one single layer of dress displaced the numerous layers the nineteenth century men and women used to wear. The Transhumanist Artist takes this evolution to another level: It is about redesigning the body and adjusting it to new demands. Therefore, the Extropic Art Manifesto of Transhumanist Arts starts with: "I am the architect of my existence." Transhumanist Artists are "active participants" in their "own evolution from human to posthuman." Transhumanism is the transient state of humanity to become posthuman. Transhumanist Artists "shape the image – the design and the essence" of what they are becoming. Not only do they actively work on and with their own body, but also on their mind.

Transhumanist Art is highly influenced by technological and scientific developments. It is focused on designing new bodies and new types of senses. Obviously, it is Science Fiction coming true. What used to be imagined has finally become possible. Art theories develop and change as the world evolves. Now that our lives and, thus, our perspectives are highly defined by technology and globalization, this influences art. Artists use computers, Virtual Realities, robotics, etc. But this is modern art per se. What defines Transhumanist Art is that the artist's own body is his/her work. The Transhumanist Artist is an Automorph, a constant refinement of the body, the senses, and the mind. Therefore, the Transhumanist Artist is committed to his/her individual growth, seeking to overcome all biological, psychological, technological and moral limits. The ultimate aim is to integrate art with technology, life and science.

Today, technology is included in art, which can be seen in computer graphics, digital installations etc. However, there is still a big taboo, which has long been dealt with in Science Fiction literature and movies and which finally becomes realizable: The technological alteration of the body. Technology will ultimately infiltrate not only in our daily processes but into our very mode of living, into our way of thinking and into our bodies. Transhumanist Artists are the first to break this taboo now that fiction becomes true. What is accepted as medical treatment (e.g. implants, prostheses, in-vitro fertilization, laser treatment, etc.) is often perceived as morally bad when used for 'fun'. If some people would get x-ray vision, a third hand, or chips that activate unused parts of the brain this would soon raise heavy discussion about the equality of man as guaranteed in most of the world's constitutions. These scenarios will happen in the future and need to be discussed now. Transhumanist Artists play with these possibilities, test them out but as artists they are allowed to do so. Transhumanist Art follows the path of modern art per se: Art is increasingly integrated into everyday life. Once the artistic and aesthetic sphere is left and the social coexistence is entered, questions will arise. Therefore, Transhumanist Art is a valuable movement that works to communicate Transhumanist ideas and values to a broad audience.

Let me give you an example. Technology is highly embraced and evolving fast in science and medicine. It is already possible to make the blind see and to enable paralyzed people who could communicate merely with their eyelids to have advanced communication possibility via a brain-computer interface. Furthermore, genetic engineering will soon allow us to explore genetic defects early and correct them. Diseases will be cured before they start. Life will be of a higher standard, healthier and longer. Now imagine a healthy person who attaches a third hand to his body. Stelarc, a Transhumanist Artist, did so. Since this is art, it might appear weird but it is accepted. Then imagine a healthy person getting a brain update so he knows everything about law and becomes a perfect lawyer. This is fiction and could be seen in the TV show *Angel*. However, what we know as Science Fiction now finally becomes realized. Someday someone might be able to get an update like this. If that person were an artist that would be fine because art is a context that (usually) does not threaten society. However, if an average person gets this upgrade many ethical and moral questions will arise. There is hardly any discussion about these future scenarios that will ultimately evolve. Transhumanist Artists embrace and accept technology, they are willing to alter and upgrade their body. Not only do they show us what the future will be about, they also want to raise awareness and discussion. In the transitional state of Transhumanism, a state our society is currently stepping into, there will always be those people who embrace technology and those who don't. There lies the real danger of a two-class society. What we need to become aware of and what we have to face directly is that the technological alteration of society will advance and become even more central in our lives than today.

Utopia?

Traditionally, art is illusion. But, as Herbert Marcuse states in his essay *Art in the one-dimensional society* (1967), the status of art has changed since Romanticism. Art itself can help create a new reality. Marcuse states that a new type of human being is necessary to develop a new society. This is the role of the artist, because the artist shows us what could be. He portrays the needs and longings of the people. Let's take a look at the Dandy who partially developed out of the romantic anti-hero. The Dandy aesthetizes his self and his surroundings. Therefore, he not only creates art but is a piece of art himself. He confronts the world outside, which seems to him ugly, barbarian, dirty, loud and evil, with sensual art. This art invites the participant to enter into a different stream of reality. As we can see in the art of the Fin de Siècle, the so-called ivory tower-artists, this ultimately led to self-estrangement and alienation from the world.

Transhumanist Artists are the people that Marcuse was waiting for. The progress of science and technology continually reminds us that we failed to create a free society, based on the traditional aesthetic of the beautiful, the good and the true. We depend more and more on technology, Marcuse said that art – as technology – could be in the position to be the leading force in creating a new society based on aesthetic values that are per se non-aggressive. And this is exactly what Transhumanist Artists aim for, especially the sub branch of extropian art, which raises values such as liberty, autonomy, peace and optimism. This is the basis for creating the post-human world. Art has gone a long way from being an exclusive masterpiece affordable merely for a few distinguished people to now being part of everyday life. Just take a look at the electronic devices industry. It is not only about function, but to a great deal also about design. Aesthetics has become a part of modern technology. As the technologization of society increases, so will the aesthetic shape of this society. Transhumanist Artists use technology as their artistic tool, give it form as well as shape it to their needs. They embrace technology in a positive way, they use it to enhance life, make it more beautiful and valuable. This society, according to Marcuse, needs beauty for its existence and this beauty will affect everything people desire. Beauty disables aggression and destruction. However, Marcuse disregards that destruction does possess a beauty in itself. This is exactly what Baudelaire realized: There is beauty in the ugly or in the bizarre. This is where Marcuse and Transhumanists, especially Extropians, fail. It is naïve to assume a world without aggression and distress is possible. It is creditable to aim for a society where everybody is at peace with oneself and the world. However, there are always power-seekers, be it individuals, parties or companies.

Transhumanist Artists

Finally, I want to give you some examples of Transhumanist Artists. Doing this research, I was not satisfied with what I found. It seems that the theory about Transhumanist Art is more developed than the practical outcomes. This may be due to the fact that the technological alteration of society is running smoothly. Many ideas that exist in the minds of artists and scientists need more elaborate technologies in order to be realized. However, the position of Transhumanist Artists as visionaries and mediators of these ideas has to be stressed.

Natasha Vita-More developed the so-called Primo 3M+, a prototype of the future body. It transcends the known images of the body: the classical human body, the cyborg, the robot, the transcendent being. Primo 3M+ is the first conception of the posthuman body, although it may appear as an upgrade of the human body. Primo is a piece of conceptual art and was developed in conjunction with a scientific team including Transhumanists like Max More, Marvin Minsky, Vernor Vinge and Hans Moravec. Many of the intended upgrades are merely conceptualized and will be put into action once technology and the cultural understanding allow this. Due to replaceable genes, this body does not age and can be upgraded. It features enhanced senses, an advanced metabrain, a nano-engineered spinal communication system which runs under the guidance of networked Artificial Intelligence, a 24-hour remote Net relay system and multiple gender options. Primo 3M+ is endued with smart skin that can repair itself and features tone and texture changeability. Nanobots communicate with the brain, e.g. to transmit sensory data. Furthermore, it is able to identify toxins in the environment and protects from sun radiation. The heart can be protected, shut down and replaced by a cardiac turbine.

Natasha Vita-More's concern is that ongoing development does bring about new technologies but that aesthetics plays a major role in our culture. Therefore, the Primo body does not look like a robot or a cyborg but rather smooth and human-oriented. She views the artist as a mediator, communicating ideas. The technological development, especially Artificial Intelligence, super longevity, genetic engineering, nanotechnology, etc. give way to plenty of new ideas. Certainly, these things need to be thought about, scenarios must be created, ideas have to be spun. Art is, and has always been, a great medium to express ideas such as these.

Stelarc is probably the most convincing Transhumanist Artist for he actually integrates technology into his own body. Stelarc developed a third hand, a stomach sculpture and other machine-human connections. Biological restraints are transcended. Just to mention some of his works: The *Third Hand* is attached to the right arm and controlled by EMG signals from the abdominal and leg muscles. The *Stomach Sculpture* was swallowed and 'worked' in his stomach; it glows, it makes sound, it opens and closes, it enlarges and contracts. It's a piece of art located inside the body and thus the function of the skin to divide the inner self from the outside world is annihilated. The body is hollow to a certain extent and serves as an exhibition site. His *Fractal Flesh* performance featured his body, which was equipped with a computer-interfaced muscle-stimulation system. Participants in different cities could ping his body through the Internet. The incoming information flow caused his body to move, independent of Stelarc's will. Stelarc states that the human body is obsolete, given the intense information environment we live in. Since our philosophies are connected to the way we perceive the world – that is through our senses – we may need to find new ways of perception now that we face the information age. Our body is already facing problems such as jetlag and nightshifts. Space colonization, which might indeed be very far away, will ultimately confront the body with zero gravity, etc. The dichotomy of man and woman will be replaced with that of the human being and the machine. These days, people can already create alternate, digital personas through the Internet and there, the body is not merely obsolete, it's already gone.

Kevin Warwick implanted a device into the median nerves of his left arm. Thus, he linked his nervous system directly to a computer. However, he is not an artist but a scientist. His aim is to use this system with disabled people. The blurry connection between art and science is recognized by the fact that his implants are exhibited in the Science Museums of London and Naples.

Transhumanist Artists are as provocative as Dandies were. Both groups claim the right and full control over their bodies. They transcend borders – be they mental or physical. It can be argued that Dandies primarily transcended mental borders, for many things were taboo in nineteenth century life. The Dandy dared to break with the etiquette and say the truth. This often made him *persona non grata*. He dared to enjoy the beauty of ugliness. He dared to think about the beauty and aesthetic of a crime. This is why many Dandies were writers and why their works were often heavily discussed. To be able to transcend mental and physical borders today, one thing is needed: technology. However, Transhumanist Artists are *not* the new Dandies. But: they have the potential to make a similar impact on culture and society. Both are a small elite of darers and vanguards of a coming era. Dandies were the first to fully enjoy and diversify their individuality. A Dandy first and foremost is an I, an egotist, even an egomaniac. The shaping of the self was the ultimate sense of the life of a dandy. He made himself a piece of art. What would he say of the possibility of extending his bodily functions through technology?

List of References

Bostrom, Nick. 2003. *The Transhumanist FAQ. A General Introduction. Version 2.1.* http://transhumanism.org/resources/FAQv21.pdf

Bostrom, Nick. 2005. *A History of Transhumanist Thought.* http://jetpress.org/volume14/bostrom.pdf

Danius, Sara. 2002. *The Senses of Modernism. Technology, Perception, and Aesthetics.* Ithaca/London: Cornell University Press.

Marcuse, Herbert. 2000. *Kunst und Befreiung. Hgg. und mit einem Vorwort versehen von Peter-Erwin Jansen.* Lüneburg: zu Kampen.

Mill, John Stuart. 1991. *Über die Freiheit. Essay.* Leipzig/Weimar: Kiepenheuer.

Pudelek, Jan-Peter. 2000. *Der Begriff der Technikästhetik und ihr Ursprung in der Poetik des 18. Jahrhunderts.* Würzburg: Königshausen & Neumann.

Vita-More, Natasha. 2002. *Transhumanist Arts Statement.* http://www.transhumanist.biz/transhumanistartsmanifesto.htm

Vita-More, Natasha. 2003. *Extropic Art Manifesto of Transhumanist Arts.* http://www.transhumanist.biz/extropic.htm

Wolter-Abele, Andrea. 1996. *How Science and Technology Changed Art.* History Today: 46: 11.

Zweite, Armin/Krystof, Doris/Spieler, Reinhard (Eds.). 2000. *Ich ist etwas Anderes. Kunst am Ende des 20. Jahrhunderts.* Global Art Rheinland 2000. Köln: DuMont.

Art in the Age of Biotechnology

INGEBORG REICHLE

Scientific images in contemporary art

In recent years, many artists have taken as their theme the effects of research and development in modern biotechnology. Contemporary artists give diverse responses when faced with recent scientific and technological advances and the production of compelling images in science, which seem to suggest an almost artistic endeavour. As visual experts, artists translate and reassume societal or scientific issues into a visual language and conduct a visual exploration of and into other representational and signifying practices, such as molecular biology, because many artists share with scientists from these fields a common interest in life itself. Artists today try to decode 'scientific' images through the linking of art and the images of the life sciences, to find a new way of reading them. With the aid of an *iconography of images from science* (Shearer 1996: 64-69), an attempt is being made by artists to decipher the cultural codes that these images additionally transport as well as making them recognizable as a space where other fields of knowledge and areas of culture may also be inscribed. (See Nelkin/Lindee 1995; Haraway 1998)

Fig. 1: Suzanne Anker, *Zoosemiotics* (1993). View of the installation in the exhibition *Devices of Wonder. From the World in a Box to Images on a Screen* at the J. Paul Getty Museum in Los Angeles 2001.

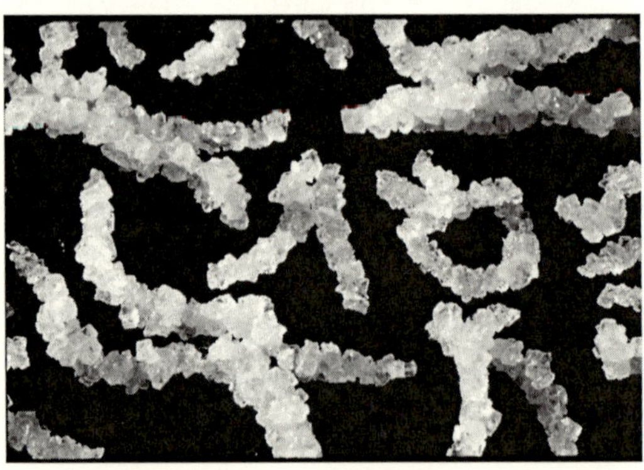

Fig. 2: Suzanne Anker, *Sugar Daddy: The Genetics of Oedipus* (1992).

In her works *Zoosemiotics* (1993) (fig. 1) and *Sugar Daddy: The Genetics of Oedipus*, from 1992 (fig. 2), the New York based artist Suzanne Anker, takes up, for example, the visualization of chromosomes (Anker 1996, 2001; Anker/Nelkin 2004). In her 1993 installation *Zoosemiotics (Primates)*, which could be seen in the 2001 exhibition *Devices of Wonder: From the World in a Box to Images on a Screen*, at the J. Paul Getty Museum in Los Angeles, Anker, again turning to visualizations of the chromosomes of various species, crosses her own visual language with that of genetics. Anker's renderings of chromosomes, enlarged and sculpted in bronze, are careful arranged on one of the walls of the gallery in eight rows, one above the other, and on the other walls in an irregular circular pattern. On a delicate pedestal set out from the wall we see a glass filled with water. Viewed through the curves of the glass, the sculptured chromosome pairs installed on the wall appear entirely distorted. The intention here is not to depict the diversity and forms of chromosomes, but rather to instruct the eye in the simple optical technique of enlargement using a water-filled glass. The production here of an optical distortion serves to demonstrate the artificiality of scientific images and their dependence upon the optical media and conventions of perception associated with their respective time and age. In her 1992 installation *Sugar Daddy: The Genetics of Oedipus*, Anker again plays with human perception. Shimmering blue velvet, draped in heavy folds, serves as the background for pieces of crystal sugar laid out to form pairs of chromosomes. Only on closer inspection does it become clear that this is not a scientific illustration.

By the end of the 1980s, the New York artist Steve Miller had also turned to the new human images in the natural sciences and neurobiology and was no longer producing traditional portraits. Miller instead arranged for DNA to be extracted from bodily samples taken from his subjects; the chromosomes were then scientifically visualized and captured by Miller on canvas, as in the *Genetic Portrait of Isabel Goldsmith* (1993) (fig. 3). In earlier works such as *Self Portrait Black* (1993) (fig. 4), Miller's work dealt with images of the organic interior of the human body:

> In these portraits, the sitters' identity is no longer limited to outward appearance, but viewed through medical images, such as x-ray, MRI, sonogram, EKG, and CAT scans. Rather than being a depiction, these new portraits focus on identification using internal vistas and abstract symbols of medical nomenclature.[1]

The traditional genre of portrait painting, which lives from the tension between the image and its representation, is replaced in the examples given here by the presumably objective image produced by the technically advanced visualization processes used in the life sciences: presented now as a *vera icon*.

1 http://www.geneart.org/miller-steve.htm (20.07.2002).

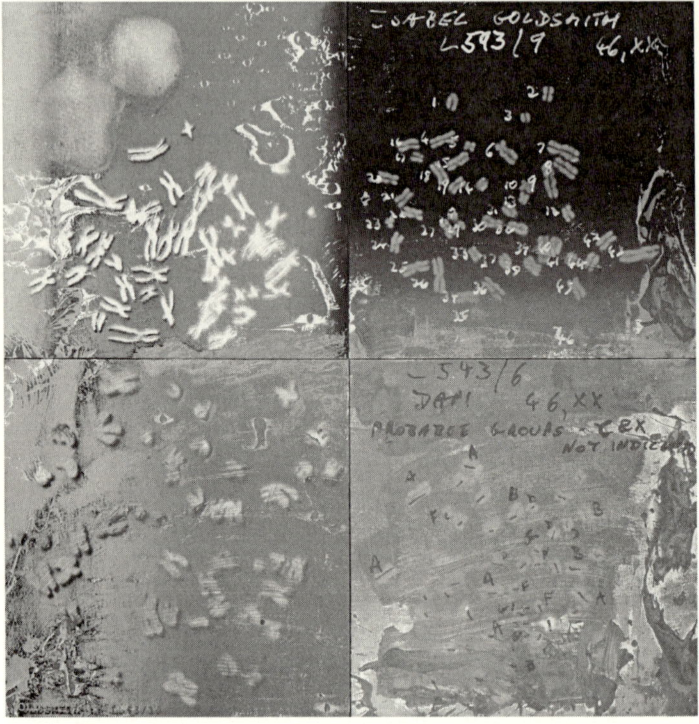

Fig. 3: Steve Miller, *Genetic Portrait of Isabel Goldsmith* (1993).

In their work, artists such as Anker and Miller examine the representational context of the respective experimental processes and the various visual preparations – DNA, for example –, which reveal more about the investigative approach of the experimenter and the circumstances of the matter than about the "matter itself". Even the highly dimensional digital worlds of the sciences remain forever loaded with cultural associations and values. Suzanne Anker therefore views the visual language of the life sciences as enhanced by advanced techniques of image processing – not as "objective" and "neutral", but rather as a socially influential force, shaping the development of the *individuum* within society, biopolitics and genetics.

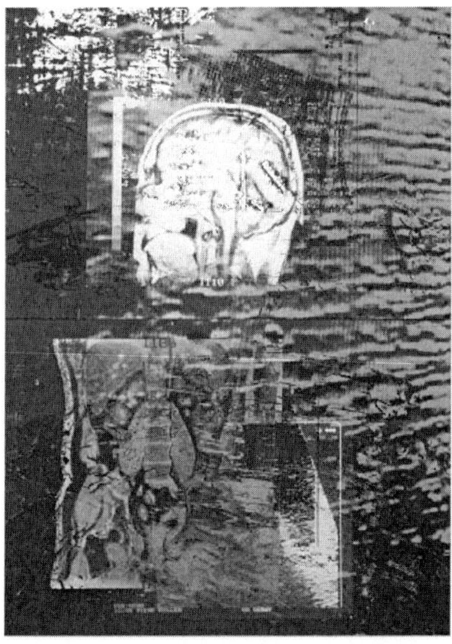

Fig. 4: Steve Miller, *Self Portrait Black* (1993)

Molecular vision has increasingly dominated the assumptions and methods of the biological sciences. Reducing life itself to molecules, it has displaced the visceral references that had once defined the authenticity of the body and the authority of traditional biology as a descriptive science. Despite the complexity of life, this vision implies that we are but a sequence of nucleic acids, a "code script" of information. This transformation of biology from organism to code and/or text parallels developments in art. Artists are adapting images revealed through hightechnology apparatus, and their pictorial and sculptural products have shifted toward the abstract. They have recognized in genetic iconography an underlying narrative that resonates with familiar forms and issues in the history of art. (Anker/Nelin 2004: 19)

In contemporary art today, we also see approaches that reveal the complex relationship between art and science, especially in the use of controversial technologies such as genetic engineering and tissue engineering (see Reichle 2005). In the last two decades we have seen a number of artists leave the traditional artistic playground to work instead in scientific contexts such as the laboratories of molecular biologists and then explore or intervene in the laboratory practices while working with the same materials and technologies as scientists do. The use of biological materials by artists ranges from tissue engineering to stem-cell technologies and even transgenic animals, a phenomenon that raises ethical questions with regard to both scientific and artistic endeavours. These new approaches in art differ dramatically from those approaches which explore art and genetics through the use of traditional media. Artists create new 'life forms', i.e. new organisms which are to a greater or lesser extent artificial entities rather than 'natural' organisms. Many artists today use transgenic organisms in their works, addressing the perpetuation of evolution by humans through the creation of novel organisms according to aesthetic criteria, processes which the advent of recombinant DNA technology has now made possible. Some years ago the Paris-based art theorist Frank Popper introduced the term *Techno-Science-Art* to describe a form of art that is situated between art, science and technology)[2]. This new term, which places 'technoscience' in the dominant position, seems to be a suitable meta-term for describing these emerging new art forms. The term *technoscience* was introduced by both Bruno Latour (Latour 1995: 21, 46) and Donna Haraway[3] to describe the effects of the enormous transformations in the production of knowledge in the life sciences since the beginning of the twentieth century. According to Latour and Haraway, these transformations in science will lead to a redefinition of nature and science. As a consequence, the term *natural sciences* will no longer seem adequate and should be replaced by the term *technoscience* (Weber 2003).

Genetic Engineering in Contemporary Art

Artists turning today to the technical production of transgenic organisms, hybrids and other technofacts, have apparently touched a raw nerve with the modern life sciences, and this is turning the transfer of technofacts from the scientific laboratory to the art world into a precarious ordeal. The Brazilian media artist and theorist, Eduardo Kac, now based at the Art and Technology Department of the Art Institute Chicago, operates at the interface of art and genetic engineering in his projects *GFP K-9* (1998), a bioluminescent dog, *GFP Bunny* (2000) (Kac 2001), a green-glowing rabbit, and the installation *Genesis* (1998-1999)[4].

2 See Popper 1987: pp. 301–302 and Sakane 1998: 227. For the term "Techno-Science-Art" see recently the interview of Joseph Nechvatal with Frank Popper (Nechvatal 2004).

3 See for Haraways definition of the term *Technoscience*: Haraway 1995: 105.

4 See for the art of Eduardo Kac: Kac 2004.

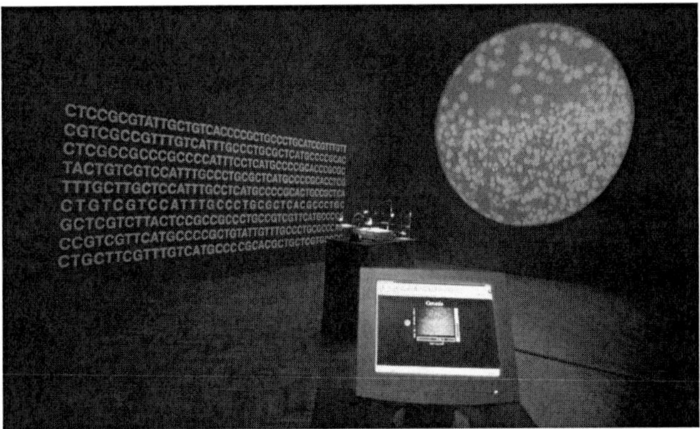

Fig. 5: Eduardo Kac, *Genesis* (1999). View of the installation at the Ars Electronica in Linz in 1999.

With these works, Kac puts up a new art form for debate: the concept of Transgenic Art (Kac 1998). By creating transgenic animals and integrating them domestically and socially, it is Kac's declared intention to draw attention to the cultural effects and implications of a technology that is not accessible visually and bring these to the public's attention for debate. Using biotechnology, Kac transfers synthetic genes to organisms and natural genes from one species to another. Projected is the creation of original, unique organisms. In his installation *Genesis* (fig. 5), Kac attempts to make biological processes and technological procedures visible, which for years now have been standard practice in research laboratories (Kac 1999).

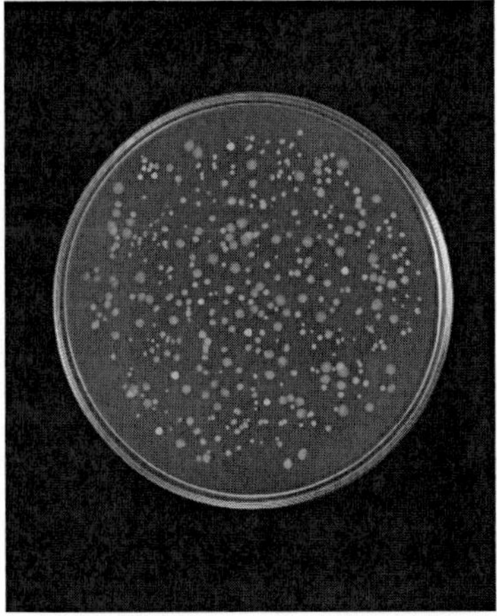

Fig. 6: Eduardo Kac, *Genesis* (1999). View of the petridish.

In a dark room, a brightly illuminated petri dish stands on a pedestal. A video camera, which is positioned above it, projects an oversize image of the dish onto the wall. Ultraviolet light falls onto the petri dish (fig. 6) and the intensity of the light can be controlled by the visitor via a computer. This can be done either in the gallery or via the Internet. In this way the users can influence the processes of replication and interaction of the bacteria in the petri dish and observe these in the magnified projection on the wall or in the Internet – processes, which normally can only be seen under a microscope. Thus, the role of the observer is enhanced to that of active participant who is able to intervene in the processes and influence the course of the work's presentation. The focus of the installation is a synthetic gene created by Kac, a so-called "artist's gene". Kac has transferred this process, which would normally take place only within a laboratory, to an art gallery. With his Transgenic Art, Kac wishes to draw attention to the cultural implications of biotechnology and its possibilities for transforming and manipulating life.

Still, the laboratory methods used to manufacture these "artist's genes" are not in any way new. For already three decades, genetic engineering techniques from the field of molecular biology have made possible the technical reproduction or even new production of life at the molecular level. These organisms, until now non-existent in the natural human world, no longer resemble any natural evolutionary architecture and reinforce the transformation of biology lab organisms into *epistemic objects* (Amann 1994). Molecular biology, as well as other fields in the life sciences, to a large extent construct and design the objects of their research today themselves, thereby producing technological artefacts which owe their existence to the culture of experiment and the expanding technological systems of the laboratory. At the same time these organisms in the laboratory now often have an epistemological status in terms of knowledge models that merely serve as representational models. In this way the technofacts of the 'third nature' have, to a large extent, currently replaced life forms of the first nature as the reference objects of the laboratory (Ritsert 1996: 338). Reports of experimental results as well as the discourse of research organisations are therefore primarily focused on these manufactured, epistemic objects, whose modelling takes place within the immense science complex and the physical infrastructure of the laboratory. Such an implementation of model realities without a reference makes possible a controlled technical manipulation of the processes of life, which then leads to a denaturalisation or artificiality of the object under investigation.

The development of recombinant DNA technology in the 1970s led to a fundamental change in the way molecular structures and processes of living organisms could be made available for scientific experimentation. With the production of transgenic organisms, molecular biology moved beyond the current borders of species and subspecies that are a result of millions of years of evolutionary change, thereby shaking up the existing system of scientific classification.

From an epistemological perspective, this new access to organisms represents a break with previous methods and approaches in molecular biology: Macromolecules themselves became manipulative tools of recombinant DNA technology and thus were transformed into *technological entities*. The nature of these is such that they are no longer distinguishable from the processes in which they intervene, and in the molecular biology lab they begin to resemble industrial production systems becoming, in effect, *molecular machines* (Knorr Cetina 2002: 199). As a consequence of this development, the organism acquires the status of a technological object; the organism or even the molecule itself becomes a laboratory (Rheinberger 1997: 661). Thus molecular biology, as a central domain of biology and the life sciences, finds itself on the way to becoming a science that not only handles, dissects, processes, analyses and modifies its subjects – life forms and the parts thereof – , but rather constructs these henceforth in a fundamentally new sense, as *technofacts*, which can no longer be described as biological objects of a 'natural nature' (Amann 1994: 25). This construction, however, does not correspond to an understanding of the production of matter as a form of 'creation' in the sense of the bringing forth or generation of life, but is rather to be seen as a process of transformation and conversion of matter.

Art in the Age of Biotechnology

The transfer of scientifically produced transgenic organisms from the laboratory into the artist's space in the last few years has led to passionate debates which tend to focus less on the status of such objects as works of art, but rather much more on ethical debates about the limits of manipulation by the natural sciences of the unadulterated natural world and its economisation by business. At the same time there has been reoccurring criticism of the 'artistic' production of living organisms according to aesthetic criteria and without any considerations of use or purpose, as opposed to the work of scientists in fields such as molecular genetics or cell biology. Art was seen as transforming such life forms without legitimate purpose or reason into aesthetic artefacts, wanting thereby to rewrite the story of Creation for its own outrageous purposes. It thus became clear that the public is not yet ready to accept 'glowing dogs' and 'glowing rabbits', which are viewed as eerie and monstrous hybrid life forms not belonging – not permitted to belong – to the creature world, where their presence would lead to disarray within the traditional, ontological orders. With regards to the production of new hybrid forms in art, it seems to be less a debate about the acceptance of new art forms or shifting borders in the art world itself; but more significantly, the negotiation processes of the forces shaping society, forces which can lead to the construction of very specific life forms and worlds, thereby excluding others. Living things that are manipulated and modified in laboratories for specific scientific or economic purposes, will, to a certain extent, be accepted, but not, however, in day-to-day life. This all the more so since in the course of the *mechanisation of the living* it is becoming ever more difficult to determine what is still 'nature' and what is already technology, what can be regarded as real and what is imaginary; the certainties of the daily world have already been severely shaken.

While more traditional epistemological viewpoints, focused on the idea of the organic, continue to persist in the old 'humanistic' connotation of nature, regarding nature as static, abiding, and more or less endowed with inalienable properties, and while postmodern epistemology continues to concentrate on deconstructing the accompanying classical humanistic categories, the biosciences ceased operating with this humanistically understood idea of nature some time ago. New art forms emerging from the lab show the world how precarious the category of 'nature' appears today and how great the fear is that the results obtained in the laboratory with artificially created technofacts and epistemic objects will, in the *Age of Technoscience*, as a rule, be applied to other organisms and eventually humans. These fears, in light of the tremendous speed with which the technosciences are developing, are well justified. Furthermore, on account of the increasing amalgamisation of technology, industry, and science today, one can barely distinguish between the technical, social, economic and political factors that are responsible. The extent of the current ubiquitous 'scientification' and mechanisation leads furthermore to the situation that technology will become increasingly constitutive for social structures and processes – a situation which, according to recent scientific research, will lead to a fundamental transformation of the constitutive social structures (Ropohl 1991: 184).

List of References

Amann, Klaus. 1994. "Menschen, Mäuse und Fliegen. Eine wissenssoziologische Analyse der Transformation von Organismen in epistemische Objekte." *Zeitschrift für Soziologie* 23.1: 22-40.

Anker, Suzanne. 1996. "Cellular Archaeology". *Art Journal.* Contemporary Art and the Genetic Code 55.1: 33.

Anker, Suzanne. 2001. "Gene Culture. Molecular Metaphor in Visual Art." *Leonardo* 33.5: 371-375.

Anker, Suzanne and Dorothy Nelkin. 2004. *The Molecular Gaze. Art in the Genetic Age.* New York: Cold Spring Harbor Laboratory Press.

Kac, Eduardo. 1998. "Transgenic Art". *Leonardo Electronic Almanac* 6.11.

Kac, Eduardo. 1999. "Genesis". In O.K Centrum für Gegenwartskunst, ed. *Gail Wight "Spike" Eduardo Kac "Genesis".* Ars Electronica. Cyberarts 99. Linz, p. 50-59.

Kac, Eduardo. 2001. "Bio Art: Proteins, Transgenics, and Biobots". In Stocker, Gerfried and Christine Schöpf, eds. *Takeover. Who's Doing the Art of Tomorrow. Wer macht die Kunst von morgen.* Vienna/New York: Springer, pp. 118-124.

Kac, Eduardo. 2004. *Telepresence and Bio Art: Networking Humans, Rabbits and Robots.* Ann Arbor: University of Michigan Press.

Knorr Cetina, Karin. 2002. *Wissenskulturen. Ein Vergleich naturwissenschaftlicher Wissensformen.* Frankfurt/M.: Suhrkamp.

Latour, Bruno. 1995. *Wir sind nie modern gewesen. Versuch einer symmetrischen Anthropologie.* Berlin: Akademie Verlag.

Hammer, Carmen and Immanuel Stieß, eds. *Donna Haraway: Die Neuerfindung der Natur. Primaten, Cyborgs und Frauen.* Frankfurt/M./New York: Campus.

Haraway, Donna. 1998. "Deanimation: Maps and Portraits of Life Itself". In Jones, Caroline A. and Peter Galison, eds. *Picturing Science – Producing Art.* London/New York: Routledge, pp. 181-207.

Steve Miller: http://www.geneart.org/miller-steve.htm (20.07.2002).

Joseph Nechvatal. 2004. "Origins of Virtualism: An Interview with Frank Popper Conducted by Joseph Nechvatal." *CAA Art Journal* 62.1: 62-77.

Nelkin, Dorothy and Susan Lindee. 1995. *The DNA Mystique: The Gene as a Cultural Icon.* New York: Freeman.

Popper, Frank. 1987. "Techno-Science-Art: the Next Step." *Leonardo* 20.4: 301-302.

Ingeborg Reichle. 2005. *Kunst aus dem Labor. Zum Verhältnis von Kunst und Wissenschaft im Zeitalter der Technoscience.* Vienna/New York: Springer.

Rheinberger, Hans-Jörg. 1997. "Kurze Geschichte der Molekularbiologie." In Jahn, Ilse et al eds. *Geschichte der Biologie. Theorien, Methoden, Institu-*

tionen, Kurzbiographien. Heidelberg/Berlin: Spektrum Akademischer Verlag, pp. 642-752.

Ritsert, Jürgen. 1996. *Einführung in die Logik der Sozialwissenschaften*. Münster: Westfälisches Dampfboot.

Ropohl, Günter. 1991. *Technologische Aufklärung: Beiträge zur Technikphilosophie*. Frankfurt/M.: Suhrkamp.

Sakane, Itsuo. 1998. "The Historical Background of Science-Art and Its Potential Future Impact." In Sommerer, Christa and Laurent Mignonneau, eds. *Art@Science*. New York: Springer, pp. 227-234.

Shearer, Rhonda Roland. 1996. "Real or Ideal? DNA Iconography in a New Fractal Era." *Art Journal. Contemporary Art and the Genetic Code* 55.1: 64-69.

Weber, Jutta. 2003. *Umkämpfte Bedeutungen. Naturkonzepte im Zeitalter der Technoscience*. Frankfurt/M./New York: Campus.

Cultural Aspects of Interface Design[1]

LAURENT MIGNONNEAU/CHRISTA SOMMERER

We are artists working on the creation of interactive computer installations since 1991 for which we design metaphoric, emotional, natural, intuitive and multi-modal interfaces. The interactive experiences we create are situated between art, design, entertainment and edutainment. When creating our interactive systems we often develop novel interface technologies that match conceptual and metaphoric content with technically novel interface solutions. While our main focus is to design interactive systems for the art context, our interactive or immersive systems also often find use in edutainment and in mobile communications areas. The following article summarizes some of our key concepts for our interface designs and presents some of our interactive technologies in more detail.

Introduction

Human-computer interaction (= HCI) is a discipline concerned with the design, evaluation and implementation of interactive computing systems for human use, including the study of major phenomena surrounding this theme. Human-computer interaction is concerned with the joint performance of tasks by humans and machines; the structure of communication between human and machine; human capabilities to use machines (including how easy it is to learn how to use interfaces); algorithms and programming of the interface itself; engineering concerns that arise in designing and building interfaces; the process of specification, design and implementation of interfaces; and design trade-offs. Human-computer interaction thus has science, engineering and design aspects.

1 This paper is largely based on a longer version in Mignonneau/Sommerer 2005.

Because human-computer interaction studies a human and a machine in communication, it draws from supporting knowledge on both the machine and the human side. On the machine side, techniques in computer graphics, operating systems, programming languages and development environments are relevant. On the human side, communication theory, graphic and industrial design disciplines, linguistics, social sciences, cognitive psychology and human performance are relevant. And, of course, human aspects such as emotions and feelings become relevant as well.

In main stream HCI science and commercially available interfaces, the prevalent human-computer interface of today is still the mouse-keyboard-screen based interface (desktop metaphor). Bill Buxton observes that "despite all of the technological changes, I would argue that there has been no significant progress in the conceptual design of the personal computer since 1982" (Buxton 2001). Buxton backs his observation by looking at the design of the first GUI, the Xerox Star 8010 Workstation, introduced in 1982 (Smith/Irby/Kimball 1983). When we compare this workstation with currently available systems, not much has actually changed in the design of these systems and computers available nowadays (after almost 20 years) still look much the same as back then.

> But Buxton also notes that "despite the increasing reliance on technology in our society, in my view, the key to designing a different future is to focus less on technology and engineering, and far more on the humanities and the design arts. This is not a paradox. Technology certainly is a catalyst and will play an important role in what is to come. However, the deep issues holding back progress are more social and behavioral than technological. The skills of the engineer alone are simply not adequate to anticipate, much less address, the relevant issues facing us today. Hence, fields such as sociology, anthropology, psychology and industrial design must be at least equal partners with engineering and technology in framing how we think about, design and manage our future." (Buxton 2001)

Buxton has long been one of the most prominent critics of commercially available interfaces and their techno-centric design. Buxton proposes a more human centric design (Buxton 1994), which reflects the importance of usage and activity rather than technology. Clearly the social component, as described by Buxton, is one of the key competencies artists can contribute to the process of interface design. For a long time artists and designers have been aware of the emotional and intuitive quality of objects and situations and since the late 1980s and early 1990s media artists, including ourselves, have designed interactive systems that aim to match emotional and metaphoric content with actual physical objects, or images and sounds employed in these systems (Dinkla 1997).

Emotional, Metaphoric, Natural and Intuitive Interfaces

Before describing some of our interactive artworks in detail, let us first look at our conceptual considerations and principles for designing these systems.

Emotional and Metaphoric Interfaces

Artists and designers have always been skilled in applying metaphors when designing systems, objects or works of art. Metaphors can evoke certain sensations or emotions in the spectators, feelings that can often not be described with words alone. The power of metaphors is in the fact that they tap into cultural, historic and emotional knowledge that we humans have built up in the course of our lives. Touching, for example an object that looks like a cat (even if it is in fabric or plastic), will evoke a nice, warm and cozy feeling or emotions of personal attachment and care might be triggered. Through our daily interactions with objects or even beings that we touch, manipulate, look at, perceive or interact with, we have developed a rich intuitive knowledge of how these things work and what kind of emotions and sensations are being attached to them.

To design successful interfaces that make use of this emotional and metaphoric knowledge, it is necessary to ask a few questions before designing a system. These questions are:

- Which emotion or sensation do I want to convey?
- What do I want the user to feel when experiencing this system?
- What is the cultural and historic background of the user?
- What is the emotional and metaphoric knowledge already available to the user?
- What kind of object or interface can convey this desired emotion or sensation?
- If not available, how can I design a new interface that evokes this sensation?
- Which interface would feel most natural and most intuitive to the user?

David Rokeby, a well known media artist, also noted that "user interfaces are also a kind of belief system, carrying and reinforcing our assumptions about the way things are. It is for this reason that we must increase our awareness of the ways that the interface carries these beliefs as hidden content. [...] It is also useful to realize that effective interfaces are usually intuitive precisely because they tap into existing stereotypes for their metaphors. [...] A metaphoric interface borrows clichés from the culture but then reflects them back and reinforces them." (Rokeby 1997) Rokeby's "Very Nervous System" is an example of a natural and intuitive interface, where the user's body actually becomes the interface to the virtual space, as described in literature (Rokeby 1997). As the user moves about in the interaction space, he/she starts to learn how to use his/her body for triggering and playing sounds and music. All this feels very easy and intuitive to the user, almost playful, and users will not even realize that they are being tracked and that they have become part of a computerized system.

Using this immanent, intuitive and emotional knowledge, we became interested in exploring the power of emotional and metaphoric interfaces in 1991.

Interactive Plant Growing – living plants as metaphoric and natural interface

The first emotional and metaphoric interface we designed was a living plant in 1992. In this system, real plants are the interfaces and users can touch these real plants to create artificial plants on a computer screen. One of the first interactive computer art installations to use a natural interface instead of the then-common devices such as joysticks, mouse, trackers or other technical interfaces is our installation *Interactive Plant Growing* (1992) (Sommerer/Mignonneau 1993). In this installation, living plants function as the interface between the human user and the artwork. By touching or merely approaching the real plants, users engage in a dialogue with the real plants in the installation. The electrical potential differences (voltage) between the user's body and the real plant is captured by the plant and interpreted as electrical signals that determine how the corresponding virtual 3D plants (which look similar to the real plants) grow on the projection screen. By modifying the distance between the user's hands and the plant, the user can stop, continue, deform and rotate the virtual plant, as well as develop new plants and new combinations of plants. As the growth algorithms are programmed to allow maximum flexibility by taking every voltage value from the user's interaction into account, the resulting plants on screen are always new and different, creating a complex combined image that depends on the viewer-plant interaction and the voltage values generated by that user-plant interaction. When users place their hands towards the real plant, voltage values are generated that become higher the closer the hand is towards the plant. We use 5 different distance levels to control the rotation of the virtual plants, their color values, their place where they grow on screen as well as on/off growth value. Figure 1 shows a user as he touches real plants to produce a collective image of virtual plants on a projection screen.

Fig. 1 shows a user as he interacts with real plants to grow artificial plants on the screen.

Users who experienced this system have often reported that they suddenly realized that plants are actually living beings and that they were astonished that plants really "feel something." When thus touching a real plant and seeing the effect of this touch translated into a graphical form on a screen, users are suddenly reminded of that immanent intuition they already had about plants.

Natural and Intuitive Interfaces

Closely linked to designing emotional and metaphoric interfaces is the concept of natural and intuitive interfaces. By this we mean interfaces that feel very easy and natural in their use, without the user having to go through a lengthy learning process when he/she wants to interact with this system. Natural interfaces are, for example, gesture-, speech-, touch-, vision-, and smell-based interactions or basically actions and sensations that refer to our daily life experience. Using, for example, living plants as an interface not only provides a new, emotionally charged and unusual connection between computers and living beings, but also poses the questions of what a plant is, how we perceive it and how we interact with it when we touch it or approach it. Natural interfaces also circumvent the annoyance of wearing unpleasant devices before entering virtual space (= unencumbered interaction). The natural interfaces we have used and developed include living plants, light, camera detection systems, a "3-D Video Key" system, a multimodal scanning device and a window touch screen combined with speech input, all described in-depth in literature (Mignonneau/Sommerer 2005).

Phototropy – intuitive interaction with artificial insects using a lamp interface

An installation where users could intuitively interact with artificial life creatures is called *Phototropy* (Sommerer/Mignonneau 1995), which we created in 1995. Here, an in-house light detection system measures the position and intensity of a spot light shone through a flashlight onto a large projection screen. As the user of the system moves the light spot onto different parts of the screen, virtual insects appear and follow the light's beam; the user can "feed" the creatures with light or eventually kill them if he or she provides too much of it. Figure 2 shows an overview of the *Phototropy* system with its four light detectors located on the four corners of the projection screen. The actual position of the flashlight's beam is communicated to the virtual creatures, which in turn change their behavior patterns according to the light intensity of the light spot. This system is very intuitive and natural in its use, as everyone knows how to switch on a flashlight and how to shine light onto the screen. The system needs virtually no explanation and users can become increasingly skilled in creating new creatures when interacting with the system for a longer period of time.

Fig. 2: A visitor interacting with artificial insects by holding and moving a flash light interface to feed or kill virtual insects.

Non-linear, multi-layered and multi-modal interaction

We also believe that interaction in interactive systems should not be linear but instead feel like a journey. The more one engages in interaction, the more one should learn about it and the more one should be able to explore it. We call this non-linear interaction as it is not pre-scripted and predictable, but instead develops as users interact with the system.
The interaction path in our systems should also be multi-layered meaning that the interaction feedback should be simpler at the beginning and become increasingly complex when users interact with the system further in order to continuously discover new levels of interaction experiences.

And finally, a last cornerstone in designing our systems is the design of multi-modal interaction experiences that combine several senses, such as vision, sound, touch and smell. We have developed multi-modal interfaces since around 1994 and some of these systems will be described in the following sections as well.

Over the past years, multi-modal interaction has, in fact, become a mainstream research trend in HCI. It defines multi-modality as the combination of multiple input modalities to provide the user with a richer set of interactions compared to traditional uni-modal interfaces. The combination of input modalities can be divided into six basic types: complementarity, redundancy, equivalence, specialization, concurrency and transfer (Martin 1998). Examples of such multimodal interaction system include Waibel and Vo's series of input modes that include speech, pen-based gestures, eye tracking, lip reading, handwriting recognition and face recognition for applications such as text editing and calendar management (Waibel/Duchnowski/Manke 1995; Sommerer/Mignonneau 1997).

A-Volve – non-linear, evolutionary interaction with artificial creatures in a pool of water

Created in 1994, *A-Volve* is an interactive computer installation where visitors can interact with artificial creatures that live, mate and evolve in a water-filled glass pool. An in-house touch screen GUI interface is used for designing creatures. When a user draws a two-dimensional side view and a section view of any possible shape onto the touch screen's GUI, the user generates data that are used to calculate a three-dimensional shape. The x, y and z parameters for side view, section view and speed of the drawing process, are then used to calculate a 3-D form that becomes "alive" in the water of the pool. The x, y and z parameters become part of the creature's genetic code. The genetic code determines the behavior and fitness of a three-dimensional creature. As soon as it has been created, the creature starts to swim in the real water of the pool. Details about the creature's behavior and the complex interactions that arise among creatures can be found in literature (Sommerer/Mignonneau 1997).

Creatures in *A-Volve* not only interact with each other, but also react to the visitors' hands in the water. An in-house camera detection system is used to measure the users' hand positions: these data are then communicated to the artificial creatures, which instantaneously react to the users' gestures; creatures can, for example, stop to move once they are caught by the user's hand, or become afraid when being touched too often.

Figure 3 shows users as they interact with the creatures in the pool. The system is multi-modal as it combines touch-based interaction of the touch screen GUI, a gesture-based interface for hand tracking and a sound-based interaction (the creatures make also sound depending on their body shape). By using real water in the pool and by linking the design of the virtual fish-like creatures to the pool metaphor, *A-Volve* is clearly also a very metaphoric interface as users are often reminded of the emergence of life or to life-like situations in Darwinian systems ("the survival of the fittest"). The interaction development in *A-Volve* is also non-linear and multi-layered as it depends on the users skill and the system's internal evolutionary algorithms to create increasingly complex interaction patterns.

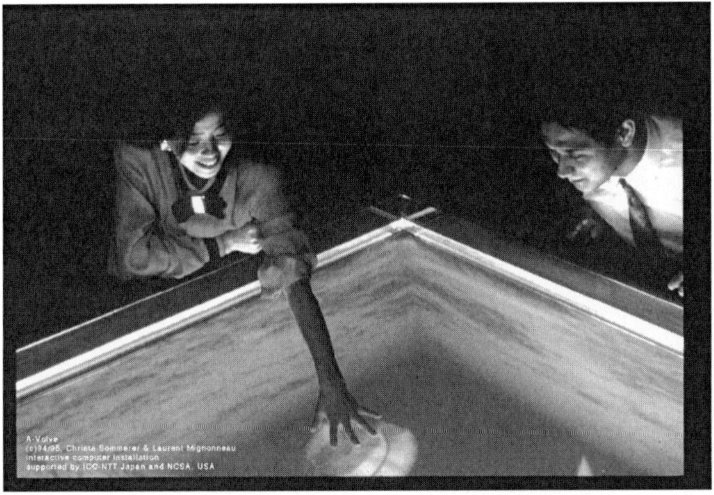

Fig. 3: Users as they interact with the *A-Volve's* artificial life creatures in the pool.

Riding the Net – Intuitive and Multi-modal Interaction with Internet Data

Until now, interaction with data from the Internet has been mostly restricted to the use of the common mouse-keyboard interface (=desktop metaphor). On the other hand, future applications for entertainment, edutainment and interactive art that involve the Internet call for more intuitive and more playful interaction experiences. To create richer, more stimulating and more intuitive information spaces involving the Internet, we created several interactive systems that propose novel and entertaining ways to browse the Internet through multi-modal and immersive interactions.

In 2000, we created an interactive web-based image retrieval system called *Riding the Net.* This system was first presented at the Siggraph 2001 Emerging Technologies exhibition (Sommerer/Mignonneau/Lopez-Guliver 2001). In *Riding the Net,* users can use speech communication to retrieve images from the Internet, watch these images as they stream by on an interactive window touch screen and can also touch these images with their hands. Two users can interact in this system simultaneously and, while communicating with each other, their conversation is supported and visualized in real-time through images streamed from the Internet. The system functions as a brain storming tool and allows intuitive and playful interaction with the enormous wealth of image and sound data from the Internet.

Fig. 4: Users as they interact through speech and touch with the *Riding the Net* system.

Figure 4 shows two users as they use the *Riding the Net* system by speaking into headset microphones to generate keywords that are used to download images from the Internet and by using their hands on the interactive screen to hold and interact with image icons derived from the Internet. A detailed description of this system and its multi-modal interface as well as the image search algorithm is provided in literature (Lopez-Gulliver/Sommerer/Mignonneau 2002).

Summary and Outlook

For the above described interactive systems we have developed various types of interfaces technologies that do not use off-the-shelf standard interface technology, but instead aim to explore novel and perhaps slightly unusual concepts of interaction. By matching the metaphoric and emotional content of the system with the technological solution we designed interfaces that aim to keep the interaction for the audience as intuitive, natural, multi-layered and multi-modal as possible (Mignonneau/Sommerer 2005), perhaps fulfilling Buxton's vision of a more human centric design (Buxton 1994) in HCI.

We see interactive installations as an important medium to bridge the gap between purely artistic and purely technological applications and we aim to create interactive systems that can fulfil artistic, educational and entertainment purposes. Over the years, we have worked on the convergence of art and science (Sommerer/Mignonneau 1998) and art and technology and by designing interactive systems that convey scientific principles to an art audience and vice versa artistic content to a scientific audience, we hope to further contribute to the renaissance of art-science collaborations.

List of References

Buxton, W. 1994. "Human skills in interface design." In: L.W. MacDonald and J. Vince (Eds.). *Interacting with virtual environments*. New York: Wiley, pp. 1-12.

Buxton, W. 2001. "Less is More (More or Less): Some Thoughts on the Design of Computers and the Future," In: P. Denning (Ed.), *The Invisible Future: The seamless integration of technology in everyday life*. New York: McGraw Hill.

Dinkla, S. 1997. *Pioniere Interaktiver Kunst*. Karlsruhe: Cantz Verlag.

Martin, J. C. 1998. "TYCOON: Theoretical Framework and Software Tools for Multimodal Interfaces." In: *Intelligence and Multimodality in Multimedia interfaces*. (ed.) John Lee, AAAI Press.

Mignonneau, L. and C. Sommerer. 2005. "Designing Emotional, Metaphoric, Natural and Intuitive Interfaces for Interactive Art, Edutainment and Mobile Communications" In: *COMPUTERS & GRAPHICS: An International Journal of Systems & Applications in Computer Graphic*. Elsevier, 2005, pp. 837-851.

Rokeby, D. 1997. "Constructing Experience," In: *Digital Illusions*, C. Dodsworth (Ed.). New York: Addison-Wesely.

Smith, D.C., Irby, C., Kimball, R., Verplank, W. and Harslem, E. 1983. "Designing the Star User Interface." In: P. Degano & E. Sandewall (Eds), *Integrated Interactive Computing Systems*. Amsterdam: North-Holland, pp. 297-313. Originally appeared in Byte, 7(4), April 1982, pp. 242-282.

Sommerer, C. and Mignonneau, L. 1993. "Interactive Plant Growing," In: *Visual Proceedings of the Siggraph '93 Conference.* ACM Siggraph, pp. 164-165.

Sommerer, C. and Mignonneau, L. 1995. "Phototropy," In *Oltre il villaggio globale – Beyond the Global Village*, M.G. Mattei, (Ed.). Milan: Electra Edition, pp. 134-135.

Sommerer, C. and Mignonneau, L. 1997. "Interacting with Artificial Life: A-Volve," In: *Complexity Journal.* New York: Wiley, Vol. 2, No. 6, pp. 13-21.

Sommerer, C. and Mignonneau, L. (eds.) 1998. *Art @ Science.* Wien/New York: Springer Verlag.

Sommerer, C., Mignonneau, L. and Lopez-Gulliver, R. 2001. "Riding the Net: a Novel, Intuitive and Entertaining Tool to Browse the Internet," In: *Siggraph 2001 Conference Proceedings*, ACM Siggraph, p. 133.

Lopez-Gulliver, R., Sommerer, C. and Mignonneau, L. 2002. "Interfacing the Web: Multi-modal and Immersive Interaction with the Internet" In: *VSMM 2002 – 7th International Conference on Virtual Systems and MultiMedia*, Conference Proceedings (Korea), pp. 753-764.

Waibel, A., Vo, M. T., Duchnowski, P. and Manke. S. 1995. "Multimodal Interfaces." In: *Artificial Intelligence Review*, Special Volume on Integration of Natural Language and Vision Processing, Mc Kevitt, P. (Ed.), pp. 299-319.

Imaging Reality. On the Performativity of Imagination

MARC ZIEGLER

This article focuses on some aspects in the relation of imagination, technique, society and subjectivity. Over the course of the last few decades, new approaches were initiated dealing with a reappraisal of the role of imagination in its relation to technology.

Current forms of subjectivity arise in the cross-over points of society and technique. These forms are endowed with an aesthetical as well as socio-technical character: Hermeneutical and creative practices merge with technical mediated modes of imaging the world. Before this background I suggest to regard imagination as an enfolding performative force in the indeterminate centre between technique and society. Furthermore, the productivity of imagination is owed to a radical negativity that withdraws from any technological grasps. This leads to the desideratum to project a post-hermeneutical concept of a techno-scientific overdetermined imagination.

Reflexivity, indeterminacy and the sense of possibilities

Since Marx' comment that natural limits recede in proportion as industry advances (Marx 1976), industrial production, science and technology can be outlined as one reflexive medial event or rather as a *medium*: In a (historical) process, reaching from feudal tool-technology over modern industry to post- or radical-modern cybernetic networks, an increasing technical complexion is taking place. Technology proves more and more to be a constitutive moment in the construction not only of *social* life, but also of *organic* life insofar as it is possible to subsume every existing system under the concept of technique that, like the infusoria or an alga, is able to identify, store and manipulate relevant information for its survival in order to derive specific forms of behaviour from regularities (Lyotard 2001: 23). It is Lyotard who reminds us not to forget that we humans are an invention of technique rather than technique being an invention of humans (ibid.).

Epistemologically regarded as a medium, technique should no longer be described in terms of traditional action theory. Technology has lost (or even never had) the sole form as an object, that opposes human subjectivity and, to keep the analysis in a terminology of means and ends, conducts the analysis on an overly simplified level.

Furthermore, the analysis of technique as a medium elaborates a significant shift by focusing on the enabling-character of the medium insofar as the system of means marks the room of possibilities in the use of means and ends. Tracing a sense of possibilities ("Möglichkeitssinn" – Robert Musil) becomes more important than an analysis of a however given "reality". Traces of (intentional) action can only be abductively developed in a "real virtuality" of technical systems, which constitute worlds of action ("Handlungswelten") on their own (Hubig 2005: 61f.).

Our every-day life is in nearly every single strand the result of specific technical arrangements. Technology, therefore, has lost its main-character as a conglomeration of tools and artefacts. Nevertheless, the thesis of technology as a (fatal) determining totality[1] also misses the point: "[T]he essence of technology is by no means any technological" (Heidegger 1977: 4). To regard technique as a medium emphasises the paradox that a boundless technique is universal and (*because* of its universality) indeterminate, or, that the *Internität* of the medium, the being-amid of our hermeneutical relations in a technical sphere, corresponds with an *Entzogenheit* of a representable centre: the medium always remains in the back of the different processes of action (Gamm 2004: 158ff.).

Technical acting shares in so far the same attitude as language: Both are irrevocable inscribed in our self-relations and figure as the opening as well as the closing force of an ever specific horizon of current hermeneutical exegesis. Attempts to understand language or technique from a mere instrumental point of view constantly undergo this *performative* moment in the structure of language or technique (Hetzel 2005: 280).

1 As an example see Ellul 1970.

Technoscience: New fields of indifference

Another approach to interpret the amalgamation of society, politics, economics and science is linked with the concept of *technoscience*. The term *technoscience* has its origin in the theoretical works of Bruno Latour (Latour and Woolgar 1979, Latour 1987, Latour 1993) and Donna Haraway (Haraway 1991, 1997). At first glance, the main theoretic intention of using this term is to draw attention to a current transformation within the culture of science (Nordmann 2005): It refers to changes in the cultural self-conception of science and the work of the scientists in regard to their researching practice. Traditionally, the scientific process of methodical accumulation of experiment-based and falsifiable knowledge is separated from the process of a technological application. Therefore, scientific knowledge about the laws of nature can ideal-typically be regarded as the precondition of a technical intervention which takes place later on in reality. Theoretical knowledge and its practical use are clearly parted from each other. With regard to the researching practices of technoscience, the significant shift consists in the merging of these two spheres: The researching practices in, for example, nanotechnology or bionics, make it no longer possible to keep up this distinction. Much more: The annihilation of this distinction is the starting point and programme of these new forms of research and technology (Nordmann 2005: 215).

This apparent indifference to the self-conception of the technology-based scientific research involves a radical different view of the classical object of modern science: nature. Nature no longer plays the role of a resistant objective reference that is endowed with universal, natural, human/subject-indifferent and timeless laws. Furthermore, nature is becoming a mere epistemological factor, something that is seen to be technical at its core.[2] In this turn against an ontological perspective, the techno-scientific grasp on nature also emphasises the rediscovering of a natural productivity that disappeared from sight in the modern physical and Cartesian view on nature as a mere mechanical structured *res extensa*. The current techno-scientific view on nature regards nature as eminently lively or at least as eminently productive.[3] Determinations

2 For example see following quotation from the NSTC: "Even an early instance of nanotechnology like catalysis really is young compared to *nature's own nanotechnology*, which emerged billions of years ago when molecules began organizing into the complex structures that could support life". (National Science and Technology Council 1999: 5). Emphasis marked by me, M.Z. The quotation was found in (Nordmann 2005).

3 The prominence and growing importance of the term "Life-Science" may help to make this thesis plausible. More to the core of the argument: Ge-

like the Greek distinction between what is naturally produced or set by nature itself without any other influence (*physei*) and what is produced or set by man (*thesei*), or the scholastic distinction between nature as a creative-productive force (*natura naturans*) and nature as a product (*natura naturata*) can now be reread (or: misread) under the assumption of a post-metaphysical, non-ontological, non-teleological and radical secularised concept of nature, in which its basic productive structure is signed with technology. The overwhelming productivity of nature returns in a historical and scientific situation where nature itself is merging with its traditional opposite: technology. But the technological productivity that is ascribed to nature does not announce the creation of an omnipotent god or the idealistic sense of a cosmological closure of an (anonymous) entelechy that is developing itself towards its inherent final causes. Furthermore, regarding the productive nature under the sign of technology, nature becomes disqualified in a universal sense. Again indifference is taking place. The scientific view of nature as a technology or as an engineering artwork not only negates the difference between nature and technology, it also levels the distinction between organic and inorganic nature. Nature, on the whole, becomes fungible. Both organic and inorganic nature turns out to be (re-)designed in terms of system-, communication-, and information theory which allows the transfer of organic as well as inorganic nature into quantifiable open and alterable functions, parameters and components that again allows the potential and projected boundless recombination and operationalisation of its elements. Reciprocally, life obtains the signature of technology and, at the same time, technology obtains the character of life (for the last predications see Weber 2003: 138f.).[4]

---

netic engineering analyses the genes and their environments under the leading concept of the genetic material being a "code" and therefore the prominent bio-molecular structure in order to make "life" itself readable, to decode and to manipulate it. See (Kay 2000) and (Keller 1995). – Nanoscience also analyses the productivity of the – unenlivened – nature by focussing on the self-organisation of mineral molecular structures.

4 In German terms one talks about the reciprocal process of the *Technisierung des Lebendigen* and the *Verlebendigung der Technik*. *Technisierung* here means the process of the technical approach or interpretation of the living, whereas "living" ("das Lebendige") must be understood grammatically in the sense of a substantive. *Verlebendigung* has the meaning of "giving life to" or "becoming alive", both in the non-religious sense of a reciprocal or reflexive ascription within the relation of technology and life: Technology is becoming alive insofar as technology is able to monitor, to decode and to change the (normative unqualified and informational quantified) structure of life. Within these operations, technology and life

The turn towards inter- or transdisciplinary organized fields of research – of which life-science and nanoscience are only examples – build up on the paradigms of three powerful theories which can be seen as theories of naturalisation and symbolization at the same time: The theory of evolution, cybernetics, and the theory of systems. These theories all deal with *systemic processes* in different ways. But these structures differ fundamentally from the traditional demand for a *systematic description* of natural processes. They bargain different forms of self-reference and recursion. They are aligned to functionality. There is an irreducible instrumentalism in the core of these theories. They all share a pragmatic perspective: Their 'truth' or accuracy lies mainly in its practical applicability. Therefore, their field of application is not ontologically predetermined. So I think we can see an initial turning point in the relation of modern science and technology in the upcoming and spreading of theories of this type: The technological application is not something that stands only in an indirect relation to scientific knowledge. Furthermore the technological application is an internal moment of these theories inherent to their logical design.

The fungible-made nature is not only the field of interest of a (pure or neutral) techno-scientific research with the theoretical aim to accumulate scientific knowledge or the practical aim to manipulate the structures of life with regard to a fundamental research (free from the aim to appliance). The pragmatic orientation of current research challenges the traditional boundaries between a clearly calculated area of high specified scientific experimental work and the routines of every-day life. For example: The willingness today, to widen or to go beyond the boundaries of the laboratory is confronting us with a very different relationship of theory and praxis than under traditional laboratory conditions. To bring it to an extreme point: Today we must think of the whole world turning into a laboratory in which long-period experiments are taking place in the environment (genetically modified seeds and plants), in human and non-human bodies (practices of cloning, stem-cell research), or in political and economic strategies (economic game-theory). Most of our relations are impregnated with the feeling that we are living under experimental conditions following the logic of trial-and-error. This reaches from a more or less private experimental life-style to the "project-oriented" forms of work organization (for the latter see Boltanski and Chiapello 2003).

proceed structural isomorphic. See also (Karafyllis 2003) concerning the discussion about the merging of (laboratory made) life and technology.

Both the social aspects within the theoretical development of scientific knowledge as well as the recursive practical scientific grasps on nature, life and life-world have been described in the works of social scientists throughout the last decades as manifoldly overdetermined with industrial, economic, military, pharmaceutical, political and social interests.[5] While these studies of science focus on the social impact of scientific research by outlining a typical modern experience of a *reflexivity of knowledge*, younger publications by Latour took a significant turn within this constructivist view of the relationship between society and science. Society is now regarded as a plural hybrid consisting of human and non-human entities as well as of social and technical structures. Intending a critical *political ecology*, Latour levels traditional perspectives on the gap between humans, animals and artefacts towards a *symmetric anthropology* (Latour 1993) consisting of a *body corporate*, a *collective* or a *dispositif* of human and non-human actants[6] (Latour 1999).

Making it concrete: The calculating imagination

It seems to be the strongest intention of technoscience to penetrate the core-structures of the biological, physical, chemical and cognitive structures of our world. Human stem-cells as well as chemical molecules or the physical attitudes of atoms and electrons are considered to function as raw material for technological recombination. A paradox occurs: The fundamental changes caused by the technosciences are happening on a level far aside human perception in so far as the direct technological manipulation is taking place on a very tiny cellular and molecular level. But at the same time the technological interventions will have great and often uncontrollable consequences for human perception and human and non-human environments. *Technoscience therefore is a challenge for a new orientation within our hermeneutical understandings of ourselves*

5 The works of the Social Studies of Science (SSS) and the Science and Technology Studies (STS) have focused mainly on the (interdisciplinary) analysis in detail of these social, practical, historical, political and economic aspects as overdeterminating factors in the social construction of modern science, knowledge and technological artefacts. See the programmatic article from Pinch and Bijker (Pinch and Bijker 1987). See also Latour and Woolgar 1979, Latour 1987, Collins and Pinch 1993 and 1998. See Winner 1993 as a critical view on social constructivism of science and technology.

6 Latour (Latour 1987) and Callon (Callon 1991) and other authors use the expression *actants* in allusion and contrast to the *agent* or *actor* in classical action theory, who is endowed with a reflexive and intentional form of acting (self-) consciousness.

and the world. I would like to point out how far the imagination is involved in this challenge.

If we follow a fable from Vilém Flusser (for the following see Flusser 1999), we then are situated in an at least twofold universal process of disintegration: Both world and human consciousness are deconstructed into dot-like elements and in consequence we have to calculate and to compute these single dot-elements with the aid of technical images to form or to inform these senseless particles in order to achieve a new experience of the world as well as enable its recognition and handling (Flusser 1999: 14, 28). This is in short what Flusser calls the process of *einbilden*. Flusser calls our current form of consciousness the conscious-level of imagination. The act of imagination is a radical dialogic and inter-subjective process of turning the abstract and zero-dimensional universe of disintegrated dot-elements into a concrete surrounding of expedient and meaningful images made by technical devices and with textures that are rather mosaics than areas (Flusser 1999: 86ff.). In this fable, human imaginativeness no longer plays the Kantian role of a subject-intrinsic power of judgement. Imagination here also does not refer to the subjective act of representational thinking, magic, animistic and mythic behaviour or having fantasies (Flusser 1999: 16ff.). Flusser determinates imagination starting from the technical image (Flusser 1999: 40). Imagination has immigrated into the cybernetic and communicative structure of technical devices. They play a double-role: They build the stage as well as the medium for the enfolding scenario of the imagination. Men are linked to the machines by pressing buttons or keys (Flusser 1999: 28). They play an important role within the workings of the imagination: Flusser calls the key-pressing women and men *imaginators*, *Einbildner*. They are those who form or inform (*einbilden*) the abstract dots in order to create a concrete and meaningful image (Flusser 1999: 39ff.). The informing interaction between humans, keys and technical devices can be seen as a hermeneutical act, or stronger: as the most prominent hermeneutical act of giving sense to a world that has fallen into myriad pieces of zero-dimensional, cybernetic generated and with entropy laden information. Our current understanding of ourselves and the world is involved in an open cosmology of a universe of technical images. Flusser calls this, using a Heidegger term, our current "being-in-the-world" (Flusser 1999: 27). By analysing this being-in-the-world as a form of consciousness, he confronts us with a post-metaphysical, post-historic and post-humanistic view on human hermeneutic processes (Flusser 1999: 9f.). Flusser situates this analysis of contemporary consciousness within a genealogy of a cultural history in which every state is signed by a certain structural form of mediality as the each ultimate horizon of the hermeneutical access to world: The handling of artefacts[7] in a three-dimensional room, the two-dimensional surface of a picture, the one-dimensional linearity of a written text and now the zero-dimensional information of cybernetic systems (Flusser 1999: 10ff.).

Flusser writes the history of culture with a focus on the image. The reality of technical images thereby is the result of immense calculations of improbabilities. This equals a devaluation of an ontological fixed reality in favour of the probability calculus. While the technological structure of the technical image becomes a black box (for the user, the *Einbildner*), the technical image itself becomes a phenomenon of pure surface which functions as a projection rather than a mirror (Flusser 1999: 41ff., 57ff.).

Flussers analytical fable mixes moments of anticipation and utopian thinking. In Flussers projected "telematic society" the dialogic structured socio-technological practice of imaging seems to be an endless cognitive celebration (Flusser 1999: 164ff.). It can be seen as a current version of a utopian society of a *homo ludens* (Flusser 1999: 95ff.). The active force of imagination enables to re-concrete the deconstructed world. Imagination here is a medium of generating sense in form of communicating with images. The whole concept is based on a thinking of imagination as the ability to create images: to form and to order dispersed digital information and to concept surfaces as projection screens whereon information can appear.

Imagination beyond images

To go one step further: Imagination, understood as the ability to create images, refers to a mere negative moment. Even imagination in the sense of a synthetic function or a technical mediation holds contact to a pre-discursive, disrupted and sense-less sphere. Every concrete imagination owes its concrete appearance to an antecedence of an image-less and frame-less negativity. The negative force of imagination can therefore be seen as the power that turns the abstract into the concrete.

Imagination is essentially unreal. But, without taking into account the productive and performative force of the imagination, it would be impossible to develop a meaningful relation towards us and the world. *World* is never given to us in a direct way. Furthermore, our relation to *world* is always over-determined by the imagination that transverses a

7 *Artefact* is to be taken in the sense of *Gegenstand*, with emphasis on the fact, that a *Gegenstand* is standing in an opposition to the subject, enabling both: distance-taking and reflexivity.

naked, direct view on the things in the world. We always see things through a specific anamorphotic refraction. Or even more pointed: the specific angle of refraction is exactly this that we experience to be *our world* respectively the *things in the world.* This negative structure of imagination that can be described as the non-substantial *hypokeimenon* of subjectivity is a meandering theme in modern discurses from Herder, Hegel and Nietzsche to Bachelard, Lyotard and Žižek (see Hetzel 2006 and Mersch 2006). Lyotard calls this negativity a traceless and beyond remembrance performing perpetual process of a fracturing presence that remains out of appearance; a non-present presence, that has always ever fallen out of the orders of time and space. A continuous breaking up that systematically withdraws from any technological grasp (Lyotard 2001: 70).

With regard to this performing negativity in the work of imagination, we now can argue that Flussers narration of imagination can enfold its inspiring productivity only because it negates or masks this negativity to which it keeps intractably linked.

The outstanding task to formulate a contemporary post-hermeneutical concept of a techno-scientific overdetermined imagination exactly has to outline systematically the combination of the synthetic and negative aspects of imagination as suggested above.

List of references

Boltanski, Luc and Chiapello, Ève. 2003. *Der neue Geist des Kapitalismus.* Konstanz: UKV Verlagsgesellschaft mbH.

Callon, Michel. 1991. "Techno-economic networks and irreversibility". In John Law, ed. *A Sociology of Monsters: Essays on Power, Technology and Domination*, pp. 132-161. New York, London: Routledge.

Collins, Harry M. and Pinch, Trevor J. 1993. *The Golem. What Everybody should Know about Science.* Cambridge: Cambridge University Press.

Collins, Harry M. and Pinch, Trevor J. 1998. *The Golem at Large. What You should Know about Technology.* Cambridge: Cambridge University Press.

Ellul, Jacques. 1970. *The Technological Society.* New York: Vintage Books.

Flusser, Vilém. 1999[6]. *Ins Universum der technischen Bilder.* Göttingen: European Photography.

Gamm, Gerhard. 2004. *Der unbestimmte Mensch. Zur medialen Konstruktion von Subjektivität.* Berlin, Wien: Philo & Philo Fine Arts GmbH.

Haraway, Donna J. 1991. *Simians, Cyborgs, and Women: The Reinvention of Nature.* New York, London: Routledge.

Haraway, Donna J. 1997. *Modest_Witness@Second_Millenium? Female-Man©_Meets_Oncomouse™. Feminism and Technoscience.* New York, London: Routledge.

Heidegger, Martin. 1977. *The Question Concerning Technology, and other Essays.* Translated and with an introduction by William Lovitt. New York, London: Harper and Row.

Hetzel, Andreas. 2005. "Technik als Vermittlung und Dispositiv. Über die vielfältige Wirksamkeit der Maschinen." In Gerhard Gamm and Andreas Hetzel, eds. *Unbestimmtheitssignaturen der Technik. Eine neue Deutung der technisierten Welt.* Bielefeld: transcript Verlag, pp. 275-296.

Hetzel, Andreas. 2006. "Metapher und Einbildungskraft. Zur Darstellbarkeit des Neuen". *Dialektik. Zeitschrift für Kulturphilosophie* 1: 77-92.

Hubig, Christoph, 2005. "'Wirkliche Virtualität'. Medialitätsveränderung der Technik und der Verlust der Spuren." In Gerhard Gamm and Andreas Hetzel, eds. *Unbestimmtheitssignaturen der Technik. Eine neue Deutung der technisierten Welt.* Bielefeld: transcript Verlag, pp. 39-62.

Karafyllis, Nicole C. (ed.). 2003. *Biofakte. Versuch über den Menschen zwischen Artefakt und Lebewesen.* Paderborn: Mentis.

Kay, Lily E. 2000. *Who Wrote the Book of Life? A History of the Genetic Code.* Stanford: Stanford University Press.

Keller, Evelyn Fox. 1995. *Refiguring Life. Metaphors of Twentieth-Century Biology.* New York: Columbia University Press.

Latour, Bruno and Woolgar, Steve. 1979. *Laboratory Life: The Social Construction of Scientific Facts.* Beverly Hills: Sage Publications.

Latour, Bruno. 1987. *Science in Action: How to Follow Scientists and Engineers through Society.* Cambridge/Mass.: Harvard University Press.

Latour, Bruno. 1993. *We have never been modern.* New York, London: Harvester Wheatsheaf.

Latour, Bruno. 1999. *Pandora's Hope. Essays on the Reality of Science Studies.* Cambridge/Mass. / London: Harvard University Press.

Lyotard, Jean-François. 2001[2]. *Das Inhumane. Plaudereien über die Zeit.* Wien: Passagen-Verlag.

Marx, Karl. 1979. *Capital. A Critique of Political Economy. Vol I.* London, Harmondsworth: Penguin, "New Left Review".

Mersch, Dieter. 2006. "Imagination, Figuralität und Kreativität. Zur Frage der Bedingungen kultureller Produktivität". *Sic et Non. Zeitschrift für philosophie und kultur. im netz.* Online source [visited in December 2006]: www.sicetnon.org/modules.php?op=modload&name=PagEd&file=index @topic_id=29&page_id=476

National Science and Technology Council, Committee on Technology. The Interagency Working Group on Nanoscience, Engineering and Technology. 1999. *Nanotechnology. Shaping the World Atom by Atom.* Washington,

D.C. Online source [visited in December 2006]: www.wtec.org/loyola/nano/IWGN.Public.Brochure/IWGN.Nanotechnology.Brochure.pdf

Nordmann, Alfred. 2005. "Was ist Technowissenschaft? – Zum Wandel der Wissenschaftskultur am Beispiel von Nanoforschung und Bionik". In Torsten Rossmann and Cameron Tropea, eds. *Bionik. Aktuelle Forschungsergebnisse in Natur-, Ingenieur- und Geisteswissenschaft*. Vienna, New York: Springer, pp. 209-218.

Pinch, Trevor J. and Bijker, Wiebe E. 1987. "The Social Construction of Facts and Artifacts. Or How the Sociology of Science and the Sociology of Technology Might Benefit of Each Other". In Wiebe E. Bijker, Thomas P. Hughes, Trevor J. Pinch, ed. *The Social Construction of Technological Systems*. Cambridge/Mass.: MIT Press, pp. 17-51.

Weber, Jutta. 2003. *Umkämpfte Bedeutungen. Naturkonzepte im Zeitalter der Technoscience*. Frankfurt/M., New York: Campus.

Winner, Langdon. 1993. "Upon Opening the Black Box and Finding it Empty: Social Constructivism and the philosophy of technology". *Science, Technology, & Human Values* 1, pp. 362-378.

Causality, Custom and the Marvellous. On Natural Science at the Beginning of Modern Times

ANDREAS KAMINSKI

Epistemic core concepts like natural law, induction, prediction or empirical falsifiability are unthinkable without causality. Modern natural science seems to be so intimately connected with the principle of causality that criticism of one could be seen as criticism of the other: 'Blind' chains of cause and effect that are without any sense; 'cold' rationality that disenchants nature as a 'constraint' causal connection.

It seems that modern natural science hardly gets away from this orientation to causality. This is not astonishing since Francis Bacon showed that the skills, abilities and the power of natural sciences depend on their causal knowledge. They are both sides of the same coin, as he pointed out: "[F]or where the cause is not known the effect cannot be produced. Nature to be commanded must be obeyed; and that which in contemplation is as the cause is in operation as the rule." (Bacon 1962 [1620]: 47). The force of the causal furor can be reconstructed by tracing the shift of meaning of what exactly philosophy is. Hobbes (re)defined it as a searching project for causalities: "Philosophy is such knowledge of effects or appearances, as we acquire by true ratiocination from the knowledge we have first of their causes or generation: And again, of such causes or generations as may be from knowing first their effects." (Hobbes 1839 [1655]: 3).

The maximum causal hope was reached in Pierre Simon de Laplace's *Théorie Analytique des Probabilités.* Laplace, a French mathematician and astronomer, gathered the assumptions which had been discussed over the last two centuries. And he drew a radical conclusion which led him to a fiction known as Laplace Demon:

> We may regard the present state of the universe as the effect of its past and the cause of its future. An intellect which at a certain moment would know all forces that set nature in motion, and all positions of all items of which nature is composed, if this intellect were also vast enough to submit these data to analysis, it would embrace in a single formula the movements of the greatest bodies of the universe and those of the tiniest atom; for such an intellect nothing would be uncertain and the future just like the past would be present before its eyes. (Laplace 1995 [1814]: 2)

Not least this radical causal scenario, mostly called strong determinism, affected the causal concept. And it dictated the (political) argumentation and contention about the causation. Complete computation of all events, this vanishing point, suggests advancing so called qualitative moments of experience against quantitative methods. It suggests regretting the 'cold' rationality of natural science that can only describe 'blind' causal relations and to criticize it as a loss of experience. Complete computation of all events is a perspective which misdirects at least partially the discussion about causality. Why? My thesis is as follows: At the beginning of modern times, causality gains evidence also as an aesthetic phenomenon. Causality persuades on an experimental stage. The German philosopher Hans Blumenberg suggested understanding rigid causal concepts as an absolute metaphor (Blumenberg 1998: 92-110).[1] This points out what often falls into oblivion: strong causal ideas pin hope on the validity of the causal principle. They are not a verified or verifiable certainty. The following offers a sketch at the beginning of modern, natural and experimental 'science'[2] in which frameworks causality appears as an aesthetic phenomenon.

1 His concept of absolute metaphor means that they involve universal assumptions which are in principle not revisable but which structure the possbilities of thinking and acting.

2 The inverted commas indicate the danger that a naive parlance of 'the beginning of modern science' involves. This leads to the hazard projecting the own present into an alien past present. Historical research arrives at the conclusion that in the 17th and 18th century there wasn't natural and experimental science in a narrower sense. Experimental research arose in contiguity with spectacular shows, alchemy, the Theatri Machinarum, religious experience and gallant science ("galanter Wissenschaft") as courtly and bourgeois fashion. Compare for the problem and the term "galante Wissenschaft" Hochadel 2003: 11-29. Also Kreißl 1999: 9-41, Nelle 2003.

Causality, Custom

It is astonishing which phenomenon at the beginning of modern times that conflicts with the claim living in a new era draws enormous attention: habit. This era (in German called 'Neuzeit') asserts to break with the traditional, the ancient beliefs, practices and forms of life. Nevertheless, the custom plays a major role in modern philosophy: habit, common things, familiar practices, usual perspectives, accepted opinions. In the works of Montaigne, Pascal, Bacon, Descartes, Hooke, Boyle and Hume habit plays a major role. The contradiction between the claim to live in a new era and the attention paid to customs can be shaken up by the following explanation: habit could not become – quasi contrapuntally – an issue not until it was shaken to the core by a strong shift in contingency. This explanation gains plausibility, as one could say, by the concept of habit itself as it was developed at that time. However, there is more to recover.

Initially, Montaigne highlighted the motive which henceforth[3] conducts speech: the power of custom. But this power is not one that presents or enters lordly and loud the stage of attention. It is really quite sovereign:

> Indeed, custom is a domineering and sneaky schoolmistress. She slips covertly and unperceived into the foot of her authority, but having by this gentle and humble beginning, with the benefit of time, fixed and established it, she then unmasks a furious and tyrannical countenance against which we have no more the courage or the power so much as to lift up our eyes. We see her, at every turn, forcing and violating the rules of nature: 'Custom is the best master of all things.' (Montaigne 1998 [1580]: 60)

The metaphors of this passage express a tension between presentation and invisibility. This metaphorical tension repeats the aforesaid tension between contingency and implicitness that coins the modern discussion of custom. Custom belongs to the group of epistemic phenomena that is not only of little manifest, but is distorted if brought up. Therefore, Montaigne strongly hassled his master although he could broach the issue of the ruling of custom only in mode of blank conjecture. However, there are more motifs to find in Montaigne's concept of custom: stupefaction of senses, occlusion of what is special in things, bias of belief (Montaigne 1998 [1580]:60-62).[4]

---

3 Surely, the phrase is an ancient one.

4 The last point refers to natural research as Montaigne's citation of Cicero indicates: "'Is it not a shame for a natural philosopher, that is for an ob-

Montaigne stressed, as aforementioned, the enormous power of custom that also dominates rationality. However, Blaise Pascal overcalls him. He places custom step by step deeper into experience. In some kind of climax: Firstly, profane authority is reduced to trivial custom (Pascal 2005 [1662]: 25/308). Secondly, time, space and motion are identified as principles that are caused by habit: "What are our natural principles but principles of custom? A different custom will cause different natural principles." (Pascal 2005 [1662]: 125/92). Pascal carried this thought to its logical conclusion: He called the difference between nature and custom into question: "I am much afraid that nature is itself only a first custom, as custom is a second nature." (Pascal 2005 [1662]: 126/93)

So far it has merely been outlined how custom was conceptualised at the beginning of modern times. The extraordinary importance of habit has become apparent. It is custom too that leads to the connection of causality and aesthetics which is in demand. But how? This can be clarified on the basis of the following issues:

(1) *Causality is imperceptible*: In the 17th and 18th century, causality was an ambivalent, conflicting concept. On the one hand, causality is taken for granted as ubiquitous. At least in the realm of bodies, causality was considered to be without gap. But then causality is imperceptible, hence it is a major difficulty. Both ubiquitous and imperceptible causality exist partly unrelated in the same work as in, for instance Descartes. On the one hand, Descartes believed that the whole material world is ruled by the principle of causality.[5] Otherwise it can be shown that Descartes pushes the problem of imperceptible causality forward. In his *Meditationes* the problem appears in context of what is known as the wax argument. The wax, as Descartes pointed out, is only grasped by intellect. From here a no less classical scene follows. Descartes places himself at a window where he observes human beings passing by in the street. However, a question rises with full impact: By means of which criteria is it possible to make a distinction between persons and machines or brutes? As far as we can see they are human beings, so it seems that people are passing in front of the window. But maybe they are only artefacts. Beneath their clothes they could be machines. Only by means of vision is it impossible, as Descartes points out, to decide this question. Solely a look behind the scenes could provide some clarity (Descartes 1984 [1641]: 21). But as Descartes himself showed in his

server and hunter of nature, to seek testimony of the truth from minds prepossessed by custom?'" (Montaigne 1998 [1580]:60)

5 Compare Descartes *Principia Philosophia* (1985 [1644]) where he points out his methodological program: the world as a big mechanism or machine.

Discours de la Méthode, the matter is much more complicated. In this (earlier) work he raises the question: Are there criteria for a distinction between animals and people? But the premise is that animals are only mechanical machines, and insofar they are universally causally determined. People are causally determined too by their bodily parts. But they are more than mere bodies and therefore more than animals, only humans have minds (Descartes 1985 [1637]: 139-141). Hence, Descartes reasons that if people und animals would look entirely equal, the difference between them could only intellectually be discovered: Namely as a result of their prudence to act adequately in specific and complex situations. This implies that the question whether events are causally determined or not cannot be decided by perception. Causality is imperceptible. Only now is Descartes' judgment plausible: "And so something which I thought I was seeing with my eyes is in fact grasped solely by the faculty of judgement which is in my mind." (Descartes 1984 [1641]: 21)

More explicitly, David Hume articulates the imperceptibility of causality. Also Hume, as Descartes and others, affirms the universal importance of the principle.[6] Otherwise he highlights that causality cannot be watched, nor can one see a cause or an effect:

> Let an object be presented to a man of ever so strong natural reason and abilities; if that object be entirely new to him, he will not be able, by the most accurate examination of its sensible qualities, to discover any of its causes or effects. [...] No object ever discovers, by the qualities which appear to the senses, either the causes which produced it, or the effects which will arise from it; nor can our reason, unassisted by experience, ever draw any inference concerning real existence and matter of fact. (Hume 1975 [1748]: 27)

Even if no one can see a cause or an effect, as Hume points out, it is also impossible to discover them by pure thought. Only experience can discover causality. Therefore Hume means to say that experience is not identical with pure sight (or hearing or feeling). The latter guides to the next point.

(2) *Causality derives from custom*: Custom already appears on Descartes' flaneur scene as the Latin text shows: "Unde concluderem statim ceram ergo visione oculi, non solius mentis inspectione cognosci, nisi iam forte respexissem ex fenestra homines in platea transeuntes, quos etiam ipsos non minus usitate quam ceram dico me videre." (Descartes 1996 [1641]: 56) One is used, as Descartes points out, to say that he sees

6 Compare Hume 1964 [1739]: book I, part III, 23: "Why a cause is always necessary".

people. Again Hume works out what at best Descartes' argumentation suggested.[7] Hume's classical criticism of causality starts with the former argument that no one can see a cause or an effect. However, causality is experience based. But this is not a contradiction as Hume explains. So what kind of experience is meant? No cause or effect arises from a single experience. The iteration adds nothing to this experience – without the iteration itself as Hume underlines. Now a constant relation between two events emerges. Hence this iteration adds something new on perception. But this new perception caused by iteration is nothing other than custom (Hume 1964 [1739]). Custom leads to expect a defined event that follows habitually appreciating another. This custom is (mis-)interpreted in terms of causes and effects. A problematic interpretation, as Hume insists.

The derivation of causality from custom must not forget that there is a difference between causality and custom in Hume's work. They are not identical; from there Hume's persistence makes sense that it is custom not causality. But what are the differences in spite of derivation? Firstly, causality and custom appear in different time modes. An accustomed chain of events attracts no attention about the past. One event is expected immediately awakening another without any request for reasoning. Furthermore, it is, as Hume says, possible that causality is in demand in cases which are exceptional, then causes or effects are supposed, assumed and conceived hypothetically. But this implies there is causality without custom (Hume 1964 [1739]: 394-399). This search is led by means of the terms cause and effect. However, these terms don't matter in accustomed cases. Here Hume picks up Montaigne's description: "Such is the influence of custom, that, where it is strongest, it not only covers our natural ignorance, but even conceals itself, and seems not to take place, merely because it is found in the highest degree." (Hume 1975 [1748]: 28)

(3) *Custom obscures causality*: A connection between causality and custom became apparent. However, the link between causality and aesthetic is still outstanding. For this purpose we have to consider the then perception theory which apprehended custom as an ignoring phenomenon, in a way as a stupefying of human senses. On this note Montaigne writes that "custom stupefies our senses. We need not go to what is reported of the people about the cataracts of the Nile [...]. Millers, blacksmiths and armourers could never be able to live with their own noise, did it strike their ears with the same violence that it does ours." (Mon-

7 Therewith a direct or indirect influence of Descartes on Hume shall not be claimed.

taigne 1998 [1580]: 60) Bacon transforms the stupefying character into a cognition problem of how discovering causal connections in familiar things:

Let men bear in mind that hitherto they have been accustomed to do no more than refer and adapt the causes of things which rarely happen to such as happen frequently; while of those which happen frequently they never ask the cause, but take them as they are for granted. And therefore they do not investigate the causes of weight, of the rotation of heavenly bodies, of heat, cold, light, hardness, softness, rarity, density, liquidity, solidity, animation, inanimation, similarity, dissimilarity, organisation, and the like; but admitting these as self-evident and obvious, they dispute and decide on other things of less frequent and familiar occurrence. (Bacon 1962 [1620]: 106)

Furthermore, numerous falsities are believed only because people are accustomed to them, as Bacon repeats ceaselessly. Thus, custom turns out to be a cognition obstacle par excellence. Descartes writes accordingly: "[N]ot all things around us can be perceived, but in contrast those that are there most of the time can be sensed the least, and those that are always there can never be sensed." (Descartes 1989 [1630]: 29)

Intensity of Experience

Precisely at this point emerges the connection of experiment, method and aesthetical perception. Two problems are to be solved: Causality that is imperceptible has to be made observable. And custom that overlays causality and deforms perception has to be eliminated. At first seen as trivial, the experiment should solve these problems. But the question is how? As in a theater scene, the experiment should highlight causality in a prepared play. The experiment should stage and perform selected things.[8] Bacon called them "experiments of light", if they "simply serve to discover causes" (Bacon 1990 [1620]: 95).[9] They enable one to take a specific look.

8 Hereunto there is a broad discussion in the philosophy of science. For instance see works of Nancy Cartwright or the volumes of the Theatrum Scientiarum project, edited by Helmar Schramm, Ludger Schwarte, Jan Lazardzig.

9 Certainly, Bacon declines the theater metaphor in his work. He prefers court metaphors. But the court can be understood as a theater beside the theater as for instance Pascals does.

Preconditions for this examination are methodical weaning techniques. From this point of view Robert Hooke contrasts an experienced person with one that is in a foreign and strange country:

For I find it very common for Tradesmen, or such as have been much versed about any thing, to give the worst kind of Description of it for this purpose; and one that is altogether ignorant and a Stranger to it, if he be curios and inquisitive, to make the most perfect and full Description of it. And the like may be observed also in such as travel into other Countries, that they will give a better Description of the Place than such as are Natives of it; for those usually take notice of all things which because of their Newness seem strange, whereas a Native passes over those because accustom'd to them.

So Hooke advises the following weaning technique:

[A]n Observer should endeavour to look upon such Experiments and Observations that are more common, and to which he has been more accustom'd, as if they were the greatest Rarity, and to imagine himself a Person of some other Country or Calling, that he had never heard of, or seen any thing of the like before: And to this end, to consider over those Phœnomena and Effects, which being accustom'd to, he would be very apt to run over and flight, to see whether a more serious considering of them will not discover a Significancy in those things which because usual were neglected: For I am very apt to believe, that if this Course were taken we should have much greater Discoveries of Nature made than have been hitherto. (Hooke 1969 [1705]: 62)

Robert Boyle advises a similar technique: He proposes that the experimenter should imagine being a blind person who sight is restored after surgery. This would have the effect of observing everything as being an amazing spectacle (Boyle 1965 [1772]: 7).[10] But these techniques should not only disassemble custom; additionally perception gains a sensual intensity. It becomes a game, a spectacle, an adventure.

The experimental arrangement is already an extraordinary space; it displays events as vividly as possible. Both problems, that is the imperceptibility of causality and determination by custom, should be cut off. Therefore, the experimental arrangement should lead and focus the attention. For instance, it alienates the ordinariness (as Hooke demonstrates in his descriptions of observations made through a microscope. He researched familiar things like cork, a razor blade and so on and found them shockingly strange when perceived through the microscope.) Moreover, the experimental arrangement should produce (visual) con-

10 Floriran Nelle pointed these parallel techniques out. Compare Nelle 2003.

ciseness of cause and effect; as if a cause and an effect could be seen and were not only a succession of two events. The spectacular experiments with vacuum and electricity in the 17th and 18th century are exemplarily.[11] The epistemic usefulness of the animal vacuum experiments is speedily utilised. Nevertheless, countless dying animals are presented to a frightened but fascinated public.[12] In this case the experimental arrangement 'visualize' an invisible cause (the vacuum) and an invisible effect (the death) so concisely that causality seems to be reified.

The Marvellous

In the vacuum experiment, radical strangeness goes along with visual conciseness. But custom is a phenomenon that recreates quasi natural. At this point an aesthetic economy of custom, perception and spectacle emerge. Aesthetical experience conflicts with custom, as Montaigne points out when referring to an ancient author: "'There is nothing at first so grand, so admirable, which by degrees people do not regard with less admiration.'" (Montaigne 1998 [1580]: 64) At the end of the 18th century, the German writer Carl Friedrich Pockels formulates a theory of the marvellous. According to Pockels we assume a complete chain of causes and effects. But sometimes a gap or blank appears, then the marvellous fills in this gap and rebuilds the chain of cause and effect. However, the condition of the marvellous is that the filling leaves an absence; a deficiency of causal explanation remains. But this deficiency is attractive and charming by "occupying spunkily" (Pockels 1983 [1785]: 111).[13] In this case, causality emerges as a searching form the marvellous answers only partially. Attraction is attained as long as the search

11 The 18th century did not know a utility of electricity. Electricity was, as Hochadel points out (2003), a completely different matter. The list above can naturally be continued, also chronologically. For example, in the 19th century the incipient physiology performed viewing teaching experiments in their "Spectatorien". An artificial blood circuit entered the stage: A frog's heart used as 'motor' pumped up blood; the blood reaching the top, flowed down into a funnel that led it back to the frog's heart. This artificial arrangement was enlarged by a projector and projected onto a wall. The frog heart measured two metres on the wall. Compare Schmidgen 2003: 280. Or for instance: In the 20th century the brain sciences claim to visualize cerebral events as causes for mental events as effects. For a assessment of these claims compare Gehring 2004.

12 The interplay between the scene and the public is depictured impressively in Joseph Wright of Derby's painting *An Experiment on a Bird in an Air Pump* (1768).

13 Translation by the author.

does not come to completion. But custom is an antipole of the marvellous:

> A natural event does not affect us strongly because it is habitually defined in all parts, because it contains nothing particular that stimulates our curiosity, and because we often have seen and heard such. The marvellous is totally different. Here we notice a lot of new matters. A totally new scene appears for us to see, hundred of new and pleasurable images of our fancy surround us. (Pockels 1983 [1785]: 112)

Therewith an aesthetical economy of the marvellous and custom is composed. For this economy it is a general rule that "we can become accustomed to sublime and sensual objects bit by bit if we see them often enough that either have no effect on us or they only astonish us." (Pockels 1983 [1785]: 114) Causality becomes, as Boyle called it above, "the greatest Rarity".

Admirable closed Machines

The fascination with machines at the beginning of modern times, as it is documented in the machine theater ("Maschinentheater") and in the machine books, confirms my argumentation.[14] Why? Daniela Bailer Jones highlights the advantage of mechanical models:

> A mechanism is something that can be followed step by step, at least in our mind. For instance, one may understand how the wheels of a bicycle move by observing how the cyclist's feet exert a force first on one pedal and then on the other, resulting in rotating a rigid rod, i.e. the axle, that is attached to a cogwheel in its middle. The pedals drive the cogwheel over which a chain runs. This chain is connected to another cogwheel that is the axle to the back wheel of the bicycle. One movement gives rise to the next, and doing one thing causes another and thus sets a whole chain of events going. (Bailer-Jones 2005: 9)

In other words, a mechanical model visualizes a closed chain of causes and effects. The model makes observable what is normally imperceptible. Mechanical machines are also causally closed. But their closed causality is normally hidden: for the look behind a blind and for the element as a result of their complexity. Therefore, mechanical machines are am-

14 The machine concept was totally different as Hans Blumenberg (1998: 92-110) and Carl Schmitt (1985 [1938]: 62) explain. Machine was close to art, theatre and spectacle. See also Böhmes's Paper in this volume.

bivalent; and this is the reason why they are marvellous for the 17th and 18th century. They cause unknown effects challenging the assumption of closed causality; therefore they stimulate the speculation of how it is possible to produce these events. Hence they perform a blank in the chain of cause and effect, but they suggest that there is a visualizable relation, a reconstructable connection. Jan Lazardig (2006) in particular showed the importance of admirability for the machine concept. He showed that machines were involved in an ambivalent game of exposure and covering in the 17th century. In the machine books, an observer often appears in the picture. This observer refers with an indicating gesture of his forefinger to the admirable details of the machine. Lazardig also mentioned the special function of elevations in the pictures which opens a stretched game of exposing and covering. Closed causality spectacles showed and hidden – before they became common in modern times.

List of references

Bacon, Francis. 1962 [1620]. *The New Organon.* In Spedding, Ellis and Heath, eds. *The Works of Francis Bacon.* Vol. IV. Stuttgart-Bad Cannstatt: Fromann-Holzboog.

Blumenberg, Hans. 1998. *Paradigmen zu einer Metaphorologie.* Frankfurt am Main: Suhrkamp.

Bailer-Jones, Daniela. 2005. "Mechanisms Past and Present." *Philosophia naturalis* 42.1: 1-14.

Boyle, Robert. 1966 [1772]. *The Works I.* Hildesheim: Olms.

Descartes, René. 1989 [1630]. *Le Monde de René Descartes ou Traité de la Lumiere.* Weinheim: VCH.

Descartes, René. 1984 [1641]. *Meditations on First Philosophy.* In John Cottingham, Robert Stoothoff, Dugald Murdoch, eds. *The Philosophical Writings of Descartes.* Vol. II. Cambridge: University Press, pp. 1-62.

Descartes, René. 1985 [1637]. *Discourse on the Method.* In John Cottingham, Robert Stoothoff, Dugald Murdoch, eds. *The Philosophical Writings of Descartes.* Vol. I. Cambridge: University Press, pp. 109-176.

Descartes, René. 1985 [1644]. *Principles of Philosophy.* In John Cottingham, Robert Stoothoff, Dugald Murdoch, eds. *The Philosophical Writings of Descartes.* Vol. I. Cambridge: University Press, pp. 177-292.

Descartes, René. 1996 [1641]. *Meditationes De Prima Philosophia.* In *Philosophische Schriften in einem Band.* Hamburg: Meiner.

Gehring, Petra. 2004. "Es blinkt, es denkt. Die bildgebenden und die weltbildgebenden Verfahren der Neurowissenschaft." *Philosophische Rundschau* 51: 273-293.

Hobbes, Thomas. 1839 [1655]. *Elements of Philosophy. The first Section, Concerning Body.* London: John Bohn.

Hochadel, Oliver. 2003. *Öffentliche Wissenschaft. Elektrizität in der deutschen Aufklärung.* Göttingen: Wallstein.

Hooke, Robert. 1969 [1705]. *The Posthumous Works.* New York, London: Johnson.

Hume, David. 1964 [1739]. *A Treatise of Human Nature.* In Thomas Green and Thomas Grose, eds. *The philosophical Works* Vol. 1. Aalen: Scientia.

Hume, David. 1975 [1748]. *Enquiries concerning Human Understanding and concerning the Priciples of Morals.* Oxford: Clarendon Press.

Kreißl, Friedrich R. and Otto Krätz (eds). 1999. *Feuer und Flamme. Schall und Rauch. Schauexperimente und Chemiehistorisches.* Weinheim: Willey-VCH.

Laplace, Pierre Simon. 1995 [1814]. *Philosophical essay on probabilities.* New York: Springer.

Lazardig, Jan. 2006. "Die Maschine als Spektakel – Funktion und Admiration im Maschinendenken des 17. Jahrhunderts". In Helmar Schramm, Ludger Schwarte, Jan Lazardig, eds. *Instrumente in Wissenschaft und Kunst – Zur Architektonik kultureller Grenzen im 17. Jahrhundert.* Berlin, New York: dahlem universitiy press, pp. 167-193.

Montaigne, Michel de. 1998 [1580]. *Essais.* Frankfurt/M.: Eichborn.

Nelle, Florian. 2003. "Im Rausche der Dinge. Poetik des Experiments im 17. Jahrhundert". In Helmar Schramm et. al., eds. *Bühnen des Wissens. Interferenzen zwischen Wissenschaft und Kunst.* Berlin: dahlem universitiy press, pp. 140-167.

Pascal, Blaise. 2005 [1662]. *Gedanken.* Stuttgart: Reclam.

Pockels, Carl Friedrich. 1983 [1785]. "Ueber die Neigung des Menschen zum Wunderbaren." In Walter Kelly and Christoph Perels, eds. *Die deutsche Literatur. Das 18. Jahrhundert. Texte und Zeugnisse.* München: Beck, pp. 109-117.

Schmidgen, Henning. 2003. "Lebensräder, Spektatorien, Zuckungstelegraphen. Zur Archäologie des physiologischen Blicks". In Helmar Schramm et. al., eds. *Bühnen des Wissens. Interferenzen zwischen Wissenschaft und Kunst.* Berlin: dahlem universitiy press, pp. 268-299.

Schmitt, Carl. 1995 [1938]. *Der Leviathan in der Staatslehre des Thomas Hobbes: Sinn und Fehlschlag eines politischen Symbols.* Stuttgart: Klett-Cotta.

Growth of Biofacts: The Real Thing or Metaphor?

NICOLE C. KARAFYLLIS

Introduction

Technology is encroaching on nature and most of all: on life.[1] While the manipulation of living entities is already known from the ancient practice of breeding (Zirkle 1935) and is regarded as unspectacular, new technologies such as cell cultures, organ transplants, reproductive medicine and computer simulation of biological processes call into question our traditional distinctions between nature and technology. Traditionally growth is distinguishing something as living.

Aristotle said in *De anima* that whatever grows is natural and hence 'life' can be identified with nature, starting with the activity of plant soul (lat. *anima vegetativa*). Somehow in contrast he stated in his *Physica*, whatever is moved externally does *not* grow, but is considered *technē*. This antithesis corresponds to our common sense intuitions: trees, children and hair grow, whereas machines and other tools do not. In addition, self-moving automatons do not convince us that they are natural, even if they have held the illusion of having body-like features for many centuries and inspire technological visions of 'man-machines' and humanoid robots (Orland 2005). However, do these distinctions still hold today in the light of recent advances in biological and biomedical technologies, producing not artefacts, but *biofacts* (Karafyllis 2003, 2006b)?

I will argue that they do indeed still hold today, but that in recent decades the distinctions have become much more hidden than before, first of all through the design of living objects in laboratories. A laboratory is a detached sphere of changing natural entities to living proto-

1 This article contains some of the main theses I developed in my philosophical habilitation project (Karafyllis 2006a).

types, which are released as products into the public sphere afterwards. W. Krohn and J. Weyer thus call modern societies "laboratories" (Krohn/Weyer 1990). Due to this epistemic 'veil of ignorance' in society, a third status between naturalness and artificialness, which I call *biofacticity*, is made plausible.

The invisibility of what makes a difference between nature and technology is at least threefold. The first type of invisibility refers to language, because growth is renamed with technical *metaphors*. The second type can be found in the epistemic *models* of growth in Converging Technologies, as e.g. networks and model organisms, which already focus on design of life. The third type of invisibility is hidden behind a lack of *anthropological* discussion, concerning what human life actually is and, moreover, should be in a subject perspective of hybridity (Latour 2002). With the neologism 'biofact', a hermeneutic concept is developed which allows to ask for the differences between 'nature' and 'technology' in the area of the living. 'Life' thus is examined in an intermediary perspective between subject and object and is outlined by reflecting on the term 'growth'.

Metaphors of Growth and Growth Categories

Growth is the necessary presupposition for both: starting biotechnical design activities and bringing them to appearance. It functions as antecedens and mode of continuity of a self, exceeding the status of a mere material conditional. What makes 'growth' an interesting term from a philosopher's point of view is that the categories of cause and reason seem to melt into one. This means, it affects the relation of 'nature' and 'culture'. In Aristotle's works, growth is a special kind of an overall movement of the *physis* and thus reserved for the topos of nature. It includes coming into being (gr. *genesis*) and fading away (gr. *phthisis*) – and thus is strongly related to the living world. Aristotle interprets growth both quantitative (increase/decrease) and qualitative (coming into being/fading away), which can be understood as an objective versus a subjective view on 'life' in a modern reading. On a meta-level, both perspectives are connected by the category of change (gr. *alloiosis*) and by the ontological concept of substance (gr. *ousia*), which always holds its own potentials from the very beginning (gr. *arché*) of being.

Since early modern times, the concept of substance was reduced to (lat.) *materia*, which corresponds with a form open for designers to some extent. Biotechnological progress could and can make use of specific quantitative and qualitative changes of organismic growth patterns.

But this design still is embedded in organism-specific growth types and limited by the birth and death of living entities. Furthermore, growth in modern biology is dealt within evolutionary thinking.

When we look at the biological understanding of growth, we find different terms, e.g. increase (of cell volume and cell number), morphogenesis, differentiation, complexity (Karafyllis 2002). They are inspired by theories of various disciplines like physics and engineering, and biological sub-disciplines like genetics, physiology and embryology. Until recently, the humanities have employed concepts of 'nature' and 'life' that ignore the problematic notion of growth in biology, and the biological sciences continue to operate with a concept of growth which is neither unified nor comprehensive. Today it is common in Life Sciences to differentiate between 'growth' and 'development', resembling a distinction between quantitative and qualitative characteristics of life, which in reality never can be found.

Due to the recent success story of genetics, meaning a research program of analyzing and designing mainly qualitative characteristics of organisms while they develop, 'growth' is understood in a reduced manner and is by now a *dead metaphor* in Life Sciences. This means that the use of related terms like 'increase' or 'proliferation', by which growth is explained (as *explanans*), implies a common sense notion of what growth actually is; but that at the same time in epistemic contexts this notion is reduced to model and provoke processes, which always already show growth as presupposition of a continuity of being (as *explanandum*). Due to the fact that we are somehow very sure what growth is, it is not called into question for connecting 'the natural' and 'the living' in modern times. Behind this circular epistemic argumentation we find the ontological problem of how to decide, if something is actually growing and, hence, if something is actually living. Synthetic biology nourishes the doubts, if the connection of growth and life is still a necessary one. Maybe, one can assume, it could be replaced by aggregation of particles, which end up in resembling an organism. However, 'living' means a state of being which means more than just to be alive. 'To live' implies a biography, leading to a referable starting point and its spatial and temporal determination. Already Aristotle had an ontological problem with the permanent growing, but non-perceiving plants. In *De anima* he assumed plants to be 'living things' (gr. *zonta*), and not (gr.) *zoa*, i.e. real living entities like animals and humans. Nevertheless, all living things had an ontological starting point of being, the (gr.) *arché*, and moreover were united by the idea of the soul (gr. *psyché*).

According to Max Black (1968), theories of metaphors can be interpreted respectively in the sense of *substitution*, *comparison* or *interac-*

tion. For our purpose, the interactive approach is the most useful for questioning how far 'growth' refers to a 'real' living entity or is used as metaphor for any kind of continuous process with inherent possibilities of change. The thing, which is named and the word to name it, interact with a certain fluidity, depending on the explanatory context. Depending on *what* is to be explained, the metaphors function as clarifying filters. Technical and economical metaphors in biosciences thus make the idea of growth and life comprehensible for the cultural sphere, in which persons regard themselves as having control over the means and ends. They serve not only for explanation, but also for justification of envisioned ends (Pörksen 2002).

Experimental science can both name and stimulate biological growth so that only the abstract starting point of genesis remains as 'nature'. Whatever grows can equally well be understood as artificial, depending on the feature taken as characteristic of growth, e.g. increase or reproduction, and depending on technical terms, e.g. 'functions' and 'tools', which are used to describe and model cells, tissues and organs. Growing entities necessarily lack *concreteness*, which makes technological approaches to design them challenging. To postulate that something is growing can be understood in different ways. One is to use 'growing' as *alienans*, i.e. as an adjective that appears to be qualifying a subsequent description, but functions to leave open the question of whether the description applies, like e.g. 'a near victory'. The potential of a specific entity is emphasized as well as the possibility of a certain outcome, e.g. when we say that lizards can regenerate their tails by the means of growing cells. Observing that lizards have the potential of totipotent growth in some of their tissues, one can assume that they *will* actualize it, once their tail is cut. Another is to use 'growing' as substitute for 'living', referring to the continuity of process that implies the overall estimation of what living entities, and not only lizards, in general do. In the second context, background theories of evolutionary thinking become important, determining the overall growth type in which species and organisms are tokens. The instance of this type-token-relation is the reproduction of organisms and biological species, guaranteeing both identity and alterity of the process.

To sum up: the term 'growth' is used in Life Sciences in a threefold sense, building distinct categories with different theoretical backgrounds, but referring to cultural history:

- Reproduction
- Regeneration
- Permanent and persistent Growth (Process/Progression)

All three types can best be demonstrated by plants, which always have been a cultural symbol for growth. They lack a limited body with skeleton and central organizing units like heart and brain, and, after they have died, do not leave a (lat.) *corpus*, but decay. Trees are somehow an exception and were always handled separate in natural philosophy, as were carnivorous plants. Because trees leave a corpus, because they are able to overcome gravitation by growth and because their life time exceeds the average human life span, Aristotle regarded them as the highest forms of the vegetative soul. 'Reproduction' was and still is also called 'Fortpflanzung' in the German language, meaning a continuous repetition of something *planting* itself. Neither 'offspring' nor 'proliferation' emphasizes this plant characteristic of growth, which also is used on animal and man: planting a self without *being* one.

'Regeneration' is a characteristic of plants. The organisms which showed regeneration after being cut, like the polyp *Hydra* spec., were until the late 18th century regarded as plants. If we say, that something regenerates, we apply a *teleologic* understanding of growth, i.e. a growth resulting in a specific end. On the contrary, 'reproduction' in a modern sense just refers to copies of an organic self, which remains productive in order to produce more. Here the most problematic term behind is 'identity'. Finally, permanent and persistent growth is the ideal type of unlimited growth, which some trees (Oak, Sequoia) show for many hundred, sometimes a few thousands of years. As evolutionary 'tree of life', famous especially by the publications of Ernst Haeckel (Bredekamp 2005), permanent and persistent growth, which keeps being rooted in a beginning, means a genealogic view of life as a whole. It is a process which will be altered, but never is supposed to cease.

In the laboratories of the Life Sciences, these three categories of growth intermingle when 'life' is designed and ends up in biofacts – and so do their metaphorical backgrounds.

Biofacts

The term 'biofact', a neologism[2] comprised of (gr.) '*bios*' and 'artifact', refers to a being that is both natural and artificial. It is brought into existence by purposive human action, but exists by processes of growth.

2 In the meantime I found out, that the term biofact was once used by the Austrian taxidermist Bruno M. Klein in one publication (1943/1944), in order to stress that living entities produce "dead" structures like wood or limestone shells. As far as I know, this publication has completely been forgotten and thus the term has never been quoted again.

Why do we need a new term? While conventional ways of describing the artificial element in nature sharply distinguish between the natural and the artificial, the term *biofact* can account for the influence of technology on previously existing natural forms of growth (resembled in certain species), and allows for reflection on the existing borders between nature and technology, when it comes to designing practices of life. Conventional terminologies of designed living entities originate in different disciplinary and everyday contexts, ranging from agriculturally based breeding practices to the science fiction film genre: bastard, genetically manipulated organism (GMO), chimera, clone, replicate, cyborg etc. This complicates their employment as terms in a scientific context. By contrast, 'biofact' is a neutral term that can contain a wide spectrum within the two poles: natural living entities and technical artefacts. It shows that 'life' is not a secure candidate within the category of nature any more and that not only construction, but also growth is a medium for design. Human action and natural growth interfere in the act of designing, producing cloned individuals as well as biomaterials and transgenic organisms. The interesting point is that for the design process of the living objects the activity duration is *determined* by growth, not by human action. Due to the reproduction of cells, life of former living entities is always a precursor of the design process to be established.

Phenomenologically, biofacts are living beings since they grow, but their development is no longer self-determined. That means whereas the presupposition of growth always was only comprehensible with reference to ontology and metaphysics, the presupposition of development now seems to be analyzable by epistemology, e.g. in genetics and/or neurosciences. Both in 'the brain' and 'the gene' we presently find hyper-coded terms, because they fuse causes and reasons. Recollecting my previous remark on the term 'regeneration' and its inherent teleology, the new discipline of regenerative medicine, aiming at the re-growth of dysfunctional organs or organs which were not established at birth time or in early childhood, will be faced with nature's remaining autonomy in relation to both a naturally and socially structured 'environment'.

Growth as Medium and Mean: Network Modelling and Model-Organisms

Growth can be understood as *medium* and *mean* (in the sense of 'tool'). The same is true for technology (Gamm 2000; Hubig 2004). As medium, growth still is a natural process containing substantial potentiality and supporting the emanation of something living. The form of a living thing in nature is reached *by* and *through* the medium of growth. When growth becomes a mean, the form is still reached through the medium of growth, but not by it. As a mean, growth also has to be manageable enough to reach a certain end. By standardizing cultivation techniques with artificial culture media, e.g. as soil- or skin-substitutes, and mapped genetic structures of organisms, stored in gene banks as implementable 'information'[3], biosciences have come near to use growth as a tool. Nevertheless growth remains, using Martin Heidegger's distinction, both as a tool which has to be both: *ready-at-hand* ("Zuhandensein") as well as *presence-at-hand* ("Vorhandensein"). In different stages of the design process, growth is either ready-at-hand (e.g. in mapping genes of cultivated organisms, i.e. *before* the entity to be designed comes into existence) or presence-at-hand, when the growing entity already exists (within practices of breeding and growth control, e.g. the use of growth hormones). To make a growth type become presence-at-hand, its reduced substantial counterpart (a gene) has to be implemented in a nourishing and stimulating context (e.g. a denucleated egg cell). Only together can something living result.

From an epistemological point of view, there are four epistemic stages, by which biofacts are made ready-at-hand for design (for details see Karafyllis 2006a). Growth then is at first, in a mechanistic filter of perception, reduced to its material compounds and substituted by movement and functional form, and at second, brought back into reality by plantation techniques (im-, ex- and transplantation).

- Imitation
- Automation
- Simulation
- Fusion

Imitation allows focus on a certain growth state, in general the one regarded as grown-up, and the plan to design it as (gr.) *mimesis* of nature. In order to do so, the substantial growth potential of the entity has to be fixed into a material state, a 'building mass'. Automation reanimates the thing by movement of the selected automation components that can be interpreted functionally (idea of 'the organic', see Köchy 2003). Computer simulation de-materializes the growth process and suggests a complete automation of the specific growth type, which by then has lost its entity. Finally, fusion in the biotech-laboratory works with organisms as

3 In this specific sense 'medium' is used in a different way, i.e. as a mere *carrier* for information without own process capacities.

media and means, and has to prove if the modelling process can be made *real*. Real is a *living* thing only if it has *actuality*, i.e. if people dealing with it sometimes experience any kind of resistance in their practical work. This definitely is true for working with living objects in the laboratory, where only a small percentage of individuals of a model-organism are normally able to "auto-activate" the implemented genetic structures.

Evelyn Fox Keller (2005) recently pointed out how bioinformatics shape new epistemic models. The pioneer of *random scale-free networks*, Albert-László Barabási, published the idea in 1999 together with his colleagues, that there are teleologic "open" networks, which follow a *power law distribution* (Albert et al. 1999). He was inspired by the distribution of *hyperlinks* (of so-called "*hubs*") in the *World Wide Web* and used it for giving the mass of new data, collected by the Human Genome Project, an interpretative structure for the new field of proteomics. Bodies, cells and networks are treated analogously in order to model probable functions arising from protein-protein-interactions. This solves the problem of proteomics, how to find functions for genetically encoded proteins, to a certain extent. Soon after Barabási's publication, scientists working in the area of cancer research wrote about the tumor suppressor gene p53:

> One way to understand the p53 network is to compare it to the Internet. The cell, like the Internet, appears to be a 'scale-free network': a small subset of proteins are highly connected (linked) and control the activity of a large number of other proteins, whereas most proteins interact with only a few others. The proteins in this network serve as the 'nodes', and the most highly connected nodes are 'hubs'. (Vogelstein et al. 2000, quoted in Keller 2005: 1060)

This model is now international standard in proteomics (see e.g. Stelzl/Wanker 2006). The *scale-free-network*-approach combines imitation, automation and simulation, and nevertheless allows the future outcome of the living process to be imagined as "open" in a mathematical sense. *Potentiality* is substituted by *possibility*. The network is not a fixed architecture, but is regarded as generative. Life in proteomics is modelled as "interactome". However, how about growth?

The comparison with the so-called *wet ware*, i.e. the biological system in which the interactome actually functions as such, remains necessary to prove the modelled functions. In proteomics, this is often done by yeast two-hybrid-experiments. Only growth gives evidence for reality and proves the stated hypothesis about the interaction of proteins and, moreover, generates new hypotheses. By growth one can find out which protein interactions (modelled *in silico*) obviously never occur *in vivo*. The bio-informatical approach using network simulation for designing biofacts exceeds the limits of the time-consuming methods of biochemists and geneticists:

> In the well-known analogy to understanding how a car runs, biochemists disassemble the engine, transmission and body, characterize all the pieces and attempt to rebuild a working vehicle. Geneticists, by contrast, break single components, turn the key and try to determine what effect the single missing part has on the car's operation. (van Criekinge/Beyaert 1999: 2)

Genetic methods very often require a specific phenotype before they can be carried out, a so-called model-organism. Famous model organisms are the yeast *Saccharomyces cerevisiae*, the fruit fly *Drosophila melanogaster*, the mouse (*Mus musculus*) and the wall cress *Arabidopsis thaliana*: it's *the* model plant, belonging to the mustard family. For these model-organisms the genome is completely sequenced (as it is for Man since 2001), and there exist genetic libraries with different mutants, vector systems and peptides ready-at-hand. In order to prove the assumed functions of localized genes by the activity of characterized proteins, several *fusions* (some of them, e.g. mating assays, carried out by laboratory robots) have to take place. This is also true for styling model organisms towards *prototypes* for industrial production ('bio-commerce'). Fused objects still end up methodologically 'old-fashioned', in being planted into culture media, giving them the ability to grow. These plantation techniques nevertheless imply the presence of nature as it is and show that the analogy of 'implementation' and 'implantation' is a false friend.

Biofacticity and Hybridity

If we take the anthropological fact into account that humans are hybrids (Latour 2002), having both a natural and a technical essence, which makes them designers with an individual "Leib" (a German-derived term in phenomenology for the subjectivity of one's own corporeality), we can ask: In how far does biofacticity, as an epistemological concept, undermines hybridity as an anthropological concept? Thus, we read in Bruno Latour's *Politics of Nature*: "What is a subject, actually? That which resists naturalization. What is an object? That which resists subjectivation." (Latour 2004: 47) Maybe this is a bit too easy for the challenging research in the Life Sciences, and maybe this is due to the use of actor-network-theory (ANT), where the question of the beginning and end of networks, nowadays a dominant metaphor, is avoided. Cells to be designed can belong not only to either plants, animals *or* humans, but always theoretically include the application on humans. In the view of biofacticity, all biotechnologies refer to living things, i.e. objects to be managed, again including humans. Even with plants humans share more than 50% of the genetic structure, stressing phylogenetic growth as a continuous evolutionary process of gaining higher complexity. How to break free of this conclusion?

Martin Heidegger emphasized that, apart from the modern subject-object-dichotomy, the Greek term *arché* implied not only a first beginning of something to be, but also a sphere of its own *dominance* over this one being, respectively being one (Heidegger 1967; Karafyllis 2006a and b). The hybrid character of the human being as a growing entity and creative person who acts in light of self-determined ends allows us to take the fact seriously that human beings exist both within the spheres of nature and technology. One result could be a broad approach towards the idea of life, which allows for an understanding of personal growth in experience without reducing life to biological processes or functions of the genetic code.

In my article, I wanted to stress that even if growth can be modelled with technical metaphors, models and apparatus in order to reach a uniform standard of certain growth *types*, the growth *token* in order to make it real still belongs to an uncontrollable sphere of nature. Nature will not vanish, as it is often argued, but *technology* will be disguised. What living entities actually *can* 'do' is found out by provoking them 'to do' something (e.g. to fuse or to produce a certain amino acid). The potentiality of the term 'can' remains an ontological problem in the sphere of metaphysics.

Plant growth and plantation techniques often symbolize the creative act of life coming into being, e.g. already in Plato's *Timaios*. Moreover, 'culture' is derived from the Latin '*colere*' and means the planting and harvesting acts in *doing* agriculture. Quite similar at first sight, the laboratory in the Life Sciences is an epistemic unit, where growth and action, nature and technology, objects and subjects are included. However, agriculture takes place in the public sphere whereas laboratory work does not. Not just the release of knowledge (e.g. in the form of scientific images) is shaping the public view of 'nature' in a new way, but moreover the release of concrete biofacts – which still can grow and will never stop of being *not* concrete. What might be regarded as risky just shows nature's own potentials of resisting technical provocation. Where nature can resist by alternative growth types, humans can resist by action – if they want to.

List of references

Albert, Réka; Jeong, Hawoong and Barabási, Albert László. 1999. „Diameter of the World Wide Web." *Nature* 401: 130-131.

Aristoteles. 1995. *„Physik"* and *„Über die Seele."* In *Philosophische Schriften* Vol. 6. Hamburg: Meiner.

Barabási, Albert László. 2002. *Linked. The new science of networks*. Cambridge, M.A: Perseus Publishing.

Black, Max. 1968. *Models and Metaphors*. Ithaca: Cornell University.

Bredekamp, Horst. 2005. *Darwins Korallen*. Berlin: Wagenbach.

van Criekinge, Wim and Beyaert, Rudi. 1999. "Yeast Two-Hybrid: State of the Art." *Biol. Proc. Online* 2: 1-38.

Gamm, Gerhard. 2000. "Technik als Medium." In: *Nicht nichts. Studien zu einer Semantik des Unbestimmten.* Frankfurt/M.: Suhrkamp, pp. 275-287.

Heidegger, Martin. 1967. "Vom Wesen und Begriff der Φύσις. Aristoteles' Physik B, 1". In *Wegmarken*. Frankfurt/M.: Klostermann, pp. 309-371.

Hubig, Christoph. 2004. "Technik als Mittel und als Medium." In Karafyllis, Nicole C. and Haar, Tilmann, eds. *Technikphilosophie im Aufbruch. Festschrift für Günter Ropohl*. Berlin: edition sigma, pp. 95-109.

Karafyllis, Nicole C. 2002. "Zur Phänomenologie des Wachstums und seiner Grenzen in der Biologie." In Hogrebe, Wolfram, ed. *Grenzen und Grenzüberschreitungen*. Bonn: Akademie Verlag, pp. 579-590.

Karafyllis, Nicole C. ed. 2003. *Biofakte – Versuch über den Menschen zwischen Artefakt und Lebewesen*. Paderborn: Mentis.

Karafyllis, Nicole C. 2006a. *Die Phänomenologie des Wachstums. Zur Philosophie und Wissenschaftsgeschichte des produktiven Lebens zwischen den Konzepten von "Natur" und "Technik."* (Habilitationsschrift): Stuttgart University. (forthcoming: Bielefeld: transcript 2007).

Karafyllis, Nicole C. 2006b. "Biofakte – Grundlagen, Probleme, Perspektiven." *Erwägen – Wissen – Ethik* Vol. 17, Nr. 4 (main article). (in press).

Keller, Evelyn Fox. 2005. "Revisiting 'scale-free' networks." *BioEssays* 27: 1060-1068.

Klein, Bruno M. 1943/44. "Biofakt und Artefakt." *Mikrokosmos* 37: 2-21.

Köchy, Kristian. 2003. *Perspektiven des Organischen*. Paderborn: Schöningh.

Krohn, Wolfgang and Weyer, Johannes. 1990. "Die Gesellschaft als Labor – Risikotransformation und Risikokonstitution durch moderne Forschung." In Halfmann, Jost and Japp, Klaus-Peter, eds. *Riskante Entscheidungen und Katastrophenpotentiale*. Opladen: Westdeutscher Verlag, pp. 89-122.

Latour, Bruno. 2002. *Wir sind nie modern gewesen. Versuch einer symmetrischen Anthropologie*. 2. ed. Frankfurt: Fischer.

Latour, Bruno. 2004. *Politics of Nature. How to bring the Sciences into Democracy*. Cambridge, M.A: Harvard University Press.

Orland, Barbara. 2005. "Wo hören Körper auf und fängt Technik an? Historische Anmerkungen zu posthumanistischen Problemen." In: Orland, B., ed. *Artifizielle Körper – lebendige Technik. Technische Modellierungen des Körpers in historischer Perspektive*. Zürich: Chronos, pp. 9-42.

Pörksen, Uwe. 2002. "Die Umdeutung von Geschichte in Natur. Das metaphorische Kunststück der Übertragung und Rückübertragung." *Gegenworte* 9 (1).

Stelzl, Ulrich and Wanker, Erich 2006. "Proteinwechselwirkungsnetzwerke: Aufklärung der Funktion von Proteinen." *Biologie in unserer Zeit* 36: 12-13.

The Brain as Icon – Reflections on the Representation of Brain Imaging on American Television, 1984-2002

CLAUDIA WASSMANN

Since their first appearance in the lay press, pictures obtained through brain "scans" reshaped the understanding of the human psyche, of mental illness and of "normal" brain function. While colorful PET images appeared first, it was especially the virtual slice through the living human brain obtained by *functional* MRI with its color-coded hot spots that became the staple illustration of the psyche used in numerous newspaper articles and TV shows in order to explain human behavior, emotion and motivation. Here, I would like to inquire into the normative properties of brain images. I have analyzed several series on the brain, various health programs and news reports that aired on American Public Television and cable networks between 1984 and 2002. My argument is that the image of the brain as a normative instance was informed by the way brain scans were reported in highly visible TV programs on the brain, and that a gap has opened between the representation of brain imaging in the lay press and the properties brain scans acquired within the neurosciences. This gap has widened since the beginning of the new century. At this time the neurosciences adopted conceptual approaches, while the pictorial representation of the brain in the media became empty. This can be exemplified in three multi-episode series on the brain that are of particular interest because they were produced by the same New York based PBS station WNET partly in collaboration with the National Institutes of Health: *The Brain,* 1984, *The Mind*, 1988, and *The Secret Life of the Brain,* 2002.[1]

1 The NIH provided either funding and advise or NIH scientists and their research were featured in the programs.

According to a public relations officer at the National Institute of Mental Health, *The Brain,* in 1984 was the first "highly visible" documentary series on the brain that aired nationwide in the US.[2] *The Brain* encompassed eight episodes (1) The Two Brains, (2) Vision and Movement, (3) Stress and Emotion, (4) States of Mind, (5) The Enlightened Machine, (6) Learning and Memory, (7) Madness, (8) Rhythms and Drives. The episodes "Stress and Emotion" and "Madness" are particularly important with regard to the normalizing function of images. "Madness" had the explicit agenda of making a claim for the "normality" of patients, who suffer from schizophrenia, in terms of our common humanity.

For historical reasons, schizophrenic patients were among the first subjects to be studied by brain imaging techniques. In the 1950s, Seymour S. Kety the scientific director of the newly founded National Institute of Mental Health (NIMH) set up a research unit at the St. Elizabeth Hospital in Washington, DC, one of the large federal mental institutions that were the legacies of World War I and II. Kety had conducted physiological research for developing a method that would allow measuring cerebral blood flow. (Kety 1945) In his laboratory between 1950 and the late 1970s, Louis Sokoloff eventually worked out the method for measuring local changes in cerebral blood flow and brain metabolism that was then adapted for Positron-Emission-Tomography and made PET scans a practical success. (Reivich 1979) Even before PET, at the St. Elizabeth Hospital Daniel Weinberger and his colleagues built the first brain-imaging unit in a psychiatric clinic in the early 1980s. (Weinberger 1984) "Madness" was shot at the St. Elizabeth Hospital. A classic documentary, the program presented extensive footage of patients and intimate conversations with patients and their relatives. It had the clear mission to alert the population to fact that the "chemical revolution" in psychiatry had devastating social consequences for the psychiatric patients. The final commentary stated "Tens of thousands of patients who are calmed by drugs and are discharged from institutions now they are in the dark corners of our cities because they have no place to go. For them the scientific revolution has proved tragic." While the scientists explained mental illness in terms of the brain – they spoke of synapses and nerve cells, of dopamine and brain development – the human dignity of the patient and his struggle remained in the forefront. The film was set in quiet images with ample time to look and listen.

While the title sequence of *The Brain* used animation of a model of the human brain, the pictorial image of the brain appeared only infre-

2 Seed money came from the NIMH and the National Science Foundation.

quently during *Madness*. The brain appeared in the form of a black and white photograph that depicted a slice through the brain of a deceased person in a brain atlas. Therein, a scientist pointed out the hippocampus and then showed microscopic pictures of the cells in the hippocampus that made evident how very different the aspect of these cells looked in the brains of healthy people and in schizophrenic patients, nicely lined up versus chaotically disordered and pointing in different directions. Thus, the assumption was made that the disorder of the nerve cells affected the connections between the cells, and these ill-adjusted connections had something to do with the illness. They affected how the patient experienced the world. The pictorial representation of the brain was used by the *scientist* in order to explain exactly what the picture showed and to provide causal information that was pertinent to the case. The image was evidence that was presented to underscore the point.

In 1988, for instance, a CNN documentary "Peace of Mind" showed impressive looking pictorial images of brains with enlarged ventricles in the schizophrenic patient compared to its healthy sibling.[3] The commentator exclaimed, "Look at these ventricles…." (CNN 1998) Yet, the scientists never claimed that the picture *was* the substrate of madness. In their interviews and commentary they stressed the complexity of the illness. What they claimed was that the biological differences they saw in the brains of healthy and schizophrenic siblings provided sound evidence against the prevailing notion in American society that schizophrenia was caused by bad mothering. Brain imaging studies, they argued, would allow them to better understand the multiple causes of the illness and, hopefully, find better treatments in the long term. Thus, scientists first used brain images to argue for the "normality" of their patients to defend their humanity. The argument that "there are *visible physical differences*" and therefore patients deserved "to be treated like any other kind of medical disease" both by society and by physicians was made in the 1980s also with regard to depression.

History of brain imaging techniques and their representation on TV

While there was still a long way to go before they could build machines big enough to image the human body, by the mid-1980s scientists were well aware of the potential that PET and MRI techniques held for clini-

3 CNN presented arguably the best depiction and explanation of the different brain imaging techniques and their clinical relevance.

cal and research purposes.[4] The scientific phenomenon of nuclear magnetic resonance was discovered in 1946 independently by two physicists, Edward Purcell at Harvard and Felix Bloch at Stanford, who shared the Nobel Prize for their discoveries in 1952. While a Swedish scientist published medical data as early as 1955, NMR spectroscopy had its first career in chemistry where it was widely used to analyze chemical structures of solids and liquids.[5] During the 1960s and '70s Paul Ernst developed the mathematics behind the computerized analysis of NMR data, the so-called Fourier transform NMR and the 2-D NMR. In 1991, he received the Nobel Prize in Chemistry for his work. In 2003, Paul Lauterbur and Peter Mansfield shared the Nobel Prize in Physiology or Medicine for their work that made today's fMRI scans possible, which also dates back to the 1970s. Since Lauterbur's article on "Zeugmatography" appeared in *Nature* in 1973 it was only a matter of time before they designed instruments that would allow implementing magnetic resonance imaging in practice.[6] (Lauterbur 1973; Kumar 1975; Hollis 1987; Mansfield 1977) The first excellent MR images of the human brain appeared in 1980. These first MR images were static.

Television programs reported enthusiastically about these images that gave an anatomical view of the brain and showed tumors and the effects of stroke. However there was a time lapse in the reporting, and comments often associated brain imaging with science fiction. Two technological developments significantly contributed to endowing images of brain with normalizing powers. These were the development of more powerful computers, which enhanced the quality of the images obtained in brain scans. By the same token, the leap in audio-visual technology from film to video and the development of graphics enabled workstations allowed television stations to produce ever more sophisticated computer animations of brains. This showed in the second series produced by WNET in 1988, *The Mind. A Search Into The Nature Of What It Means To Be Human. The Mind* encompassed nine episodes (1) The Search for Mind, (2) Development, (3) Aging, (4) Addiction, (5) Pain and Healing, (6) Depression, (7) Language, (8) Thinking, (9) The Violent Mind. Control Data, a computer company that also provided a

4 The NIH started its PET and MRI programs almost simultaneously in the 1980s and for about 15 years invested continuously in this expensive technology. And it invested in publicizing this research i.e., through seminars for science journalists and free distribution of video material on brain imaging to the press.

5 The information about molecule structure was presented in form of graphs.

6 The first mathematically reconstructed images of the internal structure of the body in anatomical "slices" were CAT scans in 1971, based on x-rays.

"Cyber 910 3-D Graphics Workstation", sponsored the program.[7] The computerized visualization of the human brain that appeared in *The Mind* was a pioneering achievement. A 3-D brain was computed from data collected by Robert B. Livingston, professor emeritus of the Scripps Research Institute Neuroscience department.

To collect the data, a normal human brain was encased in a block of paraffin. Successive thin layers were shaved off the block, and the top of the block was photographed after each pass. The resulting frames were hand-digitized to produce approximately 100 contours spaced 1.1 mm apart. Distinct neuroanatomical structures were digitized separately, so that they could be colored and displayed individually. The contour data was converted into triangular meshes by Floyd Gillis at the Scripps Research Institute (La Jolla, CA).[8]

In 1988, this 3D-surface data of the brain and its interior structures was one of the most detailed in existence. Floyd Gillis received the 1989 *Emmy Award* nomination for Best Graphics in the Television News and Documentary category for his work on the brain animation in *The Mind.* The trend setting animations of the human brain introduced in *The Mind* were often copied and signified the workings of the brain for the next decade.

The appearance of *The Mind* was strikingly different from the series *The Brain* that aired only four years earlier. "Entertainment value" had been added. Interviews were set in exotic outdoor settings, scientists were stars such as John Searle, Jane Goodall, or Jerome Kagan, and the serious and thoughtful tone that characterized "Madness" gave way to a touch of ghee wiz. The first episode, "The Search For Mind," opened like an adventure movie with dramatic music and two researchers walking in the cave of Lascot casting search lights over the walls. Then the film turned to Jane Goodall sitting in midst of a monkey colony. The first pictorial images of brains were models of a monkey brain and the image of the larger human brain, which appeared underneath. The monkey brain then moved to the upper left corner of the screen. At that moment both, the image of the brain and the animation of the image on screen were visual novelties.

7 The James S. McDonnell Foundation, the National Science Foundation, and the John D. and Catherine T. MacArthur Foundation provided funding. Special thanks went to three NIH Institutes, NICHD, NINDS, and NIMH. Co-produced with BBC, *The Mind* was distributed in other countries i.e., Germany. The companion volume was heavily advertised and the series was offered for sale for 99 Dollars.

8 See, AFCG. Inc founded in 1987 (http://www.afcg.com/).

The first image of a human brain scan appeared twenty minutes into the program. Again, the brain was used to argue that there are *visible* "defects" (this time in the brains of autistic patients) and these "defects" underscore "that specific damage to the brain is reflected in a mind ill equipped to assign meanings and motives to others." The message of the episode was "The Mind is what the brain does." The animations used in *The Mind* depicted the brain and the nervous system in terms of electrical currents as blinking lights. One animation that was used for different purposes throughout the series showed an anatomical depiction of the human brain and of nerve fibers through which lights moved signifying, for instance, pain signals that reached the brain. The brain "lights" up; lights spread in all directions from the inner parts of the brain to the cortex. The animation was used to set visual highlights and to transport the message that "Mind is brain fully realized. Mind is a hundred billion interconnected neurons. Mind is [dramatic pause] infinity."

Computer animations of the inner parts of the brain such as basal ganglia, hypothalamus and the amygdala were supplemented with, or replaced by, brain images obtained through *functional* MRI in the mid 1990s. The first *f*MRI scans for mapping human brain function were successfully run at Massachusetts General Hospital in Boston in 1991. The ability to measure the brain's function with magnetic resonance in addition to obtaining a static anatomical picture was the key scientific reason for the explosive growth of fMRI studies in the 1990s.[9] Both, PET and fMRI use parameters related to changes in cerebral metabolism and blood flow as indicator of neural activity. The images are calculated and reconstructed on a computer screen. This yielded an intense debate questioning the validity of the new methods. (Beaulieu 2003) The controversial scientific debate was not reflected in the reporting. In the mid-1990s, the realist argument was repeated with regard to ADHD, obsessive-compulsive disorder, dyslexia and depression. Now, the slice through the brain with color-coded spots came to stand for this 'reality.' The image was shown as evidence on its own, and presented by the journalist.

While brain imaging was used during the 1990s in order to excuse socially deviant behavior or underachievement in a society of success, at the turn of the twenty-first century the picture changed again. Rather then serving to argue that "there was no reason for shame" because one could *see* in brain scans that there were "real" physiological differences (for instance in depression), the last series produced by WNET used

9 Equally essential was the increasing availability of MRI scanners to researchers. In 2002, in the US approximately 1700 MR scanners were of sufficiently high field strength to conduct *functional* imaging studies.

brain imaging to emphasize the brain's great powers to structure itself and its allegedly almost unlimited powers of renewal. In 2002, *The Secret Life of the Brain* had five episodes (1) The Baby's Brain: Wider than the Sky, (2) The Child's Brain: Syllable from Sound, (3) The Teenage Brain: A World of Their Own, (4) The Adult Brain: To Think by Feeling, (5) The Aging Brain: Through Many Lives.[10] The brain's surprising power of renewal was the dominant trope. The use of brain images balanced back the burden of responsibility to the individual who can choose whether or not to train his or her brain and keep up high performance even in old age.

Furthermore, the series featured a conspicuous shift in the scientific understanding of emotion that was brought about by brain imaging techniques. In 1984, in *The Brain*, emotions were still defined as psychosomatic manifestations of stress. "Stress and Emotion" explained that "Emotions are mediated through the brain, but the problems to study them are enormous." The one emotion that one could study was "anxiety" that resulted from "stress." Subsequently the program used the story of a couple. The wife was expecting a baby, and the husband had a highly demanding but boring job in an airport control tower and was a smoker. Embedded in this scenery lay the scientific consensus of the 1980s, convictions about stress, heart attack, ulcer, smoking, and a Freudian psyche. This was supplemented by footage from animal research showing mice that had developed ulcers when they could not control stress, and with information about brain chemistry. "Stress and Emotion" opened with a lengthy reenactment of the story of nineteenth-century railroad worker Phineas Gage. The commentary presented Gage's railroad accident as the "accidental discovery" that emotions had to do something with the brain. It is interesting to note that in 1883, the French psychologist Théodule Ribot reported on the case in *Diseases of the Will*. The case of the "American quarryman" showed that through the destruction of the frontal lobes the emotions gained the upper hand. Gage was "a child with regard to the intellect and an adult with regard to passions and emotions." (Ribot 1883: 90) While Ribot claimed that lacking in Gage was inhibition and the power of inhibition was located in the frontal lobe, a century later Antonio Damasio argued that what happened to patients like Phineas Gage was rather a lack of emotions. Because of the destruction of the nervous pathways in the frontal lobe, emotional signals could no longer be fed into the process of rational decision-

10 Sponsors were Pfizer, Metronic Foundation, Park Foundation, Dana Foundation, and National Science Foundation. An award-winning web site accompanied the series. Viewers were invited to "take an animated tour of the brain."

making. (Damasio 1994) Therefore, the patients' behavior seemed odd because it lacked the emotional value judgment that normally guided our thought.

In 2002, *The Secret Life of the Brain* devoted "The Adult Brain: To Think by Feeling" entirely to the depiction of emotion *in* the brain. "To Think by Feeling" began with great pathos. A hugely enlarged close-up of a computer animation of a brain in shiny golden tones appeared. It was somewhat flurry and Damasio's voice commented that "until now" emotion and thought were thought to be separate entities, and that was wrong: "We are not thinking machines we are feeling machines that think." Using new imaging technologies, Damasio explained, scientists "have demonstrated" that emotions have a physical place in the brain. "The adult brain is capable of the whole panoply of thought. But it is emotions that is at the heart of our thoughts. [Pause] That's who we are. We are emotional people."

The style of the program changed too. The presenter and the commentary disappeared. Only original sound was used, of patients seen on-screen and of scientists mostly as off-voices. The scientists as such were no longer featured. The image of the brain was free standing.[11] In order to make the MRI images visually interesting again, the pictures were extremely enlarged and appeared almost pixilated on purpose and artificially colored like modern art. The brain had become an icon. It was an inscrutable surface and no effort was made to explain its workings, or to explain science.

Conclusion

In summary, when the first pictures of brain obtained by the new brain imaging techniques entered the TV screen in the 1980s the scientists' excitement and the medias' excitement about these images met, albeit for different reasons. Scientists were excited about the final practical success of the technique, which they had eagerly anticipated. For the media, both the imaging technology and the brain images obtained by magnetic resonance had "news" and "entertainment" value. Subsequently, a gap opened between the temporal dynamics of media "needs" and scientific "advances." While scientists did indeed gain ever more complex knowledge about the body and the brain from the new ways to acquire, visualize and analyze functional data, the pictorial image of the brain looked basically the same, though rendered more smoothly. Sci-

11 See also Igor Babou's analysis of the French case.

ence could not produce novelties at the pace the media needed them. When brain-imaging techniques stopped being "news," the media produced ever more science fiction like animation that do not carry any real information. These images are metaphors not for science, but for socially desirable and politically correct ontological positions. They code convictions such as there is "no reason for shame," or even in old age everyone can alter his fate because the brain is endlessly plastic; there are no limits to what human beings can achieve by sheer will power.

With regard to the normalizing function of images, three major developments stand out in the reporting. During the 1980s brain images were first used to "normalize" the mentally ill, the schizophrenic patient and depression. Though scientists had already begun using new brain imaging techniques, the images used on TV for that purpose were initially photographs of real brains studied post-mortem. At about 1988 computer-generated images of brain entered the TV screen. Second, in the early 1990s a process of "normalization" of the pictorial image of brain took place. It was both a process of standardization within Science of the different ways by which images were acquired and post-processed by scientists at different universities and research institutions, as well as a process of "normalization" in terms of getting used to seeing images of brain in the public domain. Since the mid-1990s when fMRI became available and when better memory and image processing software enhanced the pictorial quality of computer-generated 3-D images, computer generated reconstructions of brains appeared more frequently on TV, in print and on the Internet. The pictorial image of the brain became free standing. Finally, since the late 1990s brain images were used in order to define the "normal" that is, processes such brain development and aging, the healthy adult and, most conspicuously, emotions in the adult brain.

Technologies of visualization grounded the understanding of the human psyche so deeply in a biological base that the scientific approach to topics as various as alcoholism, drug abuse, schizophrenia, depression, Alzheimer's disease, Parkinson, aging, obesity, trauma, memory, attention and decision making are not longer possible without referral to specific structures and networks in the brain and states of activation or deactivation represented in virtual images. Virtual images have become an icon of the mental. However, the greatest potential for "normalizing by images" that brain imaging has acquired in the neurosciences is no longer reflected in the public media because it is too complex and too conceptual for allowing depiction in a single brain image.[12] While brain imaging techniques have become an innovative tool to probe genetics in the brain, pictorial images of the brain have lost their potential to convey information in the lay press.

List of References

Babou, Igor. 2004. *Le cerveau vu par la television.* Paris: PUF.

Beaulieu, Anne. 2003. "Brains, Maps and the New Territory of Psychology." *Theory & Psychology* 13(4): 561-568.

CNN. 1988. *Peace of Mind.*

Damasio, Antonio. 1994. *Descartes' error : emotion, reason, and the human brain.* New York: Putnam.

Hollis, Donald P. 1987. *Abusing cancer science: the truth about NMR and cancer.* Chehalis, Wash: Strawberry Fields Press.

Kety, S. S. and Schmidt C.F. 1945. "The determination of cerebral blood flow in man by the use of nitrous oxide in low concentrations." *American Journal of Physiology* 143: 53-66.

Kumar A., Welti D., Ernst R. R. 1975. "NMR Fourier zeugmatography." *Journal of Magnetic Resonance* 18: 69-83.

Lauterbur, Peter. 1973. "Image formation by induced local interactions: Examples employing nuclear magnetic resonance." *Nature* 242: 190.

Mansfield Peter. 1977. "Multi-planar image formation using NMR spin echoes." *Journal of Physics C: Solid State Physics* 10: L55-58.

Ribot, Théodule. 1883. *Les maladies de la volonté.* Paris: F. Alcan.

Reivich, Martin et al. 1979. "The [18F]fluorodeoxyglucose method for the measurement of local cerebral glucose utilization in man." *Circulation Research* 44: 127-137.

Weinberger, Daniel. 1984. "Computed tomography (CT) findings in schizophrenia: Speculation on the meaning of it all." *Journal of Psychiatric Research* 18: 477-490.

12 See the recent publications by Daniel Weinberger, Micheal Egan and Ahmed Hariri.

"If There is a Risk Inside of Me, I am the First Person who Should Know About it." – Images of 'Genetic Risks' as Anticipation of the Future

ANDREA ZUR NIEDEN

Genetics can be seen as a technological anticipation of the future. Medical genetic testing for predispositions is currently used to determine the probability individuals have of developing illnesses such as breast and ovarian cancer during their lifetime. This in a way makes genetics a new form of soothsaying.

Based on my fieldwork in genetic counselling centers, I will focus on the role of visual representations and metaphors that are used to make the concept of genetic predispositions evident. As an example for my argument, I will take genetic testing and genetic counselling for the BRCA1 and BRCA2 genes (that is breast cancer 1 and 2 genes). Medicine considers these genes as significant factors in the development of breast and ovarian cancer. Currently, women considered to be 'at risk' because of their family history are encouraged to undergo genetic testing in combination with a genetic consultation in the course of a multi-centered clinical trial financed by the German Cancer Aid. Carriers of the mutations are said to be exposed to a breast cancer risk of 40 to 80% compared to 10% in the average female population.

The starting point of my analysis is that illness in general poses a dilemma to modern individuals: if enlightment is seen also as a process of the schism of body and Geist which makes the body a *thing*, an object to be controlled by the mind,[1] illness challenges this constellation. The *sub-*

1 In the words of Max Horkheimer and Theodor W. Adorno: "Erst Kultur kennt den Körper als Ding, das man besitzen kann, erst in ihr hat er sich vom Geist, dem Inbegriff der Macht und des Kommandos, als der Gegens-

ject mind finds itself committed to its body, suddenly being the body's object. Modern medicine can be understood as an attempt to re-gain control over the body, to objectify it; in the long run: to abolish natural perishability.

The geneticisation of disease, the identification of 'pathogenic genes' that we cannot alter (so far) as the main cause of a medical condition, seems to re-establish the bondage of nature in medicine, and many liberal critics warn of a new biological determinism. In contrast, others (Rose/Novas 2000 or Lemke 2005) have stressed another aspect of the New Genetics: Unlike the old eugenics which concentrated on the constraint of breeding and selection, the logic of 'genetic predispositions' tries to transform people into self-determined active subjects of their genes. The genes impose a responsibility to "do something about it", to manage them. Whoever is 'genetically at risk' is supposed to take appropriate action to prevent the actual onset of the disease, where that action may be pharmaceutic, psychotherapeutic, dietary, sporting, or where it may amount to intensified screening or prophylactic surgery of breast and ovaries. (All these options are explained to the women during the counselling sessions.) Therefore, knowing about your genes means anticipating the future in a twofold way: it means calculating the probability of becoming ill, as well as trying to control and change the future by doing everything which might "reduce the risk".

In this article I will concentrate more on the first part of the process of anticipation by analysing the visual and metaphorical representations of the gene referred to in the genetic counselling sessions and in my interviews with counsellors and women who have undergone genetic testing. Although nobody has ever seen *a gene*, there are images which create an evidence of the 'gene inside'.

Almost every day we see a three-dimensional picture of the DNA-double-helix somewhere in the media, and it seems most natural to us. The double helix, a model established by Watson and Crick only in 1953, has become so much of a cultural icon that we forget that the model is only a theory-based visualisation of something not directly observable. Nowadays all we can see of the DNA and of genes are black

tand, das tote Ding, 'corpus' unterschieden." (Adorno/Horkheimer 1969: 264). (The translation by John Cumming: "Culture defines the body as a thing which can be possessed; in culture a distinction is made between the body and the spirit, the concept of power and command, as the object, the dead thing, the '*corpus*'" (Adorno/Horkheimer 1979: 232) is mistakable because in the original "spirit" is supposed to be the subject and "body" the object.)

and white barcodes, the base sequences made visible with the methods of molecular genetics.

The scientific definition of what *a gene* is is very much disputed and has become more and more ambiguous in our days. As Fox-Keller (among others) points out, all attempts to define a gene have failed and it is even impossible to talk about a gene as a clear-cut entity (e. g. Fox-Keller 2000).[2]

Despite these confusions, medicine operates with genetic tests on 'predispositions' like the BRCA1 and BRCA2 test based on correlation-studies which are supposed to show a significant interrelation between breast and ovarian cancer and certain mutations in these genes. To make the counsellees believe in the importance of the genes, to make them become part of a high-risk-population and to initiate appropriate action on their part, it is important to make the gene evident. In my study, I analysed counselling sessions and carried out interviews with genetic counsellors and women who had undergone genetic testing. The work is based on the paradigm of qualitative social research and reconstructive research logic. This means that the goal is to reconstruct the relevant concepts of the narrators and to try to represent the meaning of what was said. For this purpose, I use a type of interpretative analysis method that combines the theoretical and methodological approaches of conversation analysis, narrative analysis, and positioning analysis.[3]

So which images and metaphors of genes or genetic risk can be found? First I give a brief impression of the counselling sessions, then I will move to some examples of how the women talk about it.

Firstly, the counsellor tells the woman her genes have something to do with her ancestors. He draws her pedigree, asking in detail about the cases of cancer in the family, the state of health of the living relatives and the cause of death of the departed. Secondly, he says those genes which the client might have gotten from her parents do something in her body which has to do with her own future. The counsellor tries to give a general introduction to the genetics of cancer, while showing pictures of the chromosomes and drawing sketches like figure 1 on a sheet.

DNA sequences are, as he tells her, hieroglyphics; the chromosomes are files or books. Inside these books are many words, combinations of the letters C, T, A, G, which stand for four different bases. As he says, you cannot see the genes, but you can make the bases visible, which has been done for the BRCA1 and BRCA2 genes. The inherited set of chromosomes from mother and father could be imagined as two differ-

2 See also Samerski 2001: 116-129 for a summary of the debate.
3 The method has been outlined in Lucius-Hoene/ Depperman 2004.

ent types of encyclopaedias: the Meyer's encyclopaedia from the father and the Brockhaus encyclopaedia from the mother,[4] two sets of slightly different information concerning the same topics.

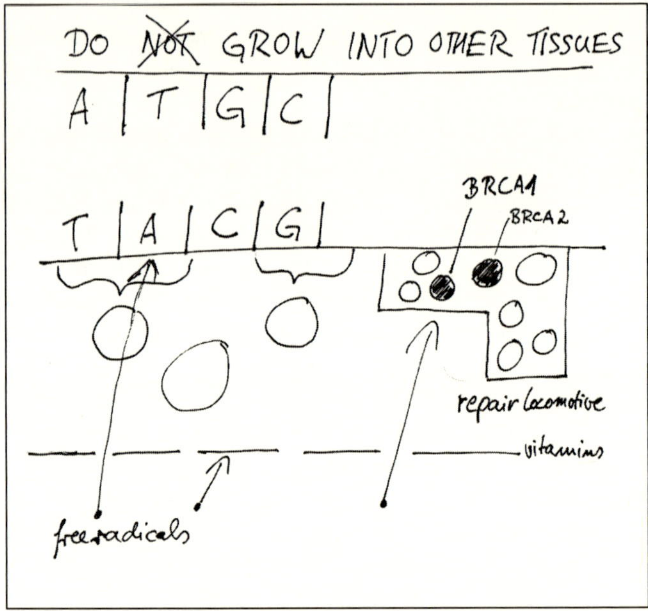

Fig. 1

Yet the inherited information in DNA is not everlasting. During our lifetime, the counsellor points out, DNA will be damaged many times by free radicals, which reduces the function of the cells. This could be imagined "as if you would stand in front of your bookshelf and shoot with a shotgun." Vitamins form a kind of protective shield against the damages, but they are not completely failsafe. The BRCA1 and BRCA2 gene are part of the so-called "repair locomotive" which normally fixes almost 100% of those damages. If you have an inherited mutation in one of these genes, the repair function is deteriorated, which might cause cancer in certain tissues. This process could be imagined like this: when the genetic information was "don't grow into other tissues!", it might be changed to "grow into other tissues!", which is a fatal change of meaning, as is obvious to everybody.

Obviously, the counsellor makes use of the discourse of DNA as a "code" or a "language" and the genes as carrying "information" that de-

4 Meyer's and Brockhaus are the most common encyclopaedias in Germany.

termines the production of a "protein", a discourse which has shaped the view of molecular biologists since the 1950s, as Lily Kay has pointed out in "Who wrote the Book of life" (2000). She traces how these concepts had a very successful career in genetics although DNA doesn't meet the criteria of being a code or a language. Instead of acknowledging the complex network of the relation between genes and proteins, the gene is perceived as the blueprint, the origin, which has a determinate "meaning".[5]

The other metaphorical concept used by the counsellor is that of war. What Susan Sontag described in her 1977 essay "Illness as metaphor" (Sontag 1977) as *the* metaphor for talking about cancer[6] and its treatment[7] is here, after the geneticisation of cancer, transferred to the molecular level. Sontag assumed in the 70s that "cancer is now in the service of a simplistic view of the world that can turn paranoid" (ib.: 69). Without speculation about the political background now, it can be at least said that the counsellor's description takes over a bit of the old cancer-paranoia.

As the sketch already implied, the coding gene does not give us the whole story required to explain the phenotype. There are other factors that influence whether you contract cancer, e. g. the amount of free radicals shooting at your "lexicon" or the amount of vitamins protecting you. The counsellor mentions a number of other relevant factors that show how little genetics can say about the disease. The test can only give a better estimation of the risk than the pedigree, as the counsellor outlines while drawing percentage-numbers on a sheet. But: the non-deterministic nature of the BRCA genes also means that you can do something to reduce the risk of cancer.

When starting my interviews with women who had undergone genetic testing, my presumption was that the genetic counselling and other popular images of genes would open a space of imagination to visualise a daily war taking place in the women's own bodies between free radicals shooting DNA, vitamins which form a protective shield and repair locomotives not doing their work, thus leading to cancer cells which form a time-bomb in the body. But interestingly most women did not speak very figuratively about the genes. To my question "do you have

5 Many other authors have criticized the notion of a one-directional relationship of cause and effect from DNA to RNA to protein to "us" (e. g. Fox-Keller 2000).

6 E. g., "invasive" cancer cells, "cancer cells "colonize" from the original tumor to far sites in the body, first setting up tiny outposts ("micrometastases")", the talk about the body's "defenses" (Sontag 1977: 64).

7 "Patients are 'bombarded' with toxic rays" (Sontag 1977: 65), etc.

any image of what the gene does in your body?" some of the women even answered that they were told about it but didn't remember. Some said they didn't *want* to imagine or refused to imagine it in a figurative way.

As one woman said, she thinks nothing of "imagining genes as (laughing) nasty things that open up their mouth and eat up the whole body or things like that"[8]

The metaphors I *did* find referred to the following concepts:

1. Cybernetics:

"how many things collude for some person or another"[9]
"about this *chain*, and if one is de*fect* then sometime you will fall *ill*"[10]

2. Cause and effect:

"w*hat* which gene ef*fects*"[11]

3. Possession:

"I *have* that gene, indeed"[12]

4. Inheritage:

"I *have* inherited it or *have not* inherited it"[13]

5. Language or code/programm:

"the DNA is not properly *read* anymore"[14]

---

8 I always add the German original quote in the references because any translation slightly changes the meaning. The stressed syllables are italicized in the transcription. My own comments during the interview are in //. "eigentlich nich, also ich halte *nichts* davon mir Gene als (lachend) böse Dinger vorzustell'n //(lacht)// die dann das Maul aufreissen und den ganzen Körper auffressen oder sonstige Dinger *nein* davon halt ich eigentlich nichts."

9 "wieviele Sachen da bei dem einen oder dem anderen zusammenspielen"

10 "mit dieser *Kette* und wenn da eins de*fekt* ist und dann irgendwann wird man dann halt *krank*"

11 "*was* welches Gen be*wirkt"*

12 "ich *habe* zwar jetzt das Gen“

13 "ich *hab* es geerbt oder ich hab es *nicht* geerbt"

14 "die DNA nich mehr richtig *ab*gelesen wird"

"I think this is ..... just a *but*ton, that can be *pres*sed or not. //mhm// isn't it? just a button, that is there, that is maybe *not* there in another person, and quite individually, and if something happens, and it is pressed, than it *starts*."[15]

6. Pregnancy:

"if I carry the gene in*side*"[16]
"because everybody carries the *ge*nes in the body, right? The cancer genes. Well, but if they break *out* by something mental, yeah, psychic"[17]

7. Outbreak

(same example)

8. Development

"what is it that happens so that the gene becomes *can*cer?"[18]

Most of these metaphors are not very exciting, even prosaic. Some refer to DNA as code or language paradigm as outlined above, but nobody takes over the whole war-story of the counsellor.

On the other hand, the quotes show that the women took over the notion of genes being clear-cut 'things'. They are things which you can posses (3.), inherit (4.), etc. And most took over the idea of one-directional cause and effect. The gene in this formulation (2.) seems to be the agent of a process in the body. The cybernetics concept (1.) seems more open: there are many factors interacting. But the first of the cybernetics quotes was part of the theory about the gene being a 'button' (5.), in which the gene also is the main cause of illness. There are other factors working together that *press* the button, (and it is not determined and uncontrollable, *why* or *by what* the button is pressed,) but then a predetermined program runs.

---

15 "Ich denke das ist einfach, das ist ein *Merk*mal, ganz normales wie soll man sagen, ein *Knöpf*chen, das ge*drückt* werden kann oder nicht. //mhm//. ne? einfach 'n *Butt*on, der da ist, der vielleicht bei jemand anderem *nicht* da ist, und ganz individuell, und wenn da irgendwas dazu kommt, und der gedrückt wird, dann geht's halt *los*"

16 "ob ich das Gen *in* mir trage"

17 "weil die *Ge*ne trägt ja jeder im Körper. ne? Die Krebsgene. //mhm// Und aber wenn sie zum *Aus*bruch kommen durch irgendwas seelisch ja, psychisch"

18 "was passiert denn damit das Gen zu K*rebs* wird?"

The other metaphorical concepts are quite common for the description of diseases (like pregnancy and outbreak or development), which indicates that the speakers do not make a big difference between a gene and something like a virus and that they see themselves as potentially ill. Again, the gene is a substance-agent in the body, a germ which – under the proper conditions – will grow and initiate the 'outbreak' of the disease. The association of pregnancy is especially common for cancer. As Susan Sontag outlines, cancer, the "disease of growth" (Sontag 1977: 12) that "works slowly, insidiously" is seen as "a demonic pregnancy" (ib.: 14). Although not very elaborate, my quotes seem to echo the old metaphors of mysterious cancer.

But as it seems, the function of the gene is not visualised by most women as a process in their own body. Symptomatically, one woman says:

> well in a way I can (laughing) imagine that. Yeah maybe not that it's *fig*uratively taking place in my body, but when I got it drawn it was shown in a simplified way and therefore, I mean, it's quite a thing what such a tiny little ge-, what such a tiny little gene-mutation can do in your body. If you imagine.[19]

Only *if* you imagine, a little thing does something big, but normally you wouldn't imagine. So far, I don't have a clear explanation of why most of the women do not visualise this as I assumed. But it may be an interesting hint that especially the women who had been told that they *had* the mutation said they didn't really want to know it in detail. Maybe they feared becoming paranoid when having too many images in mind.

But what is interesting in the aforementioned quotations and in my interviews in general, is that every woman *believes* in the genetic risk. There is no doubt about that, even for those who say they don't understand the process.

> I don't have a good biological understanding I am more a linguistic type of person. So (.) I knew the facts that were presented to me, that the risk is that high, and it...it is an hereditary affair, and actually this was *enough* for me;...[20]

19 "Ja eigentlich kann ich mir das schon (lachend) vorstellen ja vielleicht nich grad dass s so *bild*lich in meinem Körper abläuft aber, als ich d's da so aufgemalt kriegt hab dann, (.) war's ja schon in vereinfachter form (lachend) dargestellt (leise) un deswegen ich mein es is schon ziemlich heftig was so'n, (.) kleines Geh was so'ne kleine Genmutation *aus*machen kann im Körper – wenn ma' sich d's mal so vorstellt"

20 "Ich habe kein großartiges biologisches Verständnis ich bin mehr ein sprachlicher Typ. //mhm// so ich kannte die Fakten die mir dargelegt wur-

In summary, the core concept of the medical discourse is working-namely the embodiment of the risk, the notion that a statistical relation between cancer and a set of bases has something to do with the own body, the reification of this relation as a *thing* in the body. It is working as a *belief* backed by metaphorical concepts even if it is not concretely visualized in the own body. The evidence of the genetic risk inside and of the notion that this risk is a call to do something herself is taken for granted when a woman says:

> simply because I appreciate being in the know about the things that are inside of me, and if there is a risk inside of me, then I am really the first person who should know about that and ehm, so that I can deal with it, no matter *how* I deal with it. Well, basically it is my problem how I deal with such a thing,[21]

The absurdity of talking about a "risk inside of me" is a demonstrative summary of the reification of risk by the vehicle of genes.

In all the mentioned concepts of genetic risk the gene serves as a symbol of anticipating the personal future, of identifying oneself as a risk-person. But instead of a straight story about the relationship between the gene and cancer as the counsellor presented it, the women apply a variety of concepts and explanations which are often contradictory and not conclusive. This might be due to the confusion implicit in many interviews imposed by the non-deterministic nature of the gene. Perhaps for the same reason, the call for action to reduce the risk is perceived differently. While for the woman I quoted last the risk seems to be something that you can "deal with", in other quotes like the one about the "button" (5.), the risk is much more of an unpredictable *danger* and you can't influence the "outbreak". But these differences in self-management must be the subject of following articles.

List of References

Adorno, Theodor W. / Horkheimer, Max. 1979. *Dialectic of Enlightenment.* London/New York: Verso.

---

den, dass das Risiko so hoch is, und es es es ist ne erbliche Geschichte, und eigentlich das *reich*te mir; also"

21 "einfach weil ich es gut finde über die Dinge die in mir drin sind bescheid zu wissen, und wenn da in mir drin 'n Risiko is, dann bin ich eigentlich die erste die's wissen sollte, und ähm damit ich auch damit umgehen kann, ganz egal *wie* ich dann damit umgehe, dass is ja dann auch im Grunde genommen erst mal mein Problem wie ich mit so was umgehe,"

Adorno, Theodor W. / Horkheimer, Max. 1969. 'Dialektik der Aufklärung' und Schriften 1940-1950. In Horkheimer, Max. 1987. *Gesammelte Schriften. Bd. 5: 'Dialektik der Aufklärung' und Schriften 1940-1950.* Frankfurt/M.: Fischer.

Fox-Keller, Evelyn. 2000. *The Century of the Gene.* Cambridge, Mass.: Harvard University Press.

Kay, Lily. 2000. *Who wrote the book of life? A history of the genetic code.* Stanford, Calif.: Stanford University Press.

Lemke, Thomas. 2005: "From Eugenics to the government of genetic risks." In: Robin Bunton and Alan Petersen, eds. *Genetic Governance. Health, Risk and Ethics in the Biotech Era.* London und New York: Routledge, pp. 95-105.

Lucius-Hoene, Gabriele/ Deppermann, Arnulf. 2004. *Rekonstruktion narrativer Identität. Ein Arbeitsbuch zur Analyse narrativer Interviews*, Wiesbaden: VS Verlag für Sozialwissenschaften.

Rose, Nikolas/Novas, Carlos. 2000. "Genetic risk and the birth of the somatic individual." *Economy and Society* vol. 29, N. 4: 485-513.

Samerski, Silja. 2001. *"Sie müssen irgendwann 'ne Entscheidung treffen". Eine Untersuchung über die Popularisierung eines neuen Entscheidungsbegriffes in professionellen Beratungsgesprächen, dargestellt am Beispiel der genetischen Beratung.* Diss. Universität Bremen.

Sontag, Susan. 1977. *Illness as Metaphor.* New York: Farrar, Straus & Giroux.

From the Historical Continuity of the Engineering Imaginary to an Anti-Essentialist Conception of the Mechanical-Electrical Relationship

ARISTOTLE TYMPAS

Introduction

Accumulated work by historians and philosophers of science allows us to know details on how metaphor, analogy and modeling (some would add allegory and mimesis) become constitutional elements of the scientific phenomenon. I am interested here in something much less studied, namely, the place of the same elements in technology. In respect to technology, the literature is scattered and focused only on the history and philosophy of engineering models. The place of the rest of the aforementioned elements in technology – not to mention the interrelationship of metaphors, analogies, and models – remains, to my knowledge, an unexplored territory, especially for historians of technology. I will try taking a step into that territory by introducing a sample from the history of engineering analogies (Philosophers and historians of science have argued that starting from analogies may be justified by the fact that analogies are components of both metaphors and models). The first suggestion of the history of this chapter is then that analogies have been an indispensable component of not only science, but also technology.[1]

1 For an early and a more recent survey of the metaphor, model and analogy relationship in science, see Leatherdale 1974, Bailer-Jones 2002. For recent collections of essays on models, mostly scientific, see Morgan and Morrison 1999, Klein 2001, Chadarevian and Hopwood 2004. For case studies devoted to the history of engineering models, all concerning mechanical models, see Smith 1976-1977, Harley 1991-1992, Kooi 1998,

My work starts with analogies between two of the exemplar machines of eighteenth and nineteenth century capitalism, the steam engine and the dynamo. It begins with the period when the electric lighting distribution lines were first introduced. It concludes with analogies from the period when long and interconnected lines of electric power transmission had become the rule, alongside a transition from mechanical-electrical to electrical-mechanical analogies. The flow of the calculating analogy always went from the standard to the novel phenomena, which means that it was not the nature of the phenomena – mechanical or electrical – that determined the flow of the analogy. A phenomenon was more standard – and, as such, more natural – after it had become relatively familiar by the availability of a standard, mass-produced, artificial circuit that embodied it. Based on the continuities that we find in this transition, I move on to also suggest that the difference between the mechanical and the electrical are socially constructed. More specifically, I suggest that the mechanical and the electrical have been products of the expanded reproduction of a certain mode of producing nature socially, not two ontologically different states of nature.

This suggestion points to the configuration and reconfiguration of metaphors, models and analogies as normative rather than descriptive processes, interpretations of nature rather than representations of it. Following this, I further suggest that the production and use of analogies was an integral element of calculation. Indeed, the analogies that I introduce over the course of the following pages were produced and used in the context of calculating the stability of increasingly longer and interconnected electric lines. Between the 1880s, when my narrative starts, and the 1940s, when it ends, the concept 'analog' computer emerged, in interaction to a transition from mechanical-electrical to electrical-mechanical computing analogies. In this sense, the story of this chapter also involves the computer, the exemplar machine of twentieth century capitalism.

Mechanical-Electrical to Electrical Mechanical Calculating Analogies

In the 1880s, it was clear to engineers that an electrical phenomenon was not ontologically different from a mechanical phenomenon, but an expanded reproduction of it, based on the same pattern of accumulation of labor power. In the words of H. Franklin Watts, who looked at the elec-

Wright 1992. For a sample of studies devoted to the philosophy of technical analogies, see Kroes 1989, Sarlemijn and Kroes 1989.

trical from the viewpoint of the mechanical, the two differed only in the "mechanical skill" embodied in them. "It is true saying and worthy of all acceptation," stated Watts in his first of his 1887-1888 series of articles on "Practical Analogies between Mechanical and Electrical Engineering," "that an electrical engineer is about eight-tenths mechanical and two-tenths electrical." His introductory example successfully supported this argument. For Watts, "electrical knowledge" and "mechanical skill" were inseparable. He argued so by referring to the construction of a dynamo armature: to be properly balanced, so as to avoid both the electrical phenomenon "eddy" or "Foucault" currents (which increased with a bulkier supporting mechanical structure) and the mechanical phenomenon of a weak supporting structure an engineer needed both ("electrical knowledge" and "mechanical skill") (Watts 1887: 246).

With his following example, Watts elaborated on how an electrical machine was also a mechanical machine. For his comparison, he chose the machines that exemplified mechanical and electrical engineering: the representative of electrical engineering was the dynamo (which was structured around the armature); the representative of mechanical engineering was the steam engine. Watts introduced the analogy between the two by claiming that a steam engine resembles a dynamo "not only in its mechanical construction and attention necessary to operate, but also in the calculations of the theoretical performance." He acknowledged that the steam engine and the dynamo seemingly "differ greatly," as they appear to rely on reverse processes, since with a steam engine the energy of an invisible fluid is converted to visible mechanical motion whereas with the dynamo a visible mechanical motion is converted to an energy of an invisible fluid. But, by turning to history, Watts argued that there was no difference at all because the operation of both machines was actually reversible: the first steam engine, explained Watts, was used as a pumping engine. Similarly, the function of dynamo was the reverse of that of the motor. Accordingly, at this point Watts re-introduced to James Watt and Michael Faraday as inventors of machines that exemplified the same pattern (Watts 1887: 246).

For Watts, the calculations of mechanical and electrical machines were analogous. The intervening accumulation of "mechanical skill" had not changed its pattern. "The weak point of a [steam] engine," wrote Watts in the first line of his second article of the series, "may be said to be its crank-bearing, the fiction of which increases with the horse-power produced by a single crank." As we moved to horsepower of 1,000 and higher, the "lubrication of the crank becomes a very important element." "The weak point of the dynamo," added Watts, "is its commutator, the friction of which, while not necessarily increasing with the *output* of the dynamo, increases with the current to be collected." Watts was of the opinion that the difficulties with the crank, as compared to those with the commutator, were "in about the ratio of six of one to half a dozen on the other" (Watts 1887: 258).

Some of the details of the "practical analogies" between mechanical and electrical engineering of Watts displeased Rudolph M. Hunter, a mechanical engineer. In early 1888, Hunter and Watts exchanged a series of letters on the issue of practical mechanical-electrical analogies through the pages of the *Electrical World* (Hunter 1888). They were neither the only, nor the last ones to do so. When John Waddell, who was with the Royal Military College of Canada at Kingston, Ohio, published an article in order to argue that the difference of electrical potential was analogous to the pressure difference of the air, he drew the protesting response of A. W. K. Peirce. For his analogy, Watts had used standard machines. By contrast, Waddell described an analogy between the generation of electric potential in a conductor moving across a magnetic field and an air box with an "indefinitely large number of little paddles." The mechanical field in which the air box was moved was provided by two boards, which represented magnetic poles, and steel wires, which represented magnetic lines. The purpose of the analogy was to assist in computing savings in copper in dynamo (or motor) design. Peirce found this analogy "imperfect," arguing that when no current was flowing in the conductor none of the energy required to move the conductor in a magnetic field could be charged to the generation of electrical potential – the current being zero, the work would be zero at the conductor when moved in a magnetic field. During the move of the airbox in the mechanical field there was production of heat in the box. Waddell replied with a complex thought experiment – a test which he was "not in a position to make" – in order to argue that in a new machine, of which the transformer was an exemplar, heat could be developed in the secondary circuit even if it was open. I understand this debate to reveal that structural asymmetries that did not matter from a mechanical viewpoint but could become important when considered from an electrical perspective (Waddell 1894; Peirce 1894).

To be sure, the use of mechanical-electrical analogies did not start with the electric network of lighting and power. In the year 1887, on the same day when Watts started publishing his series of articles on general mechanical-electrical computing analogies in the pages of the *Electrical World*, the *Electrical Review* hosted an article on a mechanical-electrical computing analogy written by Arthur Kennelly. Watts stayed at the general level because he wrote in reference to the relatively new electric networks of lighting and power. Kennelly was, however, more specific because he wrote about the relatively old electric networks of telegraphy. As a result, unlike Watt's article, Kennelly's – entitled "On the Analogy between the Composition of Derivation in a Telegraph Circuit into a Resultant Fault, and the Composition of Gravitation on the Particles of a Rigid Body into a Center of Gravity" – included the formalization of the calculating analogy into a calculating equation (Kennelly 1887).

Mechanical-electrical analogies were central in pioneering alternating current treatises. For example, in their 1893 influential handbook on the analytical and graphical computation of alternating currents, Frederick Bedell and A. C. Crehore included an appendix on mechanical-electrical analogies (Bedell and Crehore 1893). The mechanical-electrical computing analogy between the steam engine and the dynamo machine remained fixed to the electrical engineering unconscious, to erupt in times of crisis. For example, worried about the profitable but tremendously risky acceleration of the lengthening of alternating current transmission lines, Harold W. Buck protested against those who legitimized such acceleration by advancing calculations that overplayed the profits and downplayed the risks. During a 1923 AIEE Conference paper discussion, he protested by employing an elaboration of the steam engine analogy. Given the intervening increase of transmission length, he quite properly adjusted this analogy to the perspective of the transmission component (as opposed to generation component, see Watts) of the process of the production of electric power:

> The transmission of power from a piston for instance to a flying wheel through a connecting rod is a very simple proposition, but when the connecting rod is lengthened out to such a distance that its inertia and elasticity become factors which cannot be controlled then some other method must be found. A transmission line is merely a connecting rod and in the very high-voltage lines of great length the inertia and elasticity are becoming difficult factors to handle and the papers under discussion prove it (Dellenbaugh 1923: 822).

The gradual change from mechanical-electrical to electrical-mechanical computing analogies came along the change from analogies between phenomena to analogies between circuits. Compared to the relationship between mechanical and electrical phenomena, the relationship between mechanical and electrical circuits better supported an essentialist split between the mechanical and the electrical. The two equivalent circuits appeared to be related by an analogy between two natures that were, nevertheless, different. Mediating between the electrical-mechanical circuit analogy was a mathematical function, which described both circuits.

This prepared for the completion of the conceptual transition from analogies and models to analog computers.

We can elaborate on this transition by considering a mechanical model that circulated widely among the community of electrical engineers. In 1926, S. B. Griscom, General Engineer at Westinghouse, described what he called a "mechanical analogy" to the problem of electric transmission stability (Griscom 1926). He would not refer to it as a mechanical model, and he could not, as E. W. Kimbark in 1948, refer to it as a "mechanical analogue." Kimbark, who was at the Electrical Engineering Department at Northwestern University, spoke about this model at an American Power Conference (Kimbark 1948). When Griscom was writing in 1926 amidst the peak of electrification, he avoided the concept model because it signified a past state-of-the-art computing technology. Instead, at the cost of being too general, he baptized his computing artifact a "mechanical analogy." Kimbark had a new concept that Griscom was lacking, and he could now be specific without having to resort to the term "mechanical model," which pointed to the past. His term – "mechanical analogue" – pointed to the future. For him in 1948, Griscom's 1926 "mechanical analogy" was a "material analogue." We are just a step before the concept analog computer. In the rest of his paper, Kimbark considered elaborate versions of Griscom's "mechanical analogue," appropriate for extending the study of stability to more complex networks and transient load conditions.

A detailed description of the same mechanical model was given by L. F. Woodruff in his academic textbook on transmission and by Robert D. Evans in an influential industrial textbook that was published by some Westinghouse engineers. It consisted of two rotatable units mounted on a common shaft and provided with lever arms that were connected at the outer ends by a spring. The one rotatable element was an analog of the generator and the other of the motor. Both elements were provided with means for applying torque in such a way as to stretch the spring connecting the level arms. The radial distance from pivot to any point on the spring was analogous to the line voltage at corresponding point, the length of the spring to the line reactance drop, the tension of the spring (proportional to its length) to the line current, the torque of either arm (product of the length of the arm and component of spring tension perpendicular to arm) to the active power, the product of the length of the arm and component of spring tension along the radius at any point to the reactive power, and the angle between any two points in the spring to the phase displacement at the corresponding points of the system. This mechanical model was suitable to demonstrate changes in movement that were significant from the standpoint of stability. It was proportioned so as to model both steady and transient state conditions. By addition of rotatable elements and springs, it could be extended to model complicated networks (Woodruff 1938: 181-182; Central Station Engineers of the Westinghouse and Manufacturing Company 1944).

From Evans we learn that the electrical engineers who used it still commonly called it a "mechanical model." In 1948, Kimbark would use a new concept by calling it a "mechanical analogue." The difference in the concepts used by Griscom in 1926 ("mechanical analogy") and by Kimbark in 1948 ("mechanical analogue") for the same computing artifact is suggestive about the conceptual change between the two different sub-periods of the computation-electrification relationship. Reading Griscom's 1926 article leaves us to wonder whether he had devised something material. Griscom included only a sketch of his "material analogy." It is only after reading the contribution of Evans to the influential Westinghouse electrical engineering textbook, where we see a published picture of a man holding it, that we understand that we have to do with a mechanical model. Evans provided two figures of it. Under the first, which was a picture of a man setting the devise in order to compute a certain transmission scenario, he wrote "The mechanical model." Under the second, which was a sketch of it placed next to its corresponding vector diagram and the equations that described it, he wrote "The mechanical analogy for power system stability." Evans clarified that the "mechanical analogy" could be useful even when a "mechanical model" was unavailable (Central Station Engineers of the Westinghouse and Manufacturing Company 1944).

The conceptual transition that I just outlined interacted with the shift from mechanical-electrical to electrical-mechanical analogies and the associated shift from analogies between phenomena to analogies between circuits described by the same form (the same mathematical equation). We can take a mid-point example, which show that during the second sub-period, some electrical phenomena were standardized enough to provide the analogy for non-standard mechanical phenomena. The 1931 *AIEE Transactions* hosted an article entitled "An Electric Analog of Friction: For Solution of Mechanical Systems Such as the Torsional-Vibration Damper," written by H. H. Skilling who was at Stanford University. This was no longer a paper on a mechanical-electrical analogy. It was one on an electrical-mechanical (electromechanical) analogy (Skilling 1931).

In his 1941 review article, labeled "Electrical and Mechanical Analogies," W. P. Mason started with a section entitled "Early borrowings of electrical from mechanical theory" to conclude with a section entitled "Borrowings of mechanical theory from electrical network theory" (Mason 1941). Two years later, in another review article, John Miles, who was at the California Institute of Technology reviewed an associated conceptual change. During the transition from mechanical-electrical to electrical-mechanical analogies, the electrical analogy itself was split into two in order to best adjust to the new calculating purposes: an older electromechanical analogy was conceptualized as the "direct method," and a newer one as the "inverse method" or the "mobility method" – the term "mobility" emphasized the relative flexibility of the new method. Miles also called them the "electrostatic" and the "electrodynamic" analogy (Miles 1943, 183-192).

Convinced about the superiority of the new electromechanical analogy, Miles tried to explain it:

> The fundamental imperfection of the old electrostatic analogy of velocity *across* mechanical elements being represented by current *through* electrical elements and force *through* mechanical elements being represented by voltage *across* electrical elements causes little trouble in simple systems; but it may also be said that the direct solution of these systems, sans analogies, gives even less trouble. In the case of a somewhat more complicated system, however, the newly initiated user of the old electrostatic analogy is very likely to become hopelessly confused and arrive at the most erroneous answers (Miles 1943: 184-185).

How can we understand what the advantage of the new method was so as to move on to appreciate Miles' introductory statement, according to which "the choice of analogy to be used is usually one of convenience, but that certain systems intrinsically make only one analogy possible." At first, Miles' article reads more like a confirmation of the first half of this statement than as proof of the second half. As I see it, the difference had to do with the fact that the new "electrodynamic analogy" was better for visualizing a calculation path that went from an electrical to a mechanical network whereas the old "electrostatic analogy" was better for computing the other way around. Since, by late 1940s, the dominant direction of calculations was from the electrical to the mechanical, the electrodynamic version of the analogy was becoming superior (Miles 1943: 183).

This inversion became possible by the availability of standard, relatively inexpensive electrical components. More standard electrical components meant that it was now relatively easier to construct a standard electric circuit and then use it to compute an unknown mechanical circuit. This is precisely what Gilbert D. McCann and H. E. Criner were introducing in a series of articles on calculating analogies between electrical and mechanical circuits during the second half of the 1940s. In one of these articles, McCann and Criner included a three-table figure similar to one that Miles had included in his article. The middle table included the elements of a mechanical system. The left and the right table included elements of an electrical system. The tables were similar but their titles were different. Miles labeled the left and the right table "electrostatic" and "electrodynamic" analogy perspectively; McCann and Criner "electric" and "electrical." "Physical mechanical elements," wrote McCann and Criner underneath their figure, "can be represented by either one of two analogous electrical systems." First, what was consciously noted by this sentence, was that mechanical systems were being represented by electrical (the inverse was, by then, meaningless). Second, consciously or not, through their choice of concepts, McCann and Criner had blocked the older electromechanical analogy: an analogy between elements of circuits cannot be electric-mechanical, it can only be electrical-mechanical. In other words, during the inversion from mechanical-electrical to electromechanical computing analogies, one version of the (electromechanical) computing analogy was aborted – there was not then a simple inversion, but also, a specification of the computing analogy (McCann/Criner 1945: 138).

To be sure, the concept 'model,' as we understand it, was used only for mechanical models. "The electrical-analogy method," argued McCan and Criner, "has several distinct and important advantages over the use of models or existing mechanical calculators. It is relatively inexpensive to build suitable elements to represent a wide range of other physical constants. These can readily be put in a form suitable for quick connection to represent a wide range of physical systems." In other words, the world was electrified enough so that an analogy could flow from the phenomena of a standardized electrical world to nature in order to allow the consideration of an electrified version of the nature as natural (McCann and Criner 1945, 138).

In 1887, Watts thought of the dynamo as analogous to the steam engine. He used the old exemplar of a mechanical machine (steam engine) as a model of the new exemplar of an electrical machine (dynamo). After the 1940s, the flow of this general analogy was also reversed. For example, MIT's D. C. White and A. Kusko wrote that they had developed a laboratory machine which "effectively demonstrates and supports the approach and indicates the common root that all machines basically have in a cylindrical structure with prescribed surface winding patterns." The title of their paper was "A Unified Approach to the Teaching of Electromechanical Energy Conversion." In early electrification, the unifying machine model was mechanical. By late electrification, it had become electrical. To no one's surprise, the classic engineering textbooks on analogies of the post-World War II period were about electromechanical analogies, not mechanical-electrical analogies (White/Kusko 1956: 1033).[2]

List of References

Bailer-Jones, Daniela M. 2002. "Models, Metaphors and Analogies." In Peter Machamer and Michale Silberstein, eds. *The Blackwell Guide to the Philosophy of Science*. Malden, Massachusetts: Blackwell, pp.108-127.

Bedell, Frederick and A. C. Crehore. 1893. *Alternating Currents: An Analytical and Graphical Treatment for Students and Engineers*. New York: W. J. Johnston.

Central Station Engineers of the Westinghouse and Manufacturing Company. 1944. *Electrical Transmission and Distribution Reference Book*. East Pittsburgh: Westinghouse Electric and Manufacturing Company.

Chadarevian, Soraya and Nick Hopwood, eds. 2004. *Models: The Third Dimension in Science*. Stanford, California: Stanford University Press.

Dellenbaugh, Frederick S. Jr. 1923. "Artificial Lines With Distributed Constants," *AIEE Transactions* 42: 803-823.

Griscom, S. B. 1926. "A Mechanical Analogy to the Problem of Transmission Stability." *Electric Journal* 23.5: 230-235.

Harley, Basil. 1991-1992. "The Society of Arts Model Ship Trials, 1758-1763." *Transactions of the Newcomen Society* 63: 53-71.

Hunter, Rudolph M. 1888. "Practical Analogies Between Mechanical and Electrical Engineering." *The Electrical World* (April 7): 177-178.

Hunter, Rudolph M. "Practical Analogies Between Mechanical and Electrical Engineering." *The Electrical World* (May 5): 228.

Johnson, Walter C. 1944. *Mathematical and Physical Principles of Engineering Analysis*. New York: McGraw-Hill.

Kennelly, Arthur Edwin. 1887. "On the Analogy Between the Composition of Derivations in a Telegraph Circuit into a Resultant Fault, and the Composition of Gravitation on the Particles of a Rigid Body into a Center of Gravity." *Electrical Review* 11 (November 5): 2-3.

Kimbark, Edward W. 1948. "Power System Stability," *Midwest Power Conference Proceedings* 10: 238.

Klein, Ursula, ed. 2001. *Tools and Modes of Representation in the Laboratory Sciences*. Dordrecht, The Netherlands: Kluwer.

2 For classic handbooks on engineering analogies from the 1940s, see Olson 1943, Johnson 1944, Trimmer 1950.

Kooi, B. W. 1998. "The Archer's Paradox and Modelling: A Review." *History of Technology* 20: 125-137.

Kroes, P. 1989. "Structural Analogies Between Physical Systems." *British Journal of the Philosophy of Science* 40: 145-54.

Leatherdale, W. H. 1974. *The Role of Analogy, Model, and Metaphor in Science*, Amsterdam, The Netherlands: North-Holland.

Miles, John. 1943 "Applications and Limitations of Mechanical-Electrical Analogies, New and Old." *Journal of Acoustical Society of America* 14.3: 183-192.

Mason, W. P. 1941. "Electrical and Mechanical Analogies." *Bell System Technical Journal* 20 (October): 405-414.

McCann, Gilbert D. and H. E. Criner. 1945. "Solving Complex Problems by Electrical Analogy." *Machine Design* (December): 137-142.

Morgan, Mary S. and Margaret Morrison, eds. 1999. *Models as Mediators: Perspectives on Natural and Social Science*, Cambridge, United Kingdom: Cambridge University Press.

Olson, Harry F. 1943. *Dynamical Analogies.* New York: Nostrand, 1943.

Peirce, A. W. K. 1894. "Electrical Difference of Potential: An Analogy." *Electrical World* (December 29): 671

Sarlemijn, A. and P. Kroes. 1989. "Technological Analogies and Their Logical Nature." In P. T. Durbin, ed. *Technology and Contemporary Life*. Dodrecht, The Netherlands: Reidel, pp.237-55.

Skilling, H. H. 1931. "An Electric Analog of Friction: For Solution of Mechanical Systems Such as the Torsional-Vibration Damper." *AIEE Transactions* 50 (September), pp. 1155-1158.

Smith, Dennis P. 1976-1977. "Structural Model Testing and the Design of British Railway Bridges in the Nineteenth Century." *Transactions of the Newcomen Society* 48: 73-90.

Trimmer, John Dezendorf. 1950. *Response of Physical Systems*. New York: Wiley.

Waddell, John. 1894. "Electrical Difference of Potential: An Analogy." *The Electrical World* (December 8): 589

Waddell, John. 1894. "Electrical Difference of Potential: An Analogy." *The Electrical World* (December 29): 671-672.

Watts, H. 1887. Franklin. "Practical Analogies Between Mechanical and Electrical Engineering." *The Electrical World* (November 5): 246-247.

Watts, H. 1887. "Practical Analogies Between Mechanical and Electrical Engineering." *The Electrical World* (November 12): 258.

Watts, H. 1887. "Practical Analogies Between Mechanical and Electrical Engineering." *The Electrical World* (November 19): 268.

Watts, H. 1888. "Practical Analogies Between Mechanical and Electrical Engineering." *The Electrical World* (March 17): 137.

Watts, H. 1888. "Practical Analogies Between Mechanical and Electrical Engineering." *The Electrical World* (April 7): 176-177.

Watts, H. 1888. "Practical Analogies Between Mechanical and Electrical Engineering." *The Electrical World* (April 21): 98.

Watts, H. 1888. "Practical Analogies Between Mechanical and Electrical Engineering." *The Electrical World* (May 19): 253.

White, C. and A. Kusko. 1956. "A Unified Approach to the Teaching of Electromechanical Energy Conversion." *Electrical Engineering* 75 (November): 1028-1033.

Woodruff, L. F. 1938. *Principles of Electric Power Transmission*. New York: Wiley.

Wright, Thomas. 1992. "Scale Models, Similitude and Dimensions: Aspects of Mid-Nineteenth Century Engineering Science." *Annals of Science* 49: 233-254.

Architectural Structuralism and a New Mode of Knowledge Production

MARTINA HESSLER

The spatial dimension of science and technology has been an important topic of research recently. The "spatial turn" has thus brought new perspectives to the history of science and technology. Geographies of science, sites of knowledge production as well as the relationship of architecture and science have come under close scrutiny. (e.g. Smith/Agar 1998; Galison/Thompson 1999) The work has analyzed the geography of scientific institutions, the spatial structures of buildings and how meaning has been ascribed to them. It has shown how space organized scientific and social life, how it included and excluded certain groups. Architecture and space often intend to manipulate or determine people's behavior. However, users try to describe these inscribed rules. Moreover, "architecture [...] is itself a symbolic writing of space. The very buildings where scientific inquiry was housed were often pronouncements in the language of stone, site, and plan about the place science should occupy in the wider culture." (Livingstone 2003: 38)

Taking up this work on space, architecture and science this article aims to show the close connection of urban history, the history of architecture and the history of science. In particular, it will show how the debates on cities in the 1960s and 70s merged with new architectonical concepts, but also met the needs of research institutes. In other words, I will attempt to show how scientific space was re-structured in what is termed the knowledge society, following an urban model. With important work being done on the re-structuring of space in the context of information technology and on the emergence of "new industrial spaces" (Scott), I will argue that it is not only the industrial geography that has changed. The spatial structure of industrial research areas itself has been re-organized.

My case study is the Siemens-Forschungsstadt ("science city") in Neuperlach. Neuperlach is a satellite town on the outskirts of Munich which was built in the 1960s and 1970s. The"Forschungsstadt" itself was built in the 1970s and 1980s. It is located at the edge of Neuperlach. The Siemens company established the science city in Neuperlach in order to get a stake in the market of information technology. In August 1977 the first employees moved from the inner city to the Forschungsstadt on the outskirts. By 1990 around 10,000 people were working in Neuperlach. More than half of them were involved in data processing and information technology.

What makes the Siemens-Forschungstadt a very interesting case is that it is one of the biggest structuralist architectural complexes ever built. (Schäche 1997: 86) There one can observe how contemporary concepts of architecture, criticism on the state of cities and the changing mode of knowledge production in a knowledge society were closely connected.

In what follows, I will first make some brief remarks on the trend of suburbanization among scientific institutions and the high-tech industry since the 1970s and 1980s. Second, I will show the efforts to "urbanize" these suburban research areas by focusing on my case study. I will describe the architectural design of the Siemens-Forschungsstadt and explain the main ideas of structuralism in architecture. Finally I will ask what it was that made this concept appropriate for the spatial organization of an industrial research area once knowledge came to be regarded as a more and more important production factor.

Suburbanization of scientific institutes and industrial research

Whereas traditional industries as a rule have moved to the periphery of cities since the end of the 19th century, research institutions and universities in Europe have usually remained located in the town itself for centuries because cities have commonly been regarded as the typical space for creativity and new ideas. A university or research institute on the outskirts of a city was an exception until the 20th century. The few cases in Germany, such as Berlin Dahlem or the University of Tübingen, where universities and an agglomeration of science institutes were sited outside the city at the end of the 19th century, met with resistance. However, in the 1960s these reservations no longer applied. Contrary to the criticism of the 19th century, "modern American institutions", namely the campus, now became the model for building new universities and "science cities" from the 1960s onward. They were now sited outside cities, but typically remained close to a city. To simplify complicated problems, we can state that a dramatic change can be observed in the second half of the 20th century in most European countries. Universities, scientific institutions as well as technopoles or high-tech-industries have been situated on the outskirts of cities for several reasons.[1] We can talk about a "suburbanization" of science and postmodern industry, taking place since the 1960s and 70s. In the case of the city of Munich many new sites of knowledge production emerged on the outskirts of Munich, among them the Siemens-Forschungsstadt in Neuperlach. Out there, no infrastructure, no streets existed; no public transportation was available. These sites of knowledge production were created *ex nihilo*. Of course, the relocation of scientific institutions, universities and high-tech industries to the outskirts of cities has considerable consequences for the cities themselves. This question, however, does not fall within the scope of this paper.

The key point here is that this trend towards suburbanization among high-tech industries and research institutes was accompanied by efforts to "urbanize" the new places on the outskirts once knowledge became an important resource for innovation. Since the 1970s and in particular since the 1990s, one can observe that urban planning, science policy and the administration of companies and universities have striven to "*urbanize*" these emerging suburban areas, where industries and scientific institutions are located. That was also the case with the Siemens-Forschungsstadt.

1 Less surprising, on main reason for the suburbanization of high-tech industries was the lack of space within cities and the prices for real estate were much lower on the outskirts than in the city centre. Moreover, the location on the outskirts ensured that scientific work was not impeded by circumstances like air pollution, traffic noise or vibrations caused by underground trains. Furthermore, the Siemens company was spread out across the city. As the company emphasized, this posed a problem for scientific research, which was becoming increasingly crucial for the company's success. In the 1970s the Siemens management stressed that the spatial fragmentation made interaction and communication between different departments of the company very difficult and complicated. The aim was to concentrate several departments, which were spread all over the city, in one place in order to ensure the spatial proximity of employees to enable their interaction.

Urbanization of suburban research areas

When the Siemens-Forschungsstadt was built in the 1970s, the architectural and spatial design proved to be no easy matter. The idea to locate the company's new facility at the edge of the newly built "satellite town" of Neuperlach dates back to the late 1960s. An initial design submitted by the famous architect James Stirling – a very tall building – was rejected. A design, which was submitted by the architect's office van den Broeck and Bakema was also turned down. Siemens argued that it, like Sterling's design, was still too monumental. Finally, the building department of Siemens worked in close cooperation with van den Broeck and Bakema and adapted the design. The reason for changing the design was that Siemens felt that during the planning process the world had completely changed. The company became aware of quickly changing markets, a new need for flexibility as well as the growing importance of knowledge in the innovation processes. Consequently Siemens decided it needed a new kind of concept for the planned research facility. The core of the new concept was the idea to build the new research area as a *city*. Thus we have to take a closer look at the architecture and the spatial structure of the Siemens-Forschungsstadt.

The location is organized essentially as a diagonal net structure consisting of various small buildings.[2] This layout makes it possible to enlarge the area at any time.[3] When one walks around the site, it seems complex and full of corners. From a bird's eye view, on the other hand, there is a clear and regular cross-shaped ground-plan which however is constantly interrupted in order to create courtyards, pathways and small meadows.[4] Just as townscapes are characterized by a succession of squares and open spaces, streets bordered by buildings and numerous connecting paths, the buildings on the Siemens site have been built as space-defining elements and hence allow open spaces to arise.[5] Different urban places have been built; trees and lawns as well as "streets" can be found. Every building is individual; none looks like the copy of another; which was thought to enhance the sense of being in a city.[6]

---

2 SAA-Neuperlach, Siemens München – Entwicklung auf Expansionskurs.

3 "Arbeiten in einem jungen, dynamischen Standort", in: Siemens-Zeitschrift, 2 /1987, p. 26; "Siemens in München-Perlach 1969-1985", in: Baumeister 10 /1985, pp. 55-60; Münchner Stadtanzeiger, 26. September 1978, p. 4.

4 SAA 86 / Lr 546, München-Perlach. Eine Dokumentation. 1979.

5 "Forschungs- und Verwaltungszentrum der Siemens AG, München-Perlach", in: Bauen + Wohnen 7-8, 1979.

6 "Arbeiten in einem jungen, dynamischen Standort", in: Siemens-Zeitschrift, 2 /1987, p. 26.

The Forschungsstadt was planned as a compact urban area of short distances. Thought was given to creating an urban atmosphere with the building of shops and libraries as well as a "communication centre" which is intended to support communication between the employees. In short, the objective was to establish infrastructures to enable a social life, interaction and communication. One aim was to "bridge the gap between the work space and private space."[7]

In the 1970s Siemens was not the only company to apply such a concept of bridging the gap between work space and private space. The buildings of some administrative authorities and insurance companies endeavored to do the same. An administrative building in Apeldoorn was also explicitly compared with a town. One critic spoke of the "imitation of an old-town" ("Altstadt-Imitat") and called the building an *ersatz* town for the employees ("Stadtersatz", Boskamp 1975).

Like the Siemens complex it was characterized by the addition of many small buildings, shops and "streets", etc. Unlike the Forschungsstadt, where many of these infrastructures were planned but never built, in Apeldoorn they were actually realized. (Boskamp 1975) What is striking is that the architect of the building in Apeldoorn belonged to the same group of architects as van den Broek and Bakema, namely the architectural structuralists (Joedicke/Schirmbeck 1982).

Fig. 1: Photo of Neuperlach, Siemens-Archiv, Munich.

7 Siemens'archive, Munich, legend of a photo of the science city Neuperlach.

Structuralist concepts in architecture and their critique of the functionalist city

Structuralism has not attracted much attention in the history of architecture. Moreover, those who have written about the history of structuralism tended to have great sympathies with the movement and were sometimes closely associated with its leading figures.

In the late 1950s and early 60s a critique of modernism evolved within modern architecture itself, and was focused in the group "Team X", or Team Ten (Smithon 1968). Team X organized the last congress of the CIAM in Otterlo, an event considered to be the birth of "structuralist thinking in architecture and urbanism" (Lüchinger 1981). Team X was itself part of the CIAM. It was a grouping of architects who initially identified with the tradition of functionalism (Newman 1961) but were now striving to overcome the dogmas of function separation, such as had been formulated in the Charter of Athens and had exercised great influence over many years. Bakema was among the proponents of initiatives that went beyond the CIAM and led to the foundation of Team X (Joeckicke/Schirmbeck 1982: 140). A unifying idea among the architects of the group was that architecture and town planning should be inseparable.

Jacob Berend Bakema (1914-1981), who had joined the CIAM in 1947, became a member of Team X along with Georg Candilis, Shadrach Woods, Alison and Peter Smithson and Aldo von Eyck. The group contended that the modern city was "dismembered, a collection of lifeless specialisms". They sought to counter it with a "receptivity to life in its totality" in order to make the city a "living organism" again (cited by Lüchinger 1981: 20, 22). Bakema published his ideas in a series of works, among them *Architektur – Urbanismus*. From 1948 to 1970 he collaborated with the less well known Hendrik van den Broek (1898-1978).

The terms "structure" and "structuralism" were introduced in 1966 in an essay by Kenzo Tange in which he documented his architectural, or as he put it "intellectual" development to date. He described his career in terms of a change from functionalist to structuralist thinking. In his argumentation Tange proceeded from functionalism, which requires specific rooms for specific functions, and contrasted it with a "structuralist" approach which regarded the relationship between the space and the user no longer statically and deterministically. Instead, he argued, space and its form should be kept open for possible alteration and transformation and should indeed encourage different types of utilization.

Fig. 2: Siemens-Archiv, München

Tange believed most emphatically that "a process of organization is required by the functional that unites the functional units" (Joedicke/ Schirmbeck 1982: 140). A distinct feature of structuralism therefore is the emphasis of certain principles of formation. According to Tange, what connects spaces and gives them a structure is *communication*. For Tange, horizontal connecting passages inside a building and the pathways connecting buildings are communication elements in this sense: "Giving form to these communicative activities and flows between spaces means giving structure to architectonic and urban spaces." (cited by Joedicke 1999: 141)

Referring to structuralist thought systems already employed in philosophy, linguistics and anthropology, the term structuralism as used in architecture and town planning denotes on the one hand the idea of a fundamental order, an order which constitutes the residential, the social and the urban, and on the other hand their appropriation in different individual ways. Herman Hertzberger, one of the structuralists, made explicit reference to Saussure, who differentiated between *langue* and *parole* (Lüchinger 1981: 18). For Saussure, *langue* denoted language as a system of prescribed order, a system of distinctions, a constitutive structure; *parole* on the other hand meant talk, comprising the order, its interpretation, appropriation and use. Correspondingly, structuralist buildings should, in Hertzberger's view, encourage change on the part of the users. The crucial point for all structuralists was the awareness of permanent change and mutability in the basic conditions of planning.

The relationship of order and chaos was of central importance to the structuralists. Order as a component reveals itself in the grid of structuralist construction concepts, as can be clearly seen in the Siemens-Forschungsstadt in Neuperlach, which is organized according to a strict grid-like ground-plan. "Chaos" is supposed to arise from the disruption of the grid or order. In the view of the structuralists, buildings and architectonic complexes should possess a "labyrinthine clarity". And such "labyrinthine clarity" can indeed be found at Neuperlach: the newspaper *Münchner Merkur* noted: "At first sight a bewildering labyrinth, at second sight a sophisticated system of architecture and technology."[8]

These basic considerations entail various consequences for the architectural design:

Rejection of the large format

> Things may be big only as an amalgamation of units that are in themselves small. [...] By making everything too big, too empty and thereby too far away and too unapproachable, architects become above all the manufacturers of distance and inhospitableness. (Joedicke 1990: 142).

This quotation by Herman Hertzberger from 1973 is representative of structuralism's characteristic rejection of large-scale architecture in favor of smaller units which can seem coherent and self-contained and can add up to form a larger whole (Peters 1978: 601).

"Unity of the whole and diversity of the parts," as Bakema stated.[9] This maxim may be seen as an apt description of the Forschungsstadt in Neuperlach, obeying as it does these structuralist precepts. There, small simple units are composed into a larger whole. Siemens had after all turned down the first two designs for the Forschungsstadt precisely because they were still cast in terms too grandiose.

Openness of function

Structuralism furthermore repudiated in principle the functionalist idea of constructing buildings with rooms for certain predetermined functions. Instead of this, structuralists argued that every room in a building should permit of practically every function. (Peters 1961: 602) Neither the buildings nor its rooms should have fixed functionality. The aim was to design structures without specific form, effectively incomplete, and capable of multiple interpretation; to erect structures, indeed, that are never finished (Joedicke 1990: 153).

8 "Siemens- Bau der Superlative" In Münchner Merkur, 13.10.1978, p. 13.

9 Architektur-Urbanismus. Architectengemeenschap van den Broek en Bakema. Dokumente der modernen Architektur. Hrsg. von Jürgen Joedicke. Stuttgart 1976, p. 11.

The polyvalent room was central to the designs of the structuralists. Bakema had defined his objectives by transforming Sullivan's maxim "form follows function" into "form evokes function". (Joedicke 1990: 96) In his early designs Bakema had developed the concept of the "growing building" (Joedicke 1990: 155), with many small-scale structures allowing continual modification and extension, and ensuring the space for potential alterations. This openness of function and the accompanying expectation of flexibility of use was, as I shall discuss below, a key requirement of Siemens regarding its new science city.

Giving form to the "in-between"

In addition, the structuralists emphasized the importance of public squares and spaces *between* the buildings. These were intended to enable "contact communities" to arise and were regarded as the "key to the feeling of belonging together" within a community. Residential complexes were linked by paths and courtyards that should function as "communication networks" congruent with the "network of social relationships" [10]

The user as co-constructor

The structuralists saw the unfinished state of their buildings as a challenge to the occupant. They overturned the functionalist conception of the occupant, for whom it was necessary to dictate particular functions inside their own dwellings, and instead gave the occupants or users the responsibility of making an individual and subjective interpretation of the residential space placed at their disposal.

The unity of architecture and urban planning

In declaring his program as an architect, Bakema had spoken of "architecture-urbanism". It was henceforward regarded as obsolete to make a distinction between architecture and town planning; rather they should be seen as forming an integrated whole (cited by Joedicke 1976: 6, 11). He maintained that the principles of town planning – to wit, openness to change, multifunctionality, infinite extensibility, and spaces that promote communication – should be equally valid for the architectural design of buildings: "a building like a town, a town like a building." (cited by Joedicke 1990: 159; Lüchinger 1981: 68)

10 Quoted after Lüchinger 1981: 30.

These structuralist precepts take on material form in the Forschungsstadt in Neuperlach: the combination of order and chaos, the emphasis placed on public areas and courtyards (i.e. the "in-between" given form), the emphasis placed on incompleteness, the potential to be extended, modified or differently used, and above all the concern that the architecture should enable communication.

How these concepts related to the demands of *corporate knowledge production* is a question that has not yet been addressed.

Fig. 3: Siemens-Archiv, München

Structuralism and an urban mode of knowledge production

The choice of such an urban design in the case of the Siemens-Forschungsstadt has to be seen in the context of a "knowledge society". Siemens emphasized that such an urban design corresponded to the mode of knowledge production since the 1970s when knowledge had become one of the most important resources for innovation.

Two matters were paramount for Siemens in the discussion of the spatial organization of the Forschungsstadt, and they explain why structuralist concepts seemed to be appropriate: first, the increasing need for flexibility; and second, the belief that interaction and communication constitute a necessary condition for creativity and the production of scientific-technological innovation.

The need for flexibility

The literature has observed the permanent change of markets and the increasing need for flexibility which changed the production system around the 1970s (e.g. Piore/Sabel). At this time Siemens had also observed that tasks, demands and markets had permanently changed. The company thought about an appropriate mode of production capable of reacting to these changes. In terms of spatial organization Siemens believed that the construction of several small buildings instead of one big building – as was usual before – would enable the company to react fast to changes in the market. Following the structuralist precepts and designing the Forschungsstadt as a "city" made it possible to add different buildings in various sizes, according to what was necessary in view of developments on the market.

Moreover, Siemens planned 30 % of the whole research area as "reversible spaces" which had no function at the beginning. (Baumeister 9 /79: 866) No fixed walls were built; instead it was possible to change the walls within the individual buildings in order to deal with any new tasks that might necessitate a different utilization of space. As Siemens pointed out, this adaptability of space was necessary because, when the company started the planning process of the new Forschungsstadt, it did not know what tasks the employees would be working on since the world was changing so fast. Moreover, Siemens regarded it as absolutely impossible to anticipate what be on the agenda in 5 to 10 years' time. Thus, Siemens maintained, it was not possible to make any definite plans about the utilization of space, the number of employees or tasks.[11]

> Explosive technological development and increasing specialization in research have led to new forms of cooperation within a research company. The consequence of this is new conditions for the organization and rhythm of work, which in turn have a bearing on lab design and facilities. [...] The close interconnection of science and technology means that the spatial layout of a science building must be technically flexible enough to allow rapid adaptation to changing tasks. […] This led to the evolution of a modular construction system. (SAA-Neuperlach)

Robert Venturi summarized the necessities of science buildings in a very similar way:

> This is more and more relevant in science buildings in particular and in architecture in general, to accommodate change that is more characteristically revolutionary than evolutionary and that is dynamically wide in its range: spatial, programmatic, perceptual, technical, iconographic. In our time, functional ambiguity rather than functional clarity can accommodate the potential for things not dreamt of in your philosophy (Venturi 1999: 390)

11 SAA-Neuperlach, Siemens-Forschungslaboratorien in München-Perlach.

Thus, the modular structure of the Forschungsstadt, the high number of individual buildings seemed to offer a solution in a permanently changing world. As an urban environment, which never is static, which will never be finished, which is continually re-built and constantly responding to change, science buildings were supposed to be flexible, unfinished and also able to react to changes. Structuralist architecture seemed to offer a spatial solution for this consideration.

Knowledge and creativity

Through concentration and interaction we expect the growth of ideas and of creativity. (Siemens-Mitteilungen, 11/1976: 6)

At the moment when knowledge became an important resource for economies, the urban mode of organization of "science cities" and research areas was discovered. Companies as well as politicians referred to the idea of the city as the genuine place of creativity. The central role of cities in the generation of new ideas, creativity, innovation and knowledge has been a matter of course for centuries. It is claimed that they offer the most appropriate setting for the pursuit of art, culture, scientific activity and for technological innovation. Theories and ideas are usually not generated in rural areas, but in lively, heterogeneous cities. Thereby, urban communication and personal interaction are emphasized as an essential condition for creativity. The fact that cities are traditionally defined as the typical place of innovation and creativity depends on specific features that are ascribed to the city in general, namely their characteristics as a social space, as a space of interactions, of networks and plurality. It was precisely these features to which companies and science institutions started to refer in the 1970s, and in particular in the 1990s. The Siemens' management stressed the importance of the concentration and the intercommunication of different departments, which, as the management felt certain, would lead to greater creativity: "A physician, an engineer, a mathematician – they all need incentives, communication and interaction with each other… If we concentrate their work in one place and enable their communication they will be much more creative." (Siemens-Mitteilungen, 11/1976, S. 6) These considerations were the background for the concentration of different departments in Neuperlach and for the simulation of an urban atmosphere. Communication and interaction evolved into management tools in order to enhance the output of ideas. Architecture was regarded as a medium to promote these processes.

The fact that communication and interaction between the employees was so strongly emphasized is, as I argued above, connected with knowledge becoming a decisive resource in the innovation process. Siemens employed the concept of the city in a very stylized manner. The concept of the city was used in a reduced form that focused on certain urban features such as flexibility, modularity, compactness, density, short distances, the possibility of interaction and communication and above all the idea of the city as a space of creativity. One could say that this concept of the city is a very specific one, which adapted the idea of the city to the specific needs of the company aimed at improving the output of technological-scientific innovations. The Siemens Forschungsstadt copied urban structures and tried to create an urban environment in order to stimulate innovation and creativity. A paradoxical situation may thus be observed here: while companies, universities and scientific institutions move out of cities, the idea of the city becomes a model for generating innovation.

Siemens was no exception. Particularly since the 1980s and 1990s the new guiding principles for the spatial organization of research areas, technopoles, etc. have been urbanism, interaction, face-to-face communication, interchange and networking among scientists, politicians and companies. The lack of an urban environment in the outskirts is compensated by newly constructed localities and social infrastructures such as "urban centers", squares, fountains, shops etc. An artificial urban environment is built in order to bring about an innovative and creative atmosphere. Thus, certain features of the city are copied in order to stimulate innovation processes.

Without any doubt, this "urbanization" of innovation processes is closely connected to the emergence of the high-tech industry and to the growing importance of knowledge in innovation processes. Thus the emergence of high-tech industries has changed both the spatial structure of cities as well as the organization of industrial research.

List of References

Architektur-Urbanismus. 1976. Architectengemeenschap van den Broek en Bakema. Dokumente der modernen Architektur. Hrsg. von Jürgen Joedicke. Stuttgart: Karl Krämer Verlag.

Boskamp, Inge. 1975. "Das Verwaltungsgebäude der Central Beheer in Apeldoorn." *Der Architekt*, 232-235.

Cook, Peter ed. 1991. Archigram. Basel, Bosten, Berlin: Birkhäuser.

Galison, Peter and Emily Thompson eds, 1999. *The Architecture of Science*. Cambridge: MIT Press.

Jacobs, Jan. 1961. *The Death and Life of Great American cities. The Failure of Town Planning*. New York: Random House.

Joedicke, Jürgen. 1990. *Architekturgeschichte des 20. Jahrhunderts. Von 1950 bis zur Gegenwart*. Stuttgart: Karl Krämer Verlag.

Joedicke, Jürgen and Egon Schirmbeck. 1982. *Architektur der Zukunft. Zukunft der Architektur*. Stuttgart: Karl Krämer Verlag.

Joedicke, Jürgen (ed). 1976. *Architektur-Urbanismus*. Architectengemeenschap van den Broek en Bakema. Dokumente der modernen Architektur. Stuttgart: Karl Krämer Verlag.

Livingstone, David N. 2003. *Putting Science in its Place. Geographies of Scientific Knowledge*. Chicago und London: University of Chicago Press.

Lüchinger, Arnulf. 1981. *Strukturalismus in Architektur und Städtebau*. Stuttgart: Karl Krämer Verlag.

Newman, Oscar 1961. "Preface" In Newman, Oscar, *CIAM '50 in Otterlo*. Arbeitsgruppe für die Gestaltung soziologischer und visueller Zusammenhänge. Stuttgart: Karl Krämer Verlag.

Peters, Paulhans. 1978. "Die Jahre von 1960 bis 1977" In Leonardo Benevolo, *Geschichte der Architektur des 19. und 20. Jahrhunderts*, Band 2, München: dtv.

Schäche, Wolfgang ed. 1997. *150 Jahre Architektur für Siemens*, Berlin: Mann.

Smith, Crosbie and Jon Agar eds. 1998. *Making Space for Science*. Basingstoke, London: Palgrave Macmillan.

Smithon, Alison. 1968. *team 10. primer*. Cambridge: MIT Press.

Venturi, Robert. 1999. "Thoughts on the Architecture of the Scientific Workplace: Community, Change, and Continuity" In Galison/Thompson, eds. 1999, pp. 385-398.

SAA-Neuperlach, Bauen für die Forschung. Das neue Laborgebäude der Siemens AG in München-Perlach. o.j., o.V.

Some Notes on Aesthetics in a 'Städtebau' on a Regional Scale

THOMAS SIEVERTS

The paper discusses the hitherto neglected, but nevertheless important role of aesthetics in shaping urban regions. It sketches the emancipation of 'Städtebau', working in the regional scale, from engineering, architecture and traditional urban design and describes new conditions for designing the region. It reflects on the lack of theory and on the dilemma of abstraction, on a too narrow sense of functionality and on the necessity of sensual awareness. It emphasises the meaning of aesthetics as a condition of public cooperation and ends with a plea for an extended notion of regional 'Städtebau'.[1]

'Städtebau', the regional scale and urban culture

I take the challenge of this invitation to talk at a colloquium of philosophers on technique and aesthetics to develop some considerations concerning a special problem of my profession as an urban planner: The relation of 'planning' and 'design' in influencing the shaping of urban regions. At present, the business of shaping an urban region is predominantly a matter of land market and land politics, engineering, accessibility, economics and piecemeal development on available land. Generally the role of aesthetics ends with the architectural scale or at the latest on the scale of a group of buildings or a park.

1 I learnt most about regional aesthetics by the walks and talks with Boris Sieverts. This paper is based on the discussions and results of the "Ladenburger Kolleg" by the "Gottlieb-Daimler-und-Karl-Benz-Stiftung", 2003-2005: "Mitten am Rand – Zwischenstadt. Zur Qualifizierung der verstädterten Landschaft." As an introduction into the complex research and its 12 publications, the results are summarised in Sieverts/Koch/Stein 2005).

The results of this attitude can be seen everywhere: What I call 'Zwischenstadt' and the Italians call 'citta diffusa' – those urban landscapes which are neither real landscapes nor real townscapes – these are not included in what the public understands under 'culture', though more than half the population lives and works there. It is for example significant that art generally has not invented and developed popular aesthetic dispositions, which might help 'ordinary people' to see Zwischenstadt in an aesthetic manner. And when art has begun to do so, like in some films or in advertisements, it has emphasised the boring, dark and ugly side of the agglomeration. Generally, people do not remember Zwischenstadt: With my students I exercised an experiment of the mind: I asked them to imagine in their mind their home-town without its historic core – this proved to be impossible, though the historic core covered only about 5% of the built-up area. It was however quite easy to imagine the town without its 'periphery'. This kind of environment is anaesthetic in the sense that people are only aware of it if instrumentally used, they are not aware of it with emotions. As emotions are a necessary prerequisite for attention and care, nobody therefore really cares for this kind of environment beyond its utility functions.

This state of the mind of the great silent majority of inhabitants was not always the case: There is a great and very popular tradition in pre-industrial times of designing whole landscapes: e.g. the "Gartenreich" (realm of gardens) of the Fürst Franz von Dessau, the landscape around the lakes between Berlin and Potsdam designed by Leneé, the great parks of Le Notre in France or by Olmsted in the USA. One of the last examples and a great exception to the rule is the Emscher Landschafts-Park, still in the making. But on the whole this pre-industrial tradition has vanished.

There are evident reasons for this situation: Urban Design or 'Städtebau', as we call originally the combination of urban planning and urban design in the German speaking world since more than a hundred years, is a child of the Industrial Revolution: It was created to control the evils the Industrial Revolution had caused in the cities. Städtebau helped to guide the enormous urban growth of the cities in the 19th and early 20th century and to provide the necessary technical and hygienic infrastructure. In the beginning, it was essentially a matter of engineering, and it still comprehends more than the term 'urban design', even if it came near to this term in the last decades:

In the next phase after the First World War, 'Städtebau' became an instrument of state-interventionism in the social democratic welfare state to facilitate social justice and to equalise social and cultural chances. In this period, Städtebau meant mainly publicly subsidised housing on a large scale in the form of large town extensions and new towns, but it also meant socially oriented infrastructure like schools, hospitals etc. After the Second World War 'Städtebau' was still essentially guiding urban growth, but in its last phase, from 1975 onwards it also tackled urban renewal. In the phase of the welfare state, 'Städtebau' was essentially a work of urban design on an extended architectural scale, integrating some social science in the end, as an instrument of large-scale state interventionism. With the crisis of the welfare state, Städtebau as urban design also came to a crisis, as its main purpose – the handling of growth – has seemed to disappear or has even been substituted by shrinking.

As a consequence, designing on the large scale of the urban region and taking responsibility for its sensual aesthetic qualities seems to be a field the professions of Städtebau and both urban planning and urban design as well as landscape planning and landscape architecture are not really prepared for. This fact concerns politics, administration and professional skills.

The dispersed urban region – the 'Zwischenstadt' as I call it – as we have today in Europe, is predominantly a footprint of wealth accumulated in the last fifty years. This wealth is the result of a growth in productivity based on a division of labour and cheap energy. This growth in wealth had serious consequences for the built environment: In the last 50 years, the built space occupied per inhabitant has tripled, the time for free, individual disposal has doubled, with strong demands on leisure and sports facilities and individual mobility which have increased incredibly!

Most probably this wealth will not continue to grow in the same way: Many people will have less income, many will have to work longer, and mobility will be much more expensive. This will have its effects on the urban region: It will most probably also be supported by the aging of the population and a slow reconcentration of population in old and new centres.

Some journalists hope, by this reconcentration of inhabitants, the unloved 'Zwischenstadt' will disappear, like a misfortune in urban history. But this will not be the case. It is more or less built and can only be transformed. Even with a shrinking population the global dynamics of the economy will influence the urban region further and will lead to even more dispersal: In the end, the old compact cities might appear perhaps like "erratic rocks" in an "urban sea", but not dominant any more. They will be elements among others, even if they have a special meaning and importance.

Though the majority lives and works in the 'Zwischenstadt', there is at present comparatively little public interest in urban regional problems. As mentioned at the beginning, the urban region is not counted as 'culture', there seems to be a kind of 'division of meaning': 'culture', that means the old compact city. The urban region is "a cultural no man's land": In spite of this fact, the 'Zwischenstadt' is attractive: People seem, in spite of the general dullness of the 'Zwischenstadt', to be content. The economy meets more liberal rules and less regulation and finds therefore more freedom of development than in the old city and infrastructure has a free path to develop. There is also a lack of consciousness of being part of a coherent urban region: The municipalities still continue to fight each other as in the old days before the evolution of a coherent urban region, they generally still have no 'regional feeling'. So the public climate does not favour designing on the scale of the urban region: This kind of environment is essentially anaesthetic in its character!

Some colleagues do not see a necessity for regional design, because population growth is stagnating or negative, and the economy is growing – if at all – only slowly. Even if the urban region will not disappear – they argue –, the means for a large-scale design intervention in the urban region does not exist any longer. But this is definitely not the case: The aging of the enormous urban fabric built in the last fifty years – more volume has been built in this period than in the last 5000 years altogether; – leads to 'natural' necessities of urban renewal; the structural changes in agriculture will open up new chances for large landscape design; the aging of the population needs new services, and the economy is under strong pressure to renew and adapt to the conditions of a knowledge-based economy.

To sum up: There are a sufficient number of very good reasons for skilful design interventions into the development of urban regions. And though 'Städtebau' has reached beyond the suburbs – the urban region – it has not found the right formula yet to transport its rich aesthetic tradition onto this regional scale. There are some difficulties, which are hard to overcome:

On the regional level, 'Städtebau' meets a kind of long established regional planning, based on regional science which has absolutely no aesthetic tradition and no aesthetic dimension.

'Städtebau' has lost its status as an important socio-political tool, which it was in the old social welfare state: There is no longer a responsible agent of a 'Städtebau' as a public matter.

'Städtebau' has to work with many political independent entities: There is not only no public agent responsible for the quality of an urban region as a whole, but not even a consciousness of municipalities of being part of a coherent urban region!

New conditions for designing the urban region

As we have sketched in the beginning, there are many good reasons to intervene by regional design into the more or less evolutionary and anarchic development of the urban region. We named several reasons: urban renewal, changes in agriculture, aging population and changes in the economy. I shall not discuss these 'endogenic' factors, but I shall emphasise here the arguments in connection with globalisation. From a global view the urban regions are conceived as socio-economic entities, and the global competition will be between urban regions and not between municipalities any more. The most precious 'goods' in a knowledge-based economy and in an aging society will be young people, intelligence and purchasing power. In this respect, many urban regions could do better, as the new 'creative class' of a knowledge-based economy looks for a special quality of life which many regions cannot offer yet because of cultural-aesthetic deficits. But compared with urban regions in other parts of the world the urban regions of Central Europe still have unique cultural and social qualities, and these might form a most valuable cultural capital in the global competition. Those qualities, which in the old, national economies were characterised as so called 'soft qualities', will become hard factors in the global competition of a new economy. The strategy of the IBA Emscher Park was already predominantly a cultural one in its emphasis on the large-scale design of landscapes and on good architecture, especially in the transformation of the industrial heritage into cultural institutions, and this is still a rather successful strategy.

But the qualification of the 'Zwischenstadt' is again a task, different both from designing in the historic city and in the old industrial landscape, because the hidden cultural qualities of the 'Zwischenstadt' – its aesthetics – are fundamentally different from the compact old city and from an industrial heritage. Designing on the large scale of the urban region is not just an extension of traditional design, in the sense that e.g. the town park becomes a regional park or the church tower the T.V. tower. The urban region seems to be essentially anaesthetic in character, that means, that people are not aware of their environment beyond the narrow turf that they actually use. This might have deeper reasons than just functional ones, for already about one hundred year ago the German sociologist Georg Simmel developed the theory that man in a metropolis develops an intellectual armour to protect himself against too many environmental stimuli (we shall come back to Georg Simmel later!). Therefore the conditions of design in the 'Zwischenstadt' are fundamentally different from both the traditional city and the old industrial landscape, and the traditional rules of urban design can be applied only partially: It is necessary to develop new design tools and techniques.

Design on the scale of the urban region has to handle special categories of space different the smaller urban scale. Designing space on the urban regional scale must not only aim at 'functionality' in the narrow sense of the modern movement, but it has to look for a large capacity of different meanings and functions over time. It must be 'open-minded' (Michael Waltzer) and inviting to appropriate it. Designing in this scale has more to do with (landscape) topology than with (architectural) typology, more with 'framing' than with 'painting', more with stage-building than with writing a play. Not "form follows function", but "form invites function"!

The different kinds of spaces in the urban region can be discussed productively in contrasting terms and metaphors like 'net and place', 'edge and centre', 'uniqueness and genericness' and 'participation and fight'. The metaphor of the 'net' leads e.g. to the aesthetics of flows and to a new understanding of the recent development of flow-connected' places; the metaphor of the 'edge' helps to read the urban regions as a new form of edge-producing city with different individual centres; the metaphor of 'uniqueness' directs attention to the new forms of urban nature and human initiatives at the new frontiers; and last but not least the metaphor of 'fight' sharpens the attention for new forms of political struggle and cooperation which can be observed recently in nearly all urban regions.

The special character of the 'Zwischenstadt' demands new design techniques: In the 'Zwischenstadt', we cannot speak any longer of one single form of aesthetic. At first sight we have to separate at least three different aesthetics: the classical aesthetic of conventional 'beauty', e.g. of the old city, the aesthetics of the 'prints of life', e.g. in the form of 'spontaneous appropriation', and the aesthetic of flows, e.g. in the form of transportation-networks. Looking at a it from different perspective, we can differentiate between two poles of aesthetic procedures: On the one hand, there is a thoughtful interpretation of the already existing environment, an interpretation e.g. by guided tours which help to understand the "hidden narratives of a hitherto silent" environment: Suddenly the urban region starts to 'speak'! This kind of deep understanding is a prerequisite for a design, which is rooted deeply in the situation.[2]

On the other hand, there is the concrete intervention, e.g. in form of a regional park, the shaping of a new 'address' with its own identity, the classification of a path network for slow traffic, making the urban region accessible for leisure movements, the contribution of a hitherto 'autistic' building to the public realm, just to name a few concrete forms of intervention.

In-between these two poles of a deeper understanding of the narratives of an existing environment on the one hand and the concrete, materialistic interventions on the other hand is a broad field of cultural interventions, from experiencing the urban region by the body in the form of sport, via artistic installations up to citizen workshops on possible futures of the urban region. Festivals and symbolic events can give certain points a special meaning.

'Städtebau' as design on the scale of the urban region has to work with all these different, seemingly eclectic categories of aesthetics. 'Städtebau' in this sense produces cultural events and will contribute in itself as a discipline and practice to regional culture, and is in this sense an essential part of a learning region.[3]

Beyond attracting young people, intelligence and purchasing power, there are deeper reasons that aesthetic qualities (in the wide meaning of the term) will become decisive factors in global competition: Without an aesthetic awareness of the environment, an awareness, loaded with emotions beyond function and utility, people will not develop care and responsibility for their environment! It is this dimension of design which is perhaps the most important one!

Though social conflicts cannot be solved by design, it is this deep connection of aesthetics with emotion and care, which transfer the aesthetic question into a social question!

But making regional urban aesthetics public issues is a difficult task, as it can hardly be avoided that design on the large scale of the urban region is rather abstract and long ranging. This abstractness and long timespan before realisation is one of the reason for the only limited direct interest of the public. To give large scale design the popularity it deserves, one has to complement it with concrete projects, or least to illustrate it with symbolic action in the field, which roots the long-range, abstract

2 Hille von Seggern translates Hans-Georg Gadamers term of "Verstehen" (understanding) into a productive design-method. (Seggern/Werner 2003).

3 Ursula Stein has worked intensively on the "learning urban region" and on the cultural character of planning and design procedures. Compare her forthcoming book (Stein 2006). Compare also Stein 2005.

regional design in the present. The more abstract the plan the more necessary it seems to work with real projects and simultaneously 'on' paper and 'with people'! Long-term specific strategies have to be combined with local projects and actions to provide some realistic "images" and to make them interesting for the public: The general aim should be, that the chances and potentials of the urban region will become hot public issues, as attractive as literature, theatre, fine arts or even sport! (Ursula Stein)

Some reflections on the dilemma of abstraction, function and awareness

We have described the emancipation of regional 'Städtebau' from engineering, architecture and traditional urban design to something new, which is not at all clear yet. And we have argued that culture and aesthetics of an urban regional scale will become important factors in the global competition between urban regions. We have stated that 'Städtebau' on a regional scale is not yet prepared to support culture and aesthetics in the urban region.

This has not only practical and historical reasons, there is also a lack of theoretical reflection and practical skills. We have already stated that 'Städtebau' on a regional scale has to develop a more abstract notion of space than architecture and traditional urban design without losing the sensual aesthetic dimension, which is an essential cultural quality in the life of the people. This extension of scale must lead to a far-reaching transformation of the profession.

To illustrate this transformation, I will just describe the enlargement of scales: Architecture works on the scale of 1:50 – 1:500. The traditional urban design works on the scale of 1:500 – 1:5000. The regional 'Städtebau' has to work on the scale of 1:5000 – 1-50000. This enlargement of scales is connected with a growing abstractness: The third dimension disappears at the latest with the scale 1:2000, the detailed, two-dimensional differentiation vanishes at the latest with 1:20000. This growing abstractness forces the 'Städtebau' on a regional scale to define its aims and contents much more abstractly than the architect and even the traditional urban designer. The conventional analogue sign-code of architectural presentations must be transformed into an abstract symbolic sign-code on the regional scale.

The urban region is complex, not just aesthetically, and it can be read differently: e.g. phenomenologically as a special fabric different from the traditional city, as a general structure which also intrudes and changes the old city, and last but not least as a type of urban space in steady transition. Each of these three interpretations are basic concerning urban design on a large scale, but the last interpretation of the urban region as a field in steady transition seems to be the most productive one when thinking about design, for it leads directly to the 'material' forces, as the "raw material" of design.

In this interpretation as an urban fabric in steady transition, one can emphasise the continuation of the old urban traditions, or you can, in contrast, look at the 'Zwischenstadt' as a field of new developments, as a new frontier of experiments and innovations. Both interpretations are legitimised, as the 'Zwischenstadt' is a field of 'Gleichzeitigkeit des Ungleichzeitigen' (The simultaneity of different eras). There is e.g. the strong movement of 'New Urbanism' in the first interpretation of continuity which tries to urbanise the 'Zwischenstadt' along traditional urban forms. Even if this seems to me very limited in scope, there are some good reasons for it, as there are still strong traditions which are vital and can be developed. But to me the most promising approach seems to be to work with the mutual influences of 'old' and 'new' urbanity in the sense of the second interpretation. Then you will discover the deep transformation of both old town and 'Zwischenstadt' in this era of globalisation.[4]

If you interpret old city and 'Zwischenstadt' as a regional entity in which old and new forces influence each other mutually, you can grasp the "real" socio-economic forces behind the transformation. These forces are analysed and explained first-hand by sociology and economics, and the close cooperation of urban design with these social sciences has been a standard goal in research and education for a long time. But this cooperation has become rather difficult and superficial, as both social sciences have 'lost' the category space as an independent factor more or less from their theories. Here I see a real problem of research and education, as there does not exist a good socio-economic theory of space as a necessary basis of interdisciplinary understanding. Perhaps the term and category 'capacity' may serve as a theoretical bridge and common denominator between the social sciences and design, as it can be usefully interpreted in both fields.

'Städtebau' should serve as a 'mediator', as a common platform between three-dimensional urban design with the large architectural project on the one side and the socio-economic and political-strategic planning on the other side. This function as a productive mediator does not work sufficiently in most cases yet.

4 Compare: Bölling/Christ 2005 and Bormann et al 2005. These two books document two different approaches and philosophies of regional design.

This general picture of the problematic relationship between planning and design must be modified according to the different cultural traditions on the European continent and in the Anglo-American world: In the latter world, architecture and urban planning had become separate disciplines rather early (the earliest school of planning was the 'School of Civic Design' in Liverpool, founded 1907). On the European continent, 'Städtebau' had been an unquestioned part of the discipline of architecture until the 1960s (the first faculties of urban planning in Germany were founded in 1962). It was treid to bridge The early schism between architecture and urban planning in the Anglo-American world by the discipline of 'Urban Design'. At the same time, on the European continent, the tradition of 'Städtebau' was still vital and still exists. But both traditions – 'Städtebau' and urban design – find their limits in handling the regional scale of the new urban agglomeration.

In contrast to the old town, the urban region cannot be grasped sensually by the eyes more or less as an entity, and it is used quite differently according to lifestyle, working conditions and personal interests. So there are different images and conceptions of the environment side by side, depending on lifestyle, experience and knowledge.

Urban regions are envisaged mainly instrumentally: Generally people see only those elements they use. In the mind, the spatial structure is reduced to simple patterns of orientation, the 'rest' is closed out: People experience the regional environment with an inner distance, rather without emotions. Georg Simmel wrote in 1903 'Die Großstädte und das Geistesleben': "Everything seems grey and of equal value". This 'mood', this attitude towards the environment seems to be characteristic for modern man, as Georg Simmel argued in his famous essay.

Simmel argues, that modern man in cities defends himself against the aggressive environment of the modern metropolis by filtering all impressions by his intellect, not by his emotions. He develops a kind of "psychic armour or shell" and an inner attitude, which 'dims' the differences of the environment to a kind of equal grey colour. One thing is not better than the other, all elements are 'gleichgültig' (indifferent, of equal value). This is a very good description of an anaesthetic attitude, which sees the environment without sympathy, without emotions and empathy.[5] This inner attitude could be translated as an awareness of the environment without a 'compassion of the senses' (Mitleiden der Sinne). The environment does not 'speak', it does not tell a story beyond its mere function. Or, as Kevin Lynch once remarked: 'The senses are underemployed'.

5 Compare: Sieverts 2004.

'Städtebau' as socio-aesthetic practice

Anaesthetics of a certain degree seems to be a basic condition of modern human existence in cities, and this condition obviously sets limits on a comprehensive regional design of sensual qualities. On the other hand, this means that conscious awareness directed to emotional, symbolic environmental qualities is something of special value in the urban region, because a focussed, broadly interested awareness with strong emotions, is – as we have mentioned before – a basic condition for treating the environment with care and responsibility! If something is not in the focus of awareness, it cannot be treated with care!

It is this kind of awareness, in which the different senses and emotions are integrated, which I call – in contrast to just functional or instrumental awareness – 'aisthesis' or aesthetics. Aesthetics in this context does not necessarily mean the traditional quality of beauty, but an awareness beyond the functional and instrumental realm. This kind of awareness is of basic importance not only for the visual dimensions of the environment, but for all kinds of regional projects. Even if it seems to be a modern human condition, anaesthetics is not an inherent characteristic of the regional environment itself, but a matter of attitude, of interest and convention. This conventional attitude can be broken up and transformed into an awareness of the generally hidden stimuli of an everyday environment not shaped by professional design, but by the forces of traffic, work, transport, consumer market, action and appropriation. I am convinced that it is a human desire to meet the world with open eyes and mental resonance, to listen to the narratives of our environments! If it is silent, the environment is culturally impoverished. In this sense, aesthetics belong to the basic conditions of communicating complex problems.

One could discern seven different dimensions of aesthetic attitudes and strategies, each forming a continuum from an extensive to an intensive character:

- Awareness: From anaesthetic via casual awareness to identifying.
- Learning: From understanding via experiencing to action.
- Involvement: From letting things as they are via modifying them to making them new.
- Innovation: From tradition via convention to experiment.
- Experience: From intellectual insight via sensual awareness to body experiences.
- Relation: From contemplation via interaction to appropriation.
- Movement: From the path via the threshold to the place.

Each single dimension and each degree of intensity has its own right, as it would be a wrong strategy to try to form the urban region everywhere with the same complexity and intensity: The anaesthetic or a more or less extensive casual awareness is the normal situation, the intensive awareness or even identification is rare and of special value. There will be few situations where all seven dimensions form a complex unity, but of course in most cases several of these attitudes and strategies will be combined.[6] The more the different human senses are involved, the stronger the mental impact and the deeper the memory!

Scale and 'invisibility' of the urban region need cultural translations: To realise a 'Städtebau' on a regional scale requires intensive communications and these make 'Städtebau' and its presentations an important part of public culture. By this character, 'Städtebau' as a procedure on a regional scale can transport cultural qualities beyond its functional and instrumental tasks: Public exhibitions and discussions can be interesting events, and the different steps of projects on a regional scale can be ritualised as events in their process of realisation.

The results of such strategies could be characterised as a kind of "social-aesthetics"[7], the 'real' problem, its translation into an aesthetic pattern and its public communication form together a new social-aesthetic unity. In this context, one could differentiate between three levels of reflection: The technical and functional task, the translation into a specific pattern of aesthetics and the public discourse integrating these levels of reflection.

To give you an example of my practical experience: The International Building Exhibition Emscher Park tried to work on these three levels simultaneously. One of the important technical and functional tasks was to improve some of the physical features of the Ruhr area, destroyed by 150 years of heavy industrialisation. The transformation and improvement of the degraded water systems was of special importance. To do this, it was necessary both to sketch an aesthetic vision of a new attractive landscape with new, future-oriented symbols and to install a public discourse, including festivals and events. Together, this was a large socio-aesthetic setting.

In summary, the lack of understanding between 'design' on the one hand and 'planning' on the other hand must be overcome. It will be necessary to develop new, extended forms of design, integrating the social sciences and engineering. The necessity for this integration is underlined

---

6 Compare: Hauser/Kamleithner 2006. This book discusses rather comprehensively many different aspects of the (an) aesthetic of the urban region.

7 Compare: Eckmann 2005. The book contains stimulating ideas on learning as a socio-cultural activity

by the enlarged physical dimensions of the urban agglomerations, by the extended realm of life into the scale of the region and by the political transformation from the welfare state to a new kind of enabling, encouraging and guaranteeing state, which will face a growing global competition and which must rely much more on a lively civil society than the old welfare state: Without a broadly interested public and its participation, shaping of regions will be impossible, and this public interest is connected to the aesthetic dimension!

The new forms of 'Städtebau' on a regional scale will extend the notion of urban design, which is still dominated by the tradition of architecture and landscape design. A first step, already realised at some schools, goes in the direction of landscape urbanism, but it must be extended further: The process-oriented disciplines of administration and economics must also be integrated into the notions of designing spatial strategies. For space is not just a geometric, physical entity, but has also several cultural and social dimensions: Space is produced in manifold, multidimensional social processes, and these processes must be taken into consideration and integrated into design! It is this extension of the notion of urban design which might help to develop a new 'Städtebau' for the urban region, integrating the old, valuable traditions of engineering and architecture with social geography, social sciences and political/administrative disciplines.

This extended notion of design could lead to three different groups of design-products:

- The design of methods of influencing attitudes and politics, regarding the different meanings of space.
- The design of participatory-communicative, administrative and economic procedures regarding the handling of space.
- The design of the material procedures of producing space itself in all its dimensions, including time.

The common denominator of these products might be a multidimensional socio-cultural notion of space which will open up a path of different dimensions of design perhaps in a kind of post-modern eclecticism, e.g. the instrumental, the symbolic, the playful, the communicative and the economic dimension of space. Design in this notion concerns the concrete environment as well as the paths of our understanding the environment and the way we put it into perspective: It is a complex cultural contribution, taking the new historical situation into consideration that man does not any longer transform 'nature' into 'culture', but continuously – in the sense of a reflexive modernity – old culture into new culture: Design is a perpetual transformation of a man-made world, which finds its main manifestation in the urban region.

List of References

Eckmann, Theo. 2005. *Sozialästhetik. Lernen im Begegnungsfeld von Nähe und Freiheit*. Bochum/Freiburg: Projekt.

Bölling, Lars and Wolfgang Christ. 2006. *Bilder einer Zwischenstadt. Ikonografie und Szenografie eines Urbanisierungsprozesses*. Wuppertal: Müller & Busmann.

Bormann, Oliver, Michael Koch, Astrid Schmeing, Martin Schröder and Alex Wall. 2005. *Zwischen Stadt Entwerfen*. Wuppertal: Müller & Busmann.

Seggern, Hille v., and Werner, Julia. 2003. "Verstehen oder: Wie kommt das Neue in die Welt? Studio2003 Ideen – woher nehmen?". In: *anthos* 4/03, S. 48- 57.

Hauser, Susanne and Christa Kamleithner. 2006. *Ästhetik der Agglomeration*. Wuppertal: Müller & Busmann.

Stein, Ursula. 2006. *Lernende Stadtregion. Verständigungsprozesse über Zwischenstadt*. Wuppertal: Müller & Busmann.

Stein, Ursula. 2005. *Planning with all your senses – learning to cooperate on a regional scale*. In: DISP162/2005 (download under: www.stein schultz.de)

Sieverts, Thomas. 2004. "Georg Simmels 'Die Großstädte und das Geistesleben' und die Grenzen der Ästhetisierung der Zwischenstadt". In: Georg G. Iggers et al: *Hochschule-Geschichte-Stadt*. Festschrift für Helmut Böhme. Darmstadt, p. 389-398.

Sieverts, Thomas/Michael Koch, Ursula Stein. 2005. *Zwischenstadt, inzwischen Stadt? Entdecken, begreifen verändern*. Wuppertal: Müller & Busmann.

Between Determination and Freedom: The Technological Transformation of Education and Learning

ALEXANDER UNGER

For a fairly long time the influence of technology on society and its subsystems is a major topic in the discourse of social science. A lot of research has been done to analyse this process and its consequences for society and social life. Surprisingly the educational system has received only little consideration in this context, even if on the one hand learning and education seem to be most important in a globalized and technological permeated world.[1] And on the other hand, technology, above all the computer technology, plays an increasing part in instructional and educational processes and through this more and more becomes the basis for the educational system. In this regard, the necessity of a new, technology-based learning culture is not only discussed but brought forward.

To fill this gap somewhat, this article focus on two significant trails of technological transformations that effected the educational system: the so called programmed instruction (PI) in the 1960s and the still continuing phase of E-learning that started in the mid 1990s. At first glance, both were aiming at the same goal: to boost the effectiveness of learning through a basal changeover in the process of teaching and learning. The examination of these two stages ought to display on the one hand, how the technological transformation of a social practise like teaching and learning can be realised and what consequences are associated therewith. On the other hand, it should clarify if the technological transformation of education has to be understood as determined or as a process that is technologically framed but shows certain degrees of freedom. Regarding

1 There are only few publications in this context that miss to emphasise the necessity of lifelong, self-regulated and competence-orientated learning.

this ongoing process, the article finally tries to clarify if the comparison of PI and E-learning can give hints for the shaping of the seemingly ineluctable technology-based learning culture.

Technology as media

One of the results that can be extracted from the socio-scientific discussion about technology is that the growing impact of technology on society can only be understood in its full extent if technology is treated as a ubiquity phenomenon. This means technology more and more enwombs social and cultural life, and by this gains the status of a background for society as a whole. On this account, the proposal has been made to talk about technology not only in terms of machines or (infra-)structure, but also as a medium (Gamm 2000: 275).[2] To understand technology as a medium also means that technology goes in-between and, through this, transforms and restructures social practise.[3] This function of media is reinforced by the emergence of the advanced information and communication technology (ICT). Two of the outstanding characteristics of this new technology are the abilities to simulate and proceed interactive processes and by using network technology, not only to boost the transfer and storage of information, but to establish new forms of communication.

Regarding these developments, it seems to be inescapable to broaden the definition of technology: Tcchnology, especially the ICT, is a device that can not only go in-between, but disclose technical or medial framed spaces in which social interaction can take place. In reference to the growing dissemination of this technology, it appears "that the relation

---

2 Regarding society as a whole, this development seems to lead to a state that can be described as a 'risky empowerment' by which, as social scientists like Senett have pointed out, traditional bounds and borders – often recognised as barriers for social interaction and mobility – become more flexible and fluent. While this process opens up new potentials and possibilities, it also goes along with the production of 'uncertainty', which leads on one hand to the problem how to control the development of society, and on the other to a lack of orientation and backing for the individual.

3 In the technical meaning, the medium is defined as a channel that goes in-between a sender and a receiver and allows the transfer of data which enables the extension of human communication or perception. In a more socio-scientific approach the medium does not only serve the transfer of data, but as (mass) media is closely connected with the generation of 'reality' and by this gains an influence on the development of the individual (compare Meyen 2005: 228).

between humans also as their relation to the world is largely mediated by technical media." (Fromme 2006: 1)

Virtual Spaces and Education

In respect to this background, virtual spaces, the cyberspace or the net receives a growing interest from the social sciences as a new cultural space which are accredited to produce a significant influence on social practises and the individual development. The impact of these 'new media' also seems to differ from the 'classical' media. Sherry Turkle was one of the first scientists to inquire the effect of computer mediated spaces on communication and the users self. In 'Live on the screen' (Turkle 1997) she combines the synchronic use of multiple applications, especially so called chat-rooms, with a post-modern conception of identity: Every chat-window represents a facet of the user's identity. By this and the ability of multitasking, the computer technology allows the user to act out different – even contradictary – parts of his identity, simultaneously.

In the German discussion in educational science, W. Marotzki pointed out the significance of this new cultural space in the context of education: The appropriation of virtual worlds with their special culture seems to lead to an enforcement of reflexiation and the transformation of the individual self- and world-relation – a process that is traditionally connected with "Bildung" (cp. Marotzki 2002). With the term 'Virtualitätslagerung', Marotzki also called the attention to the increasing interfusions of virtual and real spaces also as communities.

Besides these possible positive effects, the ICT still ought to be regarded as a mediatechnology in the meaning of transformation and framing: On the basis of ICT, social practises can be technologically transformed and re-implanted in the virtual space. If an already established practice is technologically reconfigured and implemented in a virtual environment, we can refer to this as a technological transformation or virtualisation.

Pedagogical Communication and the PI

In the scientific pedagogical discourse, the understanding of teaching, learning and 'Bildung' has always been closely connected with interaction in bodily presence. Even if instructional process media like the book or the blackboard inhere an important position, the 'real' pedagogical process has always been comprehended as direct (asymmetric) interaction between student and teacher. In this understanding the students' (moral) character was essentially shaped through the teachers' knowledge that was passed on in direct communication. Even if there seems to be no question that education is a form of direct social interaction, its organisation and effectiveness have always been criticized and doubted.

In the 1960s this criticism was lifted to a new level by certain cyberneticists. In their understanding, not only the organisation of education was responsible for their ineffectiveness, but the whole idea of teaching in bodily presence. With the arrival of the PI in Europe in the mid 1960s, the discussion about the pros and cons of educational-technology began. The concept of PI, which was designed on the basis of cybernetics, behaviourism and the mathematical theory of communication (Shannon/Weaver), was propagated as a new and more effective teaching-technology that ought to be able to substitute the teacher. The very basic idea of this technological approach was to re-model the interaction between learner and teacher in a way that it fit in the model of cybernetic feedback.

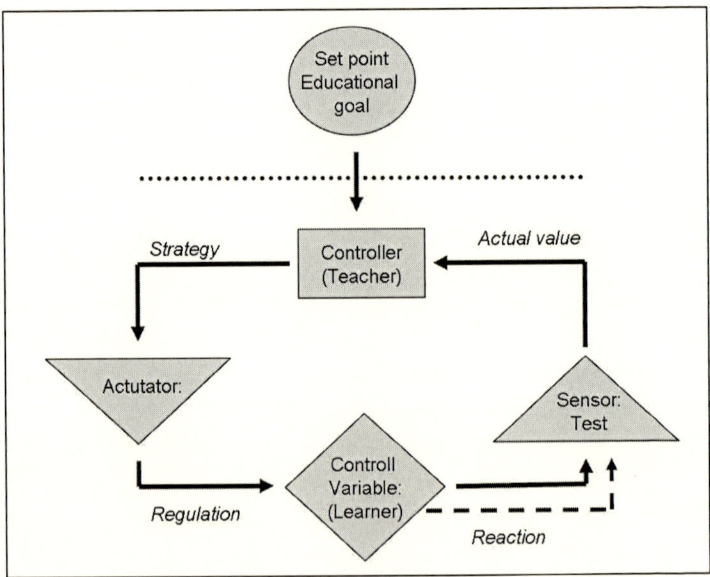

Fig. 1: The learning process as a feedback loop

Like the illustration shows, the cybernetic modelling reduces the *role* of the teacher to the *function* of a controller who uses strategies or programs to manipulate the 'target' and evaluates the successes of the operation by comparing the actual state of the learner with a predefined value. The comparison of the set point (should-be) with the actual value in short intervals allows the implementation of 'didactical branches' (see below). The cybernetic transformation of teaching and learning in bodily presence also demanded a specific structure of the learning procedure and the mediated knowledge.

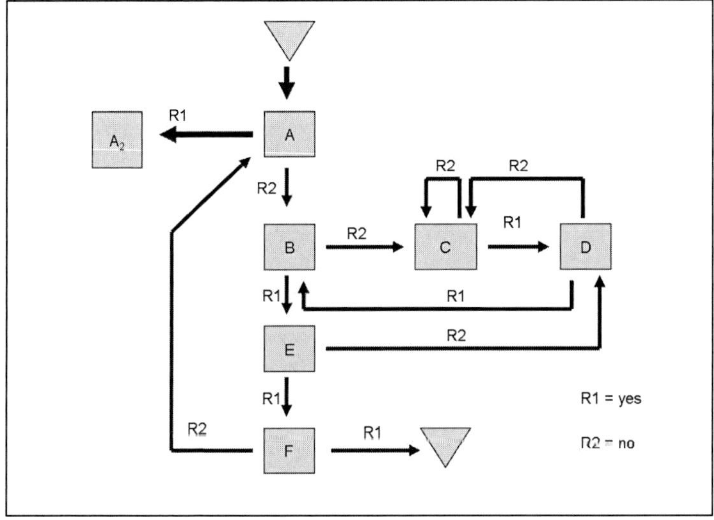

Fig. 2: Flow-chart of an learning-algorithm

Illustration 2 shows a learning-algorithm. The process of learning is presented in the form of a flow-chart and the to be imparted knowledge is organised in modules (A-F) which are combined with a test. The combination of a module to present information, a test to observe if the information has been absorbed and then the presentation of the next module is called intermitted learning. This 'mechanism' allows the regulation of the learning process on a micro level: In dependence to the test result, the learner is forwarded into different, specially prepared branches of the algorithm. When the learner fulfils or does not fulfil the required norm, a different branch is activated (R1 or R2). For example, the loop consisting of the modules C and D has the function to help a learner who is overstrained by Module B by imparting background knowledge to the content in Module B. Through this, the branching gains an instructional quality.To sum up, the transformation resulted in an analogy of pedagogical communication with an algorithm that was bound with the hope of maximising the efficiency of learning in the meaning of information transfer through permanent adjustment.[4]

What the PI aimed at was no less than the fundamental transformation of learning and teaching by converting it from human-to-human to human-maschine resp. human-programme interaction, according to the feedback-loop and the principles of modularisation and intermittent learning. In a sense, 'technology' resp. the medium wasn't longer used in the traditional way *in* the 'framework' of the instructional process, but the technology itself became the frame in which instructional processes took place. Hence the algorithm is indifferent by the fact whether the executer is human or not, and in the reverse eliminates the difference between human-controlled and machine or program controlled interaction.[5] By this a new paradigm appeared: the (technological) medium did no longer only serve to support the transfer of information, for example by illustration, but opend up a medial space for interactive learning processes without a human controller.[6]

However, the PI lacked the possibility to add new ways: Every possibility of interaction has to be anticipated and inscribed in the program. Though giving communication a high grade of accountability by eliminating 'ineffective' and unpredicted variations of the interactive process, the PI seems to undermine what it should achieve: the simulation and regulation of complex cognitive processes. While the teacher in the mode of bodily presence is on one side also concerned with stabilising and regulating the instructional 'system', on the other, he also needs the skill and experiencc to decide if a certain situation demands to alter the 'program' or even transcend it. This context-sensitive ability allows the teacher to deal with unexpected developments in the educational process that go beyond the abilities of an algorithm:[7] To sum up, anticipation

---

4 In the literal mathematical sense a learning programme is not an algorithm, because it doesn't automatically transform symbols in predefined steps, but functions as a script that waits for a low-level input (Schulmeister 1997: 27). In a broader sense, an algorithm can be understood as a complete set of commands that establish a formal system in which all possible actions respectively decisions are well-defined for all situation – and this is exactly the essence of a learning algorithm.

5 The reduction of the interaction established the possibility to delegate the reduced functions of the teacher – namely regulating, variation in the process of teaching and controlling – to a 'program' which can be executed by an 'interactive' machine resp. a computer.

6 Following this understanding Strittmaters and Niegemanns define interactivity as the ability to adapt the progression of information to the users input (compare Strittmater/Niegemann 2000).

7 According to Luhmanns (compare Luhmann/Schorr 1982) approach on education, pedagogical interaction is like all social interaction "infected"

and a program with multiple branches can't replace the human ability to act regarding a certain situation and person.[8]

Multimedia and E-learning

As has been shown, the technological transformation of education led to an extended significance of technology and technology based spaces in the context of education. In the PI, this space was designed as a causal space with the goal to determine interaction. This orientation received an increasing amount of criticism, not only from the pedagogical side but also with the further development of cognitive psychology, which showed the necessity of a more comprehensive understanding of learning. This criticism raises the question if the second virtualisation attempt, which started in the 1990's with the emergence of multimedia and E-learning, will be able to transgress the reductive paradigm of the PI.[9]

Unlike the Phase of PI, the 'multimedia revolution' includes a huge variation of different software types that integrate different functions also as concepts of learning. E-learning can be understood as a synonym for these manifold forms of advanced (multi-medial) learning software which ought to allow imparting content to groups and individuals – irrespective of time and space (cp. Minas 2002: 27). Multimedia is the label for an integrative technology based on the advanced microprocessor technology which allows the digitalisation of information, also as the integration, combination and manipulation of all types of information and (discrete and non-discrete) media in *one* technological 'environment'. In combination with network-technology, multimedia becomes a powerful tool for the processing and distribution of information and (medial) content also as new ways of computer mediated communication.

with a lack of technology. This means that social communication in its very essence is unpredictable. As Luhmann points out, the uncertainty and complexity of interaction can't be distinguished. This means social processes can't be determined, no matter what kind of 'technology' is used. The only way to handle this circumstance is to develop a technology which is based on uncertainty.

8 Finally it wasn't the transcendental argument that machines should not substitute humans and be involved in the mental and moral development which prevented the success of the PI, but its serve formal character and limitations simulating complex human communication and learning processes.

9 Meanwhile the concept of radical virtualisation changed to the more moderate of 'blended learning', that is the combination of virtual and 'real' elements to bring together the best of both 'worlds'.

In the educational context, multimedia is also understood as a new way of learning which is based on the simultaneous activation of different channels of the consciousness. This approach ought to allow a far more effective and interactive way of learning with an entertaining and motivating character. Through Multimedia, the learner seems to be delivered from the compulsion of traditional education.[10] While it is impossible to discuss all different kinds of E-learning software in this article, only a brief glance at some types can be given.[11]

Hypertext is not a learning software in the proper sense, but rather a principle for the organisation of information. Hypertext represents a non-linear structure, in which unitised content is linked in a way that allows free movement from one 'knot' to another. This type of software can be described as non-intentional because it doesn't aim at specific learning targets or integrate regulative scripts which influence the movement in the information-space, whereby the links or knots lack an instructional dimension. This 'lack' allows the user to generate individual trails through the content. Thus, Hypertext is closely connected with the idea of self-determent learning in a non-sequential setting.

Drill&pratice Software, tutorial systems and LCMS are often pooled under the label exercise-software. While these 'subtypes' coincide in their instructional approach (intermitted learning), they can be distinguished in reference to the implemented functions, interactive elements and the degree of regulation. Drill&pratice Software often exerts a strong regulative and controlling influence on the learner, and by this show a strong relation to the PI. Following the heuristic model of Baumgartner and Payr, this type of software is limited to learning in the sense of receiving and recalling information and only rarely reaches the level of applying (cp. Baumgartner/Payr 1994: 95).

Unlike drill&pratice software, LCMS aren't only capable of presenting content, but also provide an author module which allows the free creation of information modules and frames and, accordingly, the modularisation of nearly every content. In Addition to an author and presentation module, functions for the administration of courses and (the authorisation of) users are often provided, also for communication and collaborative working. The LCMS Moodle, a currently very popular open-

10 This example shows, that the introduction of Multimedia and E-learning was connected with large-scale promises that did not only aim at the substitution of the teacher, but the educational system as a whole (compare Perelman 1992).

11 Since the simulation of interaction is still one of the most important objectives of computer based learning, interactivity is one of the main criteria for the evaluation of e-learning software (compare Schulmeister 1997).

source software, shows, that the focus changed from the learning module to administration and communication.

Tutorial Systems aim in a different direction. While they also provide content that is adapted in dependence to the users input, they strive for a perfect simulation of human learning (cp. Minas 2002: 81). Therefore, intelligent tutorial systems (ITS) use methods provided by the AI to automatically create courses from a content pool in dependence to the users needs. ITS also try to analyse the cognitive process of the user while learning and permanently adapting the created course to the users cognitive development.

Some LCMS, like L3 (cp. Meder 2006), try to combine the best of these two approaches: while L3 uses an sophisticated instructional ontology and educational metadata to enable the dynamic adaptation of the provided content, all system-generated decisions and structures are transparent and the user is given the opportunity to change them. For instance, the learner can follow the system generated course, but by using a button the complete course structure is displayed and he can navigate in or between courses without any restrictions.

Simulations, however, represent a completely different type of interactivity. They simulate objects, processes or systems based on dynamical mathematical models. By interacting with the simulation, that means manipulating the parameters, the user can experience the simulated object and its behaviour in real-time. This is a completely different approach than intermitted learning which allows the virtual experience of to-be-stats without real risks. Correspondingly, simulations aim at a level of learning that goes beyond absorbing information, since they provide a free 'manipulating-space' without an adaptive functionality.

Between regulation and freedom

After recapitulating the discussed software types, we get an ambivalent picture. On the one hand, we still find a lot of software that shows a strong tendency towards intermittent learning and system regulation. On the other hand, we find a tendency to self-determent learning and non-regulated spaces. While ITS still follow the vision of a perfect learning algorithm some LCMS re-import human-to-human communication, collaboration and simulations of more or less unregulated spaces that the algorithmic approach tried to eliminate for effectiveness. Some LCMS and simulations also give the user the opportunity to manipulate or even generate their own content. This represents a form of learning that even goes beyond manipulating a simulated object or system.

From this perspective we can distinguish two possible cultures of technology-based learning which follow different understandings of interactivity. One is bound to the idea of perfect regulation that, by analysing the user behaviour, automatically generates the right instructional treatment. In contrast, Haack emphasises in his definition of interactivity the active part of the user and the degrees of freedom: The software has to offer an environment in which the user can play an active or even constructive part (cp. Haack 1997). This definition refers to an(E-)learning culture that is influenced by the idea of open, developable systems and learning as a self-determined, active process that has to be enriched with communicative and collaborative elements.

Above all, open means that all system-generated decisions are transparent and can be cancelled so that an unregulated movement in the virtual (information) space is possible. Instead of following the strategy of 'complexity-reduction' that is containing options of interaction to cleanse the virtual space from uncertainty, this approach shows a tendency towards unregulated spaces that can be explored and seem to open potential for the acquirement of advanced knowledge and competence.

Nevertheless, both cultures have to deal with specific difficulties. The algorithmic approach still struggles with a seemingly insuperable boarder concerning the formalisation of social interaction. Open learning environments, on the other hand, raise the question if exploration is a form of learning which can lead to predefined learning targets and can assure the imparting of the knowledge that is necessary for social integration and the exigencies of the information society.[12] Regarding the new defiance of the information society, we can say that the algorithmic approach is not sufficient regarding learning processes that go beyond receiving and recalling, but open spaces alone can't also solve the problem.[13]

Assuming that the future learning culture will be technology-based, we can say that all discussed forms of E-learning have their eligibility regarding certain tasks. But it also has become clear that the substitution of social interaction in the context of education is a dead end. While regulated forms of learning are essential, learning still needs human-to-

12 Opening the system also gives the user-community the chance to enhance the available knowledge, like it is practiced in Wikipedia. But on the other hand goes along with the latent danger, that standards can be undermined or even the whole system.

13 This deficit also can't simply be solved by integrating multimedia content, because it's bonded with the structure of the learning process and the degree of regulation and adaptation rather then the medial presentation. As we can see multimedia is a label that rather deflects from relevant structures, than helps to discover

human interaction and collaboration also as 'spaces of uncertainty'. From this perspective, virtual spaces seem to be quite similar to real learning settings. Both show a certain measure of regulation on one hand and liberty of movement on the other. While a certain amount of regulation seems to be indispensable for intentional processes, only a certain openness allows learning as an active and creative process through which (meta-) competences can be acquired.

In this respect, the design of virtual learning spaces seems to be an important task for educational science which demands an understanding on how virtual spaces influence learning and the individual development. As has been shown, unlike real learning settings, virtual ones are not dominated by a teacher in bodily-presence, but are pre-structured by the program code. From this point of view it seems to be an important task to balance regulation and freedom while designing virtual learning environments. Regarding the discussion of E-learning, it seems to be necessary, to integrate 'closed' also as open elements in one virtual environment. This can only be achieved by a way of coding that is based on the notion that even in intentional learning processes uncertainty is an irreplaceable component: The aim is not to completely distinguish uncertainty and complexity, but rather to generate them on the basis of a 'formal' technology and integrate them in 'ambivalent' learning setting.[14]

List of References

Baumgartner, Peter; Payer, Sabine. 1994. *Lernen mit Software. Lernen mit interaktiven Medien 1*. Insbruck: Studien Verlag.

Fromme, Johannes. 2006. "Socialisation in the age of new media." *Medienpädagogik* (online magazine) 5: 1-29. Available from: www.medien pead.com.05-1/fromme05-1.pdf [cited 1 November 2006]

Gamm, Gerhard. 2000. *Nicht Nichts. Studien zu einer Semantik des Unbestimmten*. Frankfurt/M.: Suhrkamp.

Haack, Johannes. 1997. Interaktivität als Kennzeichen von Hypermedia und Multimedia. In: Issing, Ludwig; Klimsa, Paul: *Information und Lernen mit Multimedia*. Weinheim: Beltz, pp. 150-166.

14 Regarding the concept of 'blended learning', another challenge in designing learning environments is the combination of virtual and real (in bodily presence) elements in a hybrid learning environment (compare Unger 2006).

Luhmann, Niklas. 1982. "Das Technologiedefizit der Erziehung und die Pädagogik." In: Luhmann, Niklas; Schorr, Karl: *Zwischen Technologie und Selbstreferenz*. Frankfurt a.M.: Suhrkamp, pp. 11-40.

Marotzki, Winfried. 2002: "Zur Konstitution von Subjektivität im Kontext neuer Informationstechnologien." In Marotzki, Winfried; Bauer, Wolfgang; Lippitz, Wilfried. *Weltzugänge: Virtualität – Realität – Sozialität.* Jahrbuch für Bildungs- und Erziehungsphilosophie. Hohengehren: Schneider, pp. 45-63.

Minass, Eric. 2002. *Dimensionen des E-learning. Neue Blickwinkel und Hintergründe für das Lernen mit dem Computer*. Kempten: Smart Books.

Meder, Nobert. 2006. *Web-Didaktik. Eine neue Didaktik webbasierten, vernetzten Lernens*. Bielefeld: Bertelsmann.

Meyen, Michael. 2005. "Massenmedien". In Schrob, Bernd; Hüther, Jürgen: *Grundbegriffe der Medienpädagogik*. München: kopaed, pp. 228-233.

Perelemann, Lewis. 1992. *School's out. A radical new formula fort he revitalization of america's educational system*. New York: Avon Books.

Schulmeister, Rolf. 1997. *Grundlagen hypermedialer Lernsysteme*. München: Oldenburg Verlag.

Strittmater, Peter; Niegemann, Helmut. 2000. *Lehren und Lernen mit Medien. Eine Einführung*. Darmstadt: Wissenschaftliche Buchgesellschaft.

Turkle, Sherry. 1997. *Life on the screen. London*: Orion Publishing.

Unger, Alexander. 2006: "Umgebungsanalyse: Gestaltung von virtuellen Lernumgebungen." In Sesink, Werner: *Subjekt – Raum – Technik. Beiträge zur Theorie und Gestaltung neuer Medien in der Bildung*. Münster et. al.: Lit Verlag.

(Self)Normalisation of Private Spaces – Biography, Space and Photographic Representation

KATJA STOETZER

Private spaces hold a special potential for analysing processes of adjustment and resistance to normalising effects. The inhabitant moves between the opposing poles of presenting a façade according to anticipated norms and revealing personal attitudes and biographical references while the arrangement of private spaces simultaneously serves as a guideline for potential visitors about what is accepted or tolerated – applying a normalisation of action mostly by the reception of the visual.

Introduction

Spaces for living are open to a selected public, confronting the inhabitant with specific expectations about the functional and aesthetic arrangements that serve as guidelines for the estimation of normality. Between presenting a façade according to anticipated norms and revealing very personal sides, the duality between "private" and "public" space gives way to applying different grades of "openness", especially in students residences, where space is a (very) limited resource. If access to the very private spaces is granted, personal attitudes and biographical references can be observed, serving as a repository for the self by largely excluding these normalising processes, which are typical for stage-like foreground-spaces. On the other hand, normalising effects do not refer solely to the inhabitant. The arrangement of private spaces serves as a guideline for potential visitors about what is accepted or tolerated – applying a normalisation of action mostly by the reception of the visual.

Next to aesthetical and functional techniques of producing private spaces, they transport non-narrative guidelines by visual markers. To identify them and locate the idea of different visual markers to overcome a dichotomic public-private divide in the theoretical field of sociology and biographical studies, this article will focus on students' private spaces and examine normalising effects within their production of space. Space is conceptualised as a relational model of simultaneous production by action and reception (of atmospheres, arrangements, people and social goods) according to the sociological basic concept Martina Löw (2001) developed.

"Space" as socially produced

Based on Euclidian geometry, space was understood to be independent from social action – a concept which was well passed down for generations and which has produced many irregularities with the advent of modern communication and transport systems: space seems to vanish, only the time needed to overcome specific distances seemed to be relevant.

Space conceptualised as socially produced leaves these ambiguous effects behind: Firstly, the *relations* between located human beings and social goods (materiality) substitute the absolute perspective – meaning there is not one "right" perspective to explain spatial phenomenon from –, secondly it takes the history of social processes into account as all spatial arrangements are results of prior (re)arrangements and space is no longer conceptualised as independent of social action. The relational approach (cf. Löw 2001) used here differentiates between two primary processes for the constitution of space: *spacing* as placing or moving social goods (or persons including one self) and *synthesis*, the perception of these (re)arrangements with their non-material components like atmospheres, lights or smells.

Spacing refers to the plane of action-based theories; it is the process of placing social goods, human beings or positioning primary symbolic markers to make entire ensembles of them identifiable as such. Synthesis, on the other hand, refers to structural order – by processed of perception, remembrance or imagination human beings or social goods are grouped to spatial arrangements, to spaces; but not to arbitrary ones. The constitution or production of space is pre-structured by class-, cultural- and gender specific habitus and is influenced by spatial concepts accepted in society, by the place the production of space itself occurs (e.g. the materiality found in situ limits the possibilities of (re)arrangement), by biographical elements and spatial structures like legal boundaries. The latter is quite obvious since the possibilities of a landlord are quite different to the ones of a person with a short-term contract only in modifying and rearranging the spatial distribution of walls, windows etc. within a flat or house.

Both processes are interwoven and take place simultaneously, leading to complex possibilities of overlapping spaces, since the absolute perspective of the Euclidian concept was substituted by a relational one. Nevertheless, power plays a very important role in the production of space: Different groups or even individuals try to make their production and perception of space the dominant one, which depends largely on their power and the resources they can allocate, especially economic, cultural and social capital. Habitus and the provision of Bourdieu's differentiation of capital is another important factor in analysing the production of space. Neither spatial relations nor the material and non-material things found in situ can be preferred in the analysis, both of them have to be (re)considered.

Methodical Design – Photoelicitation and visual work

The students' private spaces were investigated using a newly developed method, the "request to visual work" (Stoetzer 2004): Students with different professional backgrounds were asked to take photos of their residence, their homes, by themselves. Directly afterwards the pictures were printed on the spot and should be described as a kind of self-interpretation by their creators, which was the starting point of an interview conducted with them using visual material as stimulus – a kind of photoelicitation.

Along with the photos the students took, single-shot panoramic pictures of the depicted places were taken that allow a computer-aided simulation of different perspectives later on, capturing a field of view of 360 degrees. The panoramic shots allow to identify – during the interpretation – the perspectives and arrangements that were *included* in the photos the students took by themselves – *as well as what was left out.*

This methodical approach combines the different possibilities photos can be used in the context of interviews, taking advantage of their corresponding pros:

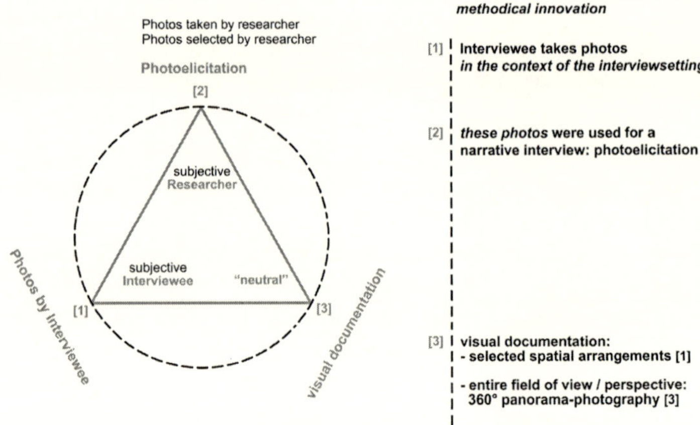

Fig. 1: Usage of Photos in context of interviews. © Katja Stoetzer

The triangle in figure 1 represents the different approaches to use photos in interviews. It differentiates between *photos taken by the interviewee [1]* – usually old ones from photo albums to call certain events to mind again, or recent photos, but used in standardised visual-content analysis (Wuggenig 1994) –, *photos taken by the researcher used for photoelicitation [2]* – they were either taken or selected for that purpose by the researcher – and *visual documentation [3]*, e.g. of items mentioned during an interview. The first two approaches are labelled "subjective" since they place special emphasis on the selection processes of the individual interviewee [1] or systematically by the researcher for the interview to conduct [2]. The dotted circle in figure 1 encompassing the triangle is unwinded on the right side, showing the use of the different approaches during the course of the interviews conducted as well as their interlinkage. Due to this complexity and the energy the interviewees expanded with taking and describing the pictures, the process is labelled "request for visual work" since it goes beyond what Harper introduced as "photo-elicitation" (Harper 2000).

Social science methodology holds a great variety of concepts about interpreting narrative or text-based data (c.f. Strauss 1996 used here), but there are only very few ideas about how to proceed with the visual. To strengthen the utilisation of visual artefacts in the social sciences as a source of its own a new approach of analysing pictorial representations, especially photos, was elaborated (Marotzki/Stoetzer 2005). It was slightly modified in its last plane to encompass biographical references in the spatial arrangements depicted, instead of the originally developed analysis of educational and cultural framing. The method of pictorial analysis is rooted in the methodological approach provided by Erwin

Panofsky (1979), which has been repeatedly revised and developed further, e.g. by Gillian Rose (2001), Ulrike Mietzer and Ulrike Pilarczyk (2003) and Ralf Bohnsack (2003).

Furthermore, integrating ideas from film-theoretical approaches, especially David Bordwell and Kristine Thompson (2004), takes the possibility to analyse the students' series of photos for implicit dramaturgical effects into account. Their differentiation between *plot* and *story* reflects very well the theoretical one between the constituting processes of the production of space, *synthesis* and *spacing* within the spatial theoretical framework elaborated by Martina Löw (Löw 2001).

The analytical model, the *4dimensional pictorial analysis (4dpa)*, operates on four planes, which refer to spatial, relational arrangements of persons and social objects depicted and their biographical references – not to absolute Euclidian geometrics augmented with a temporal frame: The first three dimensions of analysis mainly deal with a methodically controlled identification of the depicted arrangements, the *description of objects*, and their meaning variant to cultural backgrounds as well as their relational order. Analysing the relational references of the identified objects (*hypotheses and readings about the order of objects*) captures some of the elements of the spatial constituting process of synthesis, giving priority to the recipients' view (*story* according to Bordwell/Thompson). The third plane then includes the non-material aspects of spatial relations like atmospheres, lights and colouring, arrangement of fore- and background including the position of the interviewee as photographer taking the picture – *adding formal means of picture analysis*. The last level of analysis concentrates on *biographical references* found *within the spatial arrangements*.

Normalising processes within the visual

Equipped with this methodical approach, the students' series of photos have been analysed at large, starting with the interpretation of single pictures and integrating the rest of the corresponding set of photos during the last step of the method described to stabilise hypotheses or discard readings that have been proven wrong. In the final step of the interpretation of the visual, further evidence in form of analysed interviews (following Glaser and Strauss) was added and referenced against the photographic interpretation, because the visual can embody hints to biographical topics while the narrative can mirror spatial and aesthetic arrangements only partially. The constitution of space and its normalising processes will be discussed along the categories of *public and private* with respect to the students' living space as well as *order and disorder* being the second category of the production of space – both of them have been elaborated from the empirical base analysing visual and narrative data.

Starting with the latter one, *segmentation of space* occurs if visual markers like shelves, screens or curtains are used as borderlines or even barriers to separate functionally differentiated "corners" of a room. This observation is independent from the number of rooms at the interviewees' disposal, since these visual markers can simulate the segregation into different rooms by proper walls. *Overlapping spaces,* on the contrary, do not use sharp distinctions; diverse functions are more integrated into the entire ensemble – allowing different layers of functionality to overlap in relational space. This can be visually expressed by open or even removed doors between different rooms – if more than one is at their disposal – or by integrating different functions seamlessly into the entire ensemble. Figure 2 shows the division into different spaces located at the corners of the room by their usage:

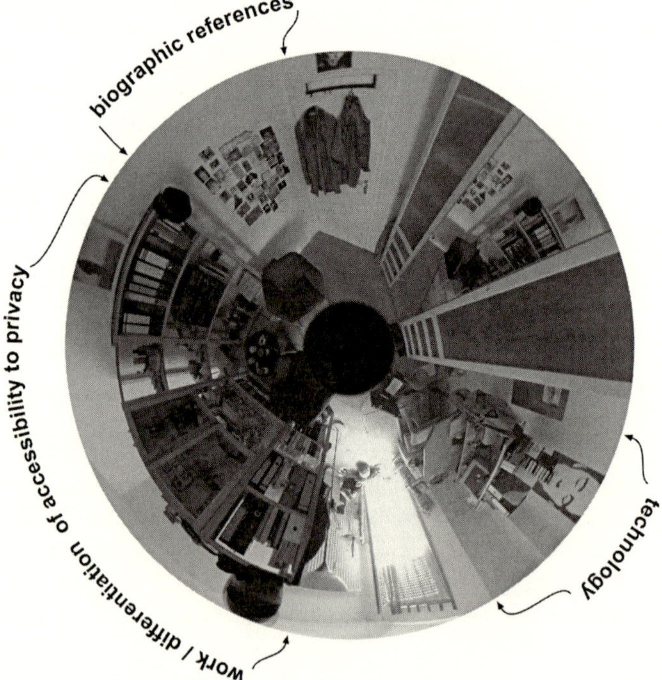

Fig. 2: Fragmentation of space – panoramic picture. Photo from the empirical base of the author's PhD-Project "Raumbiographien". © Katja Stoetzer.

At first sight this statement looks hardly surprising, like a commonplace or at least "natural" – at closer inspection, however, the question arises why this differentiation of the production of space could be "read" so easily.[1]

From the arrangement and order of social goods and persons constituting the production of space, whose grouping to ensembles corresponds with their functional aspects, guidelines are derived for just that process of synthesis: visual stimuli or marks. They are subject to processes of naturalisation, hindering them from being directly observed and presenting them as a "natural" part of the arrangement depicted. It is "normal" for ourselves to recognise immediately what kind of room we are dealing with and which functions are included or excluded with regard to social sanctions; even if hybrid constructions as multifunctional furnishing based on limited spatial resources are taken into account.

The underlying normative notions concerning order and disorder can be reconstructed methodically. Assuming that spaces are actively (re)produced as relational arrangements and order of human beings, animals or social goods specific to places, the relationship of order and disorder refers to the structural arrangement of objects on the microscopic level: They can be grouped into appropriate ensembles by action or perception in order to reduce the level of complexity of the arrangements in question – according to the observer's normative perspective: If the reduction of complexity fails, an underlying ordering principle cannot be recognised and the accommodation appears to be untidy and without order in the eyes of the beholder.

The inhabitant's subjective notions regarding the level of order of the flat occupied give insight to the constitution of spatially mediated and visually communicated agenda of normality. The permanent impression of the accommodation's state as the morning after a party can indeed be part of the inhabitants every day life (cf. figure 3) – or staged.

1 The photos shown here in greyscale due to the publication process can be viewed in colour on the author's webpage: *www.raumbiographie.de* (click on data).

Fig. 3: Order/disorder in students' private spaces – expectations about normality. Photo from the empirical base of the author's PhD-Project "Raumbiographien". © Katja Stoetzer.

Expectations about normality can also be found with regard to the existence of specific furniture or lights as well as the distribution of furnishing depending on the amount of space the inhabitant can exercise power on [number of rooms, legal form of dwelling: community, individual renting (with(out) furniture etc.)]. This is repeated on the macroscopic level by attaching functions to ensembles of goods and persons or even spaces themselves.

This level of the constitution of space corresponds with the institutionalised production of space of the theoretical model of relational space, stressing repetitive spacings that reproduce institutionalised space secured by resources. The existing structural arrangement and order is reproduced by everyday action, referencing structure and agency on each other (cf. Giddens 1997, Löw 2001). This can easily be shown by floor plans of advertised estates with the different rooms reserved for intended purposes – given a three-room-flat they are typically: living, sleeping, child.

In designing the private spaces the students therefore reproduce standardised arrangements and order. Staging a room as "consulting room" or as appropriate for "juniors" involves deliberate placing of social goods with a corresponding habitual and biographical background. The desire to create normality by reproducing images of standardised arrangements and order with respect to three- or four-room-flats is obvious. The changeableness of these notions of normality and its relatedness in cultural-historical terms becomes clear looking back in history. Insofar processes are identified, whose momentary status quo becomes manifest spatially and is reproduced by routines, but with the possibility of change.

The production of normality in students' dwellings can even be highly staged, something Thomas Düllo refers to as "Ikeaisation of the living environment" (2000 – own translation): The addiction *not* to "design" doesn't reflect any state of distress, but that the provisional arrangement was produced. Likewise the impression of the living environment as imperfect and incidental was deliberately created, with randomised arrangements and combination of furnishings dominating the rooms and creating the impression of nativeness as a top priority. To evoke such an impression is the result of an exhausting process and not only carelessness.

The normalising potential of spatial arrangements regarding the second category, the *public-private-divide*, can be traced back to the visual, too. Objects or groups of them serving as visual markers divide parts of the space depicted and take care that certain places cannot be accessed easily or unattended. Nicholas Blomley has played this through on the theoretical base of legal division of space with reference to the visual markers Carol Rose (1994) attributed with power:

> Within both the formal legal realm and popular culture, there is an expectation that property will be communicated through visual markers, such as fences, hedges, buildings, improved land, and so on. Visibility can make a difference (Blomley 2005: 282).

What Blomley describes as the public-private-divide with respect to the legal sphere and freehold property is a constitutive division within western culture: "This demarcation is constitutive of western philosophy and the empiric realm in our modern societies" (as Ulla Terlinden argues, 2002, p. 143 – own translation), going back to the division of Oikos and Polis in the Greek's ancient world. The processes of naturalisation reach back accordingly making sure this demarcation is not regarded as a human social product, but as an everyday banality that does not need any further understanding since everybody seems to know where to draw the line between the two terms.

Labelling the private with visual markers can be found in students' private spaces itself to a varying degree: Some regions within the interviewees' private spaces can be characterised as more public then others for their accessibility by peers or visitors. Other private areas were blocked intentionally or furnished that admittance of this "private privacy" is only granted to very close friends or relatives (cf. figure 4). Another technique is to take only photos of selected rooms but to leave others out that don't seem private from the interviewee's perspective, e.g. a workroom – reminding of the spatial division of labour combined with the public-private-divide at the advent of the industrialisation.

Precisely this inner differentiation is relevant: The public-private-divide operating with the same techniques of visual markers can even be found in spaces belonging ipso jure to the private sector anyway. But this inner differentiation is much more blurred and heterogeneous than the legal one due to its greater complexity: Firstly, the symbolic normalisation using visual markers doesn't have to be unambiguous. Secondly, the differentiation of public and private occurs on different levels of meaning: Next to legal concepts, psychological ones exist using proximity and distance as well as biographical backgrounds and motives of staging private spaces for distinction, allowing or denying access to more private regions (of the private space) that are regarded as authentic (cf. Löw 2002).

Anthony Giddens discusses the public-private-divide in context of a regionalisation, extending Goffman's front-back distinction critically to a broader spatial level (cf. Giddens 1997: 174ff.; Löw 2001: 41, 72, 168) with at least a connotation of privacy as a regional walling-off the individual.

Fig. 4: private and public regions within in students' private spaces[2]. Photo from the empirical base of the author's PhD-Project "Raumbiographien". © Katja Stoetzer.

2 The big shelf containing mostly files from the course of studies separates the very private space, the bed behind it, from a more public accessible re-

This regionalisation or marking of explicit private space 'within' the livingspace itself reproduces this demarcation but varies it intersubjectively: Special places implying privacy emerge – a bed, the bathtub or a 'different' place, nature – interestingly leaving the sphere of the legal dimension of privacy integrating its opponent.

Summary

Depending on the space the students can exercise power over, they choose different kinds of dividing space functionally – or allow different layers of functionality to overlap in relational space. The techniques used, visual markers, are subject to naturalisation processes, rendering them nearly invisible by their implicitness for grouping the objects depicted according to their estimated functional qualities. The normalising potential of these visual markers can be revealed with regard to order/disorder and the public-private-divide, implying a simple division that emerged from the legal sphere into the social world with its greater complexity. It can also be observed on a second level within the legal private space raising the question if there is another iteration process.

Due to more complex ways the public-private-divide is moderated in private space, it is more of a social negotiation than of a demarcation. As a result, special places are produced implying the very private self which only few have access to. Interestingly, these places can even be outside legal private space. "Normality" is a social and spatial construct secured by repeated action and synthesis of persons, animals and social goods as a relational arrangement – reproduced by iteration in a recursive form, initiated and operated by visual markers.

List of References

Blomley, N. 2005. "Flowers in the bathtub: boundary crossings at the public–private divide." *Geoforum* 36: 281-296

Bohnsack, R. 2003. *Rekonstruktive Sozialforschung.* Opladen.

Bordwell, D./Thompson, K. 2004. *Film Art. An Introduction.* New York.

Düllo, T. 2000. "Ikeaisierung der Wohnwelt." In J. Carstensen/Th. Düllo/C. Richarts-Sasse, eds. *Zimmerwelten. Wie junge Menschen heute wohnen.* Essen, pp. 92-99.

gion at the front. Figure 2 shows the same room by a 360° panoramic picture. This perspective was chosen by the interviewee.

Giddens, A. 1997. *Die Konstitution der Gesellschaft. Grundzüge einer Theorie der Strukturierung.* Frankfurt/M. / New York

Harper, D. 2000. "Fotografien als sozialwissenschaftliche Daten." In U. Flick/E. von Kardoff/I. Steinke, eds. *Qualitative Forschung. Ein Handbuch.* Reinbeck, pp. 402-416.

Imdahl, M. 1994. "Ikonik. Bilder und ihre Anschauung." In G. Boehm, ed. *Was ist ein Bild?* München, pp. 300-324.

Löw, M. 2001. *Raumsoziologie.* Frankfurt/M.

Löw, M. 2003. *Prinz Charles, Hollywood und Hongkong. Raumsoziologische Annäherung an Architektur und ihre Bilder.* Antrittsvorlesung. www.raum soziologie.de

Marotzki, W./ Stoetzer, K. 2006. "Die Geschichten hinter den Bildern. Zur Methodologie und Methodik sozialwissenschaftlicher Bildinterpretation." In B. Schäffer/H. von Felden/B. Friebertshäuser, eds. *Bild und Text. Methoden und Methodologien.* Leverkusen; in press]

Mietzner, U. /Pilarczyk U. 2003. "Methoden der Fotografieanalyse." In Y. Ehrenspeck/B. Schäffer, eds. *Film- und Fotoanalyse in der Erziehungswissenschaft. Ein Handbuch.* Opladen, pp. 19-36.

Panofsky, E. 1979. "Ikonographie und Ikonologie." In: E. Kaemmerling, ed. *Bildende Kunst als Zeichensystem.* Bd.I. Ikonographie und Ikonologie: Theorien, Entwicklung, Probleme. Köln, pp. 207-225.

Rose, C.M. 1994. *Property and Persuasion: Essays on the History, Theory and Rhetoric of Ownership.* Boulder.

Rose, G. 2001. *Visual Methodologies. An Introduction to the Interpretation of Visual Materials.* London

Stoetzer, K. 2004. "Photointerviews als synchrone Erhebung von Bildmaterial und Text." *Zeitschrift für Qualitative Bildungs-, Beratungs- und Sozialforschung* 2: 361-370.

Strauss, A. /Corbin, J. 1996. *Grounded Theory: Grundlagen Qualitativer Sozialforschung.* Weinheim.

Terlinden, U. 2002. "Räumliche Definitionsmacht und weibliche Überschreitungen. Öffentlichkeit, Privatheit und Geschlechterdifferenzierung im städtischen Raum." In M. Löw, ed. *Differenzierungen des Städtischen.* Opladen, pp. 141-156.

Wuggenig, U. 1994. "Soziale Strukturierungen der photographischen Repräsentationen der häuslichen Objektwelt. Ergebnisse einer Photobefragung." In I. Mörth/G. Fröhlich, eds. *Das symbolische Kapital der Lebensstile. Zur Kultursoziologie der Moderne nach Pierre Bourdieu.* Frankfurt/M./New York, pp. 207-228.

The Aestheticisation of Futurity. From Facts to Values – Authority to Authenticity

NIK BROWN

Introduction – The Aestheticisation of Expectations

This paper offers some reflections on the aesthetic dimensions of expectations and futures in technoscience. It highlights the place of aesthetics in what colleagues and I have called the 'dynamics of expectations' and the role of visions and imagery in the mobilisation, the construction and diffusion of expectations (Borup, Brown, Konrad and van Lente 2006).

By way of illustration, the paper explores two culturally connected (though practically contrasting) contexts in the aestheticisation of contemporary technoscience (Brown 2006). First, in the late 1990s, as controversy over GM raged throughout much of Europe and elsewhere, the consultancy firm Burson Marsteller made what was to become a highly prescient set of recommendations to a consortium of biotechnology firms. The industry's mistake, Burson Mersteller insisted, was to have been drawn into debating technical details about risk and hazard. Instead, it should get away from the 'facts' and instead offer 'symbols', a new qualitative vocabulary capable of fostering, and I quote 'hope, satisfaction, caring and self-esteem'. Within a year or so, the biotechnology firm Monsanto had rebranded itself with the logo 'Food, Health and Hope' – in benign green lettering against a leafy background. More recently, Monsanto has shifted again to a new level of future abstraction replacing the former logo with the simple message 'imagine'. Can you see where this is going yet?

In a different context, a growing number of bioscience companies are now offering new parents the opportunity to bank the cord blood of their newborn children, for future treatments both real and imagined (but mainly imagined!). The web sites and glossy brochures are replete with the teary imagery of happy parents, ideal families, perfect bouncing babies – a jarring backdrop to the implicit suggestion of shattered dreams, childhood leukaemia, bone marrow engraftment and tissue engineered prosthetics. A notable feature of many of these websites and associated advertising are the references to news and commentary on distantly related developments in tissue engineering, the treatment of rare immunological diseases and blood cancers. By these means, new parents are encouraged to imagine themselves into the role and identity of a particular kind of dutiful and responsible parenting.

Both the GM debate, together with recent developments in biobanking, have been of interest to me in thinking about various aspects of expectations (Brown 2006; Brown and Kraft 2006). But in this paper, I want to offer some remarks about shifts from the technical towards the aesthetic in the public representation of bioscience, and in the construction of new promisory relationships connecting consumers to new biological services. The paper draws on literatures in 'the dynamics of expectations', visual sociology as well as some recent accounts examining the changing governance of science, notably charting a shift from authoritative technical discourse to one based on abstract future-oriented values and ideals.

Envisioning Science and Technology Futures

For some time now, a growing number of scholars and colleagues interested in the cultural and sociological features of techno-science have paid increasing attention to the role of futures, expectations, and the imagination in innovation. Making sense of and properly interpreting science and technology brings to the fore a focus on the way 'we' engage with the future, the characteristics or dynamics of 'our' imaginings, and the bearing this has upon the present. And just as crucially, when explored in detail, these subject positions of 'we' and 'our' begin to fragment as it becomes clear that futures and imagining are highly contested and variable across different groups and communities. It is often the case that futures are far from consensual and bring into relief contrasting desires and hopes. The controversies in which biotechnology has been steeped in recent years are very much rooted in discrete longings connected to nature, rights, ethics and values. Even where such highly lauded abstractions are not at stake, there are often disagreements about whether or not various technologies will work, and to what extent they will measure up to the promises made on their behalf. Nevertheless, for all of us the future is equally removed in time and space. The future is a 'not yet', a yet to be present. This nascent latency, by definition, has a levelling effect placing all of us in a present cut off from the future. Our present then is a constant and emergent moment populated by memories of the past and imaginings about the future (Mead 1932; Koselleck 2004). To this extent, colleagues interested in technoscience write increasingly about expectations having certain attributes and dynamics, and in this sense being studyable and an important object of reflection and critique.

Amongst the most important of these dynamics is the point that expectations of technology are temporally variable, changing and mutating over time. The future, as with the present, is in constant revision and process. Just as significantly, expectations of science and technology have the now familiar pattern of early promise followed by the difficult work of proving the concept, persuading sceptics, winning allies and securing patronage. Expectations of technological innovation are manifested in alternating cycles of hype and disappointment (Geels and Smit 2000; Brown and Michael 2003) or 'hot' and 'cool' phases (Callon 1993, 1995). More usually, it simply 'takes time' for many of the contingencies and problems – both social and material – to become apparent to advocates of a new vision. Just as often, the grand ambitions associated with various 'revolutions' (stem cells, gene therapy, pharmacogenomics, etc.) eventually give way to acute crisis of legitimacy as expectations go unfulfilled (Brown 2003).

Another important observation is that the intensity of expectations is often indicative of the very early stages of innovation. When we see claims to the revolutionary breakthrough potential of say genomics, stem cells, pharmacogenetics or nanotechnology, telemedicine, this can be taken to suggest that these are very early days for those fields. And during these early moments we are often likely to see incredibly intense activity in the visualisation, picturing and imagining of the promissory potential of these kinds of sectors. It is also usually a good indication that these futures could so easily turn out incredibly differently – given the amount of contestation and competition between different imaginings. This is a point I will return to later in discussing the promissory branding (or rather 'rebranding') of biotechnology.

Plus, it is also often the case that memories of past expectations tend to rationalise the memory of past expectations according to the way things eventually turned out. Such revisionist histories tend to forget or silence contingency, transforming a resultant technology into an inevitability and hero of its own making. To this extent, the contingency of the past and the fact that things could always have turned out differently become obscured.

One visually compelling representation of the temporalities of expectations has emerged from the worlds of technology consultancy. The Gartner company's 'hype cycle' chart has, over recent years, become one of the most widely recognised and widely disseminated popular representations of the way expectations of large technological fields change over time. The chart is highly illustrative of the application of visual culture and scientific representation (ref Law, etc) to the worlds of imagination and expectation. First, like so much in the aesthetics of scientific represenation it collapses and truncates complexity. Technologies are seen to move along a path from trigger, to a peak in expectations, then plummeting into a trough of disillusionment before eventually giving rise to a range of somewhat more modest applications. This reduction in complexity gives the impression of a highly linear path dependency with little scope for being able to observe the way applications change, adapt and respond to new moments. It is also technologically deterministic, being triggered by technology and plays down the many applications that simply cease in the trough of disillusionment rather than continuing on with their journeys.

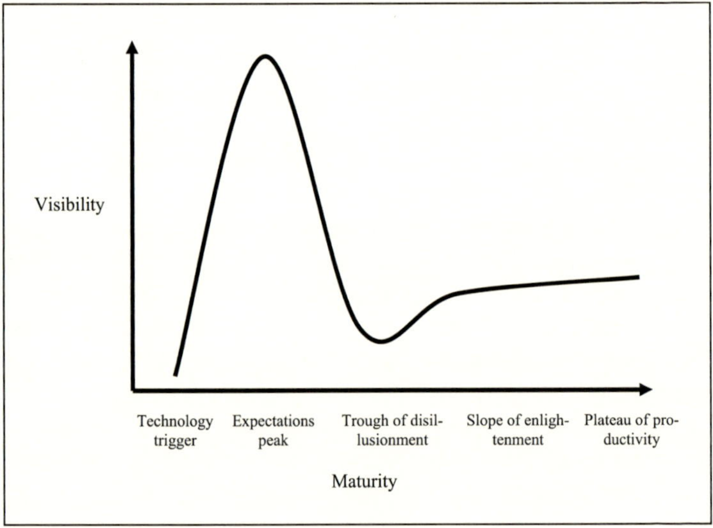

Fig. 1: The Gartner Consultancy 'Hype Cycle'

The important point here is that this is one amongst any number of images that has something to say about both the dynamics of expectations, the culture of scientific representation and the mobile propensity of aesthetics to travel long distances. There are lots of other examples of these ways of visualising technological change. And, as 'folk theories' (Rip

2003) they have a role in organising expectations and have themselves begun to contribute to the world of imagining and expectation that they are supposed to represent. Robert Merton would probably see in these some elements of self-fulfilling prophecy (1984).

Other good examples include the very many diagrammatic images that depict the futures of stem cells as the basis for a whole range of future treatments in regenerative medicine and tissue engineering. Such images populate the visual discourse of contemporary tissue engineering found in the corridors of museum display, magazine features, television documentaries, to say nothing of the thousands of websites selling the services of cord blood stem cell banks to new parents around the world. Invariably, what we are presented with in these images is the heroic stem cell as the source of a stream of arrows leading off in the direction of tissues and organs around the body including the brain, the kidneys, liver, pancreas, bone and muscle, etc. The purpose of such representations is of course to convey the incredible plasticity of stem cells, their putative ability to limitlessly regenerate the body's ailing tissues. Again, complexity and contingency are obscured. For example, the extended time frames of the stem cell vision for regenerative medicine undergo erasure through this kind of visual imagining. Each of the arrows leading from the stem cell to a particular tissue type truncates decades of complex, indeterminate and ambiguous work with little clear evidence as yet of an effortless path leading from bench to bedside.

Nevertheless, such imaginings have a performative function in shaping and structuring communities of promise and in obliging various actors to fulfil certain roles. Frans Berkhout writes of such images as central to the circulation of expectations of technology. He writes of visualisations as 'bids' or propositions which in turn create 'possibility spaces' by functioning to clear the way for certain futures to take hold. That reference to bidding is a reflection of the fact that, particularly in the context of innovation, future-making is deeply competitive and often highly antagonistic rather than mutual. Just as importantly, future visions more usually are moral in orientation with strong connection to abstract aspirational values:

> […] visions are typically moralised – effort is exerted by advocates to attach visions to widely shared values, or contrasted with undesired outcomes. This is necessary because novelty can only seem plausible if it has a chance of being widely accepted as good. If the innovation eventually does become diffused more widely, an equal and opposite process occurs in which, through a process of 'normalisation' it is emptied of moral content (2006: 309).

So the importance and relevance of visioning is important here for the reason mentioned above – that the future is only available to us through the imagination, through abstraction. The future does not have a reality status in the present – except and only in as much as those imaginings have a function in real time. What the sociologist of time, Barbara Adam, talks about as 'the future underway' in the present or 'the future in process'. And as George Herbert Mead was at pains to emphasise, the past and the future are only available to us in the present.

The connection then between expectations and aesthetics is much more than arbitrary. The imagination and the future are fundamentally integral to aesthetics, even to the point of being foundational to defining aesthetics as a disciplinary field from around the 1700s. Paul Kristeller – in *The Modern System of the Arts* (1965) – traces the first real connections between aesthetics and the imagination to Joseph Addison whose essays on the imagination first appeared in *The Spectator* in 1712. It is important to note, as Kristeller does, that as aesthetics develops as a discipline during this period, the arts become synonymous with the imagination, especially in Germany. Aesthetics is *literally* imagination, the mind's eye, envisioning. He makes note of an engraving printed in Weimar in 1769 representing the tree of the arts and sciences – putting the visual arts, poetry, music etc under the general branch of imagination.

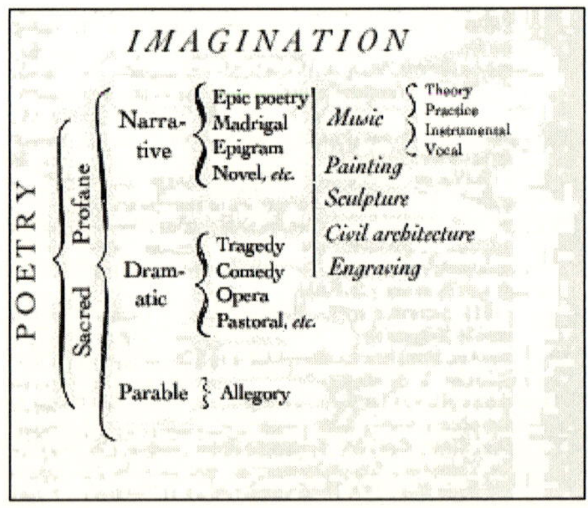

Fig. 2: The encyclopaedia of Diderot and d'Alembert 1769

The designation of aesthetics to imagination is part of an incredible jostling, pushing and shoving, as key thinkers of the period work out different and competing schemata separating one discipline from that of another. Nevertheless, aesthetics comes to be varyingly attached to the imagination, the sublime, emotions, sentimentality, beauty in work as diverse as that of Goethe, Kant, Burke and numerous others. Now, with that point in mind, I want to come to some more contemporary illustrations of the way aesthetics and expectations have been drawn closely together in science and technology. My first illustration is drawn from the branding – or rather the rebranding – of biotechnology at the close of the C20th. And my second illustration is drawn from the futuristic worlds of stem cell banking.

The aestheticisation of the biotech future

Aestheticisation plays a crucial role in the directing the futures of science and technology, both in terms of actual experimental practice and also wider cultural portrayal. The following story may be a story familiar to some, but for me it has come to represent one of my favourite recent episodes in the politics of bioscience aestheticisation (see Brown 2006). In the late 1990s, as the debate over GM reached its peak throughout much of Europe and elsewhere, a representative conglomerate of European and American biotechnology companies (EuropaBio) commissioned a consultancy report from the German public relations firm Burson Marsteller. In the worlds of industrial and environmental public relations, Burson Marsteller had become very highly regarded in managing numerous public relations disasters – like that of the Exxon-Valdez oilspill, the Bhopal tragedy and even the BSE crisis in the UK – and proved itself capable of reshaping deep-rooted perceptions of high risk sectors ranging from the nuclear to the petrochemical industries.

Controversially, their 1997 biotechnology consultancy report was leaked to Greenpeace who made the document public. What caught my eye at the time was the way the report stressed the importance of the imagination and the need for a new way of steering the debate towards intangible and aesthetic considerations of future values. In its recommendations to the EuropaBio group, Burson Marsteller made what was to become a highly prescient set of recommendations. In particular it suggested that the biotech industry get away from issues about facts, messy material squabbles about health, risk and hazard. And instead, start to offer 'symbols', a new qualitative vocabulary capable of fostering, and I quote 'hope, satisfaction, caring and self-esteem'.

A debate conducted within the highly restrictive confines of variously challenged 'truths' had in effect resulted in two things. First, it had distracted consumers, the market and policy making from being able to grasp the promissory potential of the future of GM products. The closed present tense of current and past truths had occluded the future open tense of potential and possibility, or so BM believed. And just as significantly, they recognised that given the normative positions assumed by most parties within the European controversy, all truths would be continually and unremittingly open to challenge. Science and technology studies scholars would recognise in this the ubiquitous problem of infinite epistemological regress so common to controversies involving incredibly complex proofs circulating in volatile cultural and epistemological environments. The only hope would lie in a widespread shift in the basis of the biotechnology debate from irreconcilable and untrusted truths towards the nascent and as yet deferred potential of a utopic technology to mesh with deeply held cultural values of progress, the future of nature and enlightenment advancement. A romanticised nature had become somehow diametrically opposed to technology in the debate and biotechnology would have to make a new bid for compatibility with the future of nature and life. This would depend on laying claim to future-oriented abstractions rather than messy details. In his book *Promising Progress*, Harro van Lente (1993), drawing on McGee, wrote about this kind of meaning making in terms of 'idiographs'. These are highly flexible and plastic cultural motifs or organising commitments – for example, 'justice', 'democracy', 'progress' – to which people appeal when wishing to situate a statement or an act in a much wider and more powerful context of meaning.

For BM, biotechnology had become negatively detached from a positive idiographic world and had become entrenched in a closed and circular controversy about evidence. Its first recommendation was therefore to stay away from the problem of truth describing facts as the killing fields of the bioindustries:

> Public issues of environmental and human health risk are communications killing fields for bioindustries in Europe. As a general rule, the industry cannot be expected to prevail in public opposition to adversarial voices on these issues […] a basic discipline of EuropaBio's communications strategy must be to stay off these killing fields – no matter how provocative the invitation to enter into them may be. (Leaked 1997 report from EuropaBio)

Instead, the bioindustries would have to substitute a debate about evidence for one about abstract future-oriented values representing a shift towards more aesthetic and symbolic references. Symbols and values – and especially those eliciting hope, satisfaction, caring and self-esteem – would come to represent the bioindustries escape route from messy evidential regress:

> Symbols – not logic: symbols are central to politics because they connect to emotions, not logic. Adversaries of biotechnology are highly skilled in the cultivation of symbols eliciting instant emotions of fear, rage and resentment. Bioindustries need to respond in similar terms – with symbols eliciting hope, satisfaction, caring and self-esteem. (Leaked 1997 report from EuropaBio)

Just as importantly, BM diagnosed this not just as an evidential problem but quite specifically an affective or emotional problem about the aesthetic and romantic attachments of people to the complex ideographs of nature and the rural:

> Food itself is a powerful vector if cultural – and even political – values virtually everywhere in Europe. But these values differ from country to country. And in many parts of Europe there also exists a strong corresponding emotional attachment to idealised images of rural society, farming and the countryside. (Leaked 1997 report from EuropaBio)

Following BM's report to the bioindustries representational change followed quickly. Members of EuropaBio – like Genencorp, Pfizer, Eli Lilly, Monsanto and Nestle – lost very little time in taking this massage to heart. Within a year or so, the biotechnology firm Monsanto had rebranded itself with the logo 'Food, Health and Hope' – in benign green lettering against a leafy background. More recently still, Monsanto has shifted again to a new level of future-oriented abstraction replacing the former logo with the simple message 'imagine'. In this way, Monsanto's new strap line resonates directly with the historical fusion of imagining with the disciplinary formation of C18th aesthetics.

Monsanto also set out to depict itself as a contrite organisation now much more in tune with other imaginings – particularly those of its sceptical publics and their perceived romantic longings for lost natures. And crucially, the debate became marked by the emergence of a relatively new and incredibly significant turn towards affectivity and the emotions, even at an institutional level. Monsanto's chief executive went to extraordinary lengths to portray Monsanto as having had a 'change of heart':

> Even our friends told us we could be arrogant and insensitive… We were blinded by our enthusiasm.… We missed the fact that this technology raises major issues for people – of ethics, of choice, of trust, even of democracy and globalisation. When we tried to explain the benefits, the science and the safety, we did not understand that our tone – our very approach – was arrogant. (Vidal: 2000)

This is but a small anecdote in the hopeful aestheticisation of debate in science and technology. But, as myself and Mike Michael have argued, it is part of a much broader turn in the very basis of the politics of the biosciences. In particular it represents a shift from facts and evidence (regimes of truth) towards the conduct of debates through the meta-abstractions of hope, expectations and the future. Of course, given that such futures are only available through abstraction, that future must be mediated through the evocation, performance and practice of symbols and values as a means of connecting with emotions, desires and longings. This is an illustration of the shift from 'authority to authenticity' (Brown and Michael 2002).

Modernity and science counterpoised facts against values, the emotions against rational thought, intuition and desire against direct evidence, intangible wishes against tangible proofs. What we are seeing in contemporary debates about biotechnology is evidence of a movement in the modern epistemological picture. The language of rationalistic authority is being supplemented (at the very least) it seems with a language drawn from values and aesthetics with its connections to desire, longing, the imagination and authenticity. Nussbaum (2001) goes as far as to describe this as an 'upheaval of thought' where the capabilities of rationality are proving to be ineffective at popular persuasion in the way they once were. Nussbaum describes these subtle changes as an 'intelligent responses to the perception of value' (Nussbaum 2001: x). Nussbaum's observation correctly diagnoses the rising significance of the affective in the public sphere. Within the contemporary climate, the language of rationality must therefore embody and register its modernistic opposite if credibility is to be successfully sustained.

At stake in these shifts are vast investments in future value and the economic and market reorganization of global seed and petro-chemical industries stretching back to the mid 1970s and beyond in anticipation, back then, of a future biotechnological basis to world food production (Kloppenberg 1988). The authoritative positions taken by the biotech industries – prior to their change of heart – reflects an undaunted commitment to the truth of this vast and seemingly irreversible 40-year-long restructuring. Most potential consumers of the biotech dream were for most of this period shielded from – or at least unaware of – changes taking place in the global organization of food production and manufacturing in preparation for novel GM based plants. Here then, huge temporal and spatial rifts had opened up between the expectations and hopes distributed between production and consumption. The change of heart, the shift towards shared symbols and the language of hope and promise has emerged as a means of belatedly drawing these divergent economic parties back together into a joint market agenda.

Cord Blood Futures

I now want to come to a second case that has been of interest to me where, some would argue, values and the imagination seem to prevail over that of facts. A growing number of bioscience companies are now offering new parents the opportunity to bank the cord blood of their newborn children, for present and future treatments (but mainly future promissory treatments). This example draws on work recently undertaken by myself and colleagues on blood innovation and the stem cell promise (Brown and Kraft 2006).

The web sites and glossy brochures of cord blood banks are replete with the teary imagery of happy parents, ideal families, perfect bouncing babies – a jarring backdrop to the implicit suggestion of childhood leukaemia, tissue engineered prosthetics, regenerative medicine and spinal cord injury, etc. Here, the incredible rarity of present-day uses for cord blood has not significantly deterred many thousands of new parents depositing CB thereby investing financially and corporeally in the future potential of stem cell sector.

Cord blood banking is advertised directly to parents as an investment that may, one day, prove to save the life of their newborn child. Potential 'investors' are implored to, as banks varyingly expresses it, 'put a little something away for a rainy day'; to provide 'a security blanket for your family'; by 'saving key components to future medical treatment'; or 'saving something that may conceivably save his or her life someday'. In keeping with the advertising of baby care products more generally, cord blood bank web sites are highly emotive visual and textual domains. This is a deeply suggestive form of branding and marketing that seeks to connect with new parents' sense of duty, protection and responsibility. As McCarthy and Wright note in their discussion of the way web browsing simultaneously mobilises memory, imagination and the emotions:

> Anticipation or expectation is a continuous process in experience. When visiting a company web site for the first time, we do not arrive unprejudiced. We bring expectations, possibilities, and ways of making sense that we associate with offline experience with the brand. Although anticipation suggests something that is prior to the experience, it is important to remember that it is not only prior. The sensual and emotional aspects of anticipation and our expecta-

tions of the compositional structure and spatio-temoral fabric of what follows, shapes our later experience. It is the relations between our continually revised anticipation and actuality that creates the space of experience (McCarthy and Wright 2004 – 'Technology as Experience').

Cord blood services are at once highly abstract in terms of the remote deposition of stem cells for a cash payment, and yet also highly affective and inter-subjective in the actions and experiences of parents concerned to 'do the right thing' by the future. At the centre of these debates are whether or not, and in what ways, new parents are being subject to some form of emotional manipulation during the anxieties of childbirth.

> I think in the family, certainly mothers specifically, are actually in quite an emotional state and quite vulnerable during pregnancy. (Interview with the director of a public cord blood bank – March 2005)

Whether or not manipulation is or is not taking place, what is important here is the way in which expectations are being understood and framed as affective, imaginative futures with emotional resonance. This connects with an emerging literature on the biosciences in which affectivity is seen to mediate future oriented actions in the present. Delvecchio Good writes of the 'political economy of hope' in American cancer care and research (1990; 2003). For her, it is a discourse of hope – a shared culture of images about the promise of medicine and the importance of personal and collective action in the face of potential pathology – that links together networks of clinical scientists, oncologists, patients, venture capitalists and wider public political actors. Here, health care consumers make investments in 'medical imagery', what she calls a 'possibility enterprise… culturally, and emotionally, as well as financially', and we may add corporeally in terms of the hopeful deposition of blood and cells.

> Enthusiasm for medicine's possibilities arises not necessarily from material products with therapeutic efficacy but through the production of ideas, with potential although not yet proven therapeutic efficacy (2003: 3).

More recently, work on the political economy of hope has been taken up by Novas in writing about the collaborative relationships of patient advocacy organisations and pharmaceutical innovation (2005) – he's been very interested in the branding and imagery of these expectations. Just with reference to some of the post-structuralist literature on emotions – Aesthetics easily elides with highly essentialistic and naturalistic understandings of the emotions – essentialised in terms of aesthetics being the affective properties of people – and meshes with general western assumptions about the emotions as pre-rational, pre-cognitive, embodied.

Emotions have been 'granted ultimate facticity by being located in the natural body' and therefore 'least subject to control, least constructed or learnt (hence most universal)' (Lutz and Abu-Lughod, 1990: 1).

Like many parallel enterprises in the modern biosciences, marketing relies upon creating spaces for potential consumers to imagine themselves into various future identities and to creatively engage with a cluster of values about dutiful parenting.

[…] for me it feels like clearly what they're trying to get at is they're trying to get at all your health insecurities you know and all your kind of familial health problems, so they're trying to tick as many boxes as they can it feels like…. those kind of emotional insecurities you might have about your health and your family's health… I'm kind of intuitively uncomfortable about this notion that it's emotionally coercive you know this business that you know if your kid had leukaemia would you spend £1,500? Yes of course absolutely you would in that situation. But I'm kind of intuitively suspicious and uncomfortable with the idea of feeling emotionally coerced to make a decision that I wouldn't make you know in all rationality. (Male CB parent – CDP1 March 2005).

There is a lot more to be said on the emerging economies of privatised cord blood banking (Brown and Kraft 2006; Waldby 2006) but the point here is that it illustrates the incredible mobilising capacity of the imagination, expressed in a complex aestheticisation of life science economies. That aestheticisation – and a turn towards the imagination – stems from the very fact that the usefulness of cord blood stem cells is almost purely a *future potentiality* rather than a *present reality*. It is, as Moreira and Palladino might express it, 'a regime of hope' and not of 'truth' (Moreira/Palladino 2005). Because the 'real worth' of depositing stem cells cannot be known at present, all contenders in the debate over CB banking are taking positions with respect to a not-yet-known, latent future-yet-to-be. Aesthetic imagination / future-oriented abstraction is the only recourse for a debate located in the future rather than the present.

Aestheticisation, the Future and Politics

Both of the discussions above illustrate the complex politics in which the contemporary life sciences and their future-oriented aesthetics are politically embroiled. In the context of the debate over GM we have seen a strategically important shift away from present day realism and the incredible impasse arising from evidential disputes about experimental field trials, etc. The imagination (or simply 'Imagine' as it is now expressed in Monsanto's company logo) has become a haven for an industry intractably bogged down in deadlocked controversy and debate. The future has here come to represent a means of opening up that impasse, injecting a new scope – a 'possibility space' – for open ended future potential. The imagination is after all the basis of an economy organised through 'promissory capitalisation' where latent potential is of increasingly greater importance than present value.

The divisions separating futures from presents are increasingly disintegrating. The imagination and the politics of values and aesthetics have become fundamentally integral to determining any number of recent debates in the life sciences. In parallel with this is a crumbling of the distinctions between technicist realism and cultural values. Helga Nowotny and colleagues (2001) make just this point when writing about the role of future-oriented abstractions and their articulation with concrete non-abstractions:

> Visions, images and beliefs cannot sharply be demarcated from knowledge. Far from being dangerous illusions or utopian projections, visions are a precious resource, an intangible asset that may help to provide the necessary (but typically missing) link between knowing and doing […] It is important to recognize how visions and images interact and also how wide the gap separating images from practice can become before an uncontrollable backlash is provoked (Nowotny *et al*: 232).

To draw this discussion to a close, if we are to make sense of contemporary techno-science, far greater attention now needs to be given to the articulation between temporal formations (past-present-future) as well as the disciplinary domains distributed between aesthetically mediated futures and science's present-dominated realism.

List of References

Berkhout, F. 2006. Normative Expectations in Systems Innovation, *Technology Analysis and Strategic Management*, 18, 3/4, 299-312.

Borup, M., Brown, N., Konrad, K. and van Lente. 2006. The Sociology of Expectations in Science and Technology, *Technology Analysis and Strategic Management*, 18, 3/4, 285-298.

Brown, N. and Michael, M. 2002. From authority to authenticity: governance, transparency and biotechnology, *Health, Risk and Society*, 4, 3, 259-272.

Brown, N. 2003. Hope against hype: accountability in biopasts, presents and futures, *Science Studies*, 16, 2, 3-21

Brown, N. 2006. Shifting tenses – from 'regimes of truth' to 'regimes of hope' Paper presented at 'Shifting Politics' University of Groningen. SATSU Working Paper no 30. www.york.ac.uk/org/satsu/OnLinePapers/Online Papers.htm

Brown, N. and Kraft, A. 2006. Blood Ties: Banking the Stem Cell Promise, *Technology Analysis and Strategic Management*, 18, 3/4.

Callon, Michel. 1993. Variety and irreversibility in networks of technique conception and adoption, in D. Foray and C. Freeman (eds.) *Technology and the Wealth of Nations – The Dynamics of Constructed Advantage*, London.

Callon, Michel. 1995. Technological Conception and Adoption Network: Lessons for the CTA Practitioner, in A. Rip, T. Misa and J. Schot (eds.) *Managing Technology in Society – The Approach of Constructive Technology Assessment*, London.

Geels, F.W. and Smit, W.A. 2000. Lessons from failed technology futures, in Brown et al. op cit.

Good, M.D.V., Good, B.J., Schaffer, C. and Lind, S.E. 1990. American oncology and the discourse on hope. *Culture, Medicine and Psychiatry*, 14, 59-79.

Good, M.D.V. 2003. *The Medical Imaginary and the Biotechnical Embrace – subjective experiences of clinical scientists and patients*, Russell Sage Foundation, Working paper no 20.

Kloppenburg, J. R. J. 1988. *First the Seed: The Political Economy of Plant Biotechnology, 1492-2000*, Cambridge University Press, Cambridge.

Kosellek, R. 2004. *Futures past – on the semantics o historical time,* Columbia University Press, Columbia.

Kristeller, P. 1965. *The Modern System of the Arts – Renaissance Thought and the Arts*, New York: Harper Collins.

Leaked 1997 report from EuropaBio. 1997. http://users.westnet.gr/~cgian/leak1.htm – accessed Feb 2007.

Mead, G.H. 1932. *The philosophy of the present.* Chicago: Chicago University Press.

Merton, R.K. 1984. Socially expected durations: a case study of concept formation in sociology, in W. Powell and R. Robbins (eds.), *Conflict and Consensus: a Festerchrift for L. Coser.* New York: Free Press.

Moreira, T. and Palladino P. 2005. Between truth and hope: on Parkinson's disease, neurotransplantation and the production of the 'self'. *History of the Human Sciences*, 18, 55-82.

Nowotny, H. Scott, P. and Gibbons, M. 2001. *Re-thinking Science – knowledge and the public in an age of unceraint*y, Polity Press, Cambridge.

Nussbaum, M.C. 2001. *Upheavals of Thought: the Intelligence of Emotions.* Cambridge, Cambridge University Press.

Rip, A. 2003. *Expectations about nanotechnology: patterns and dynamics*, paper prepared for the May 2003 meeting of the Expectations Network, Utrecht, NL.

Van Lente, H. 1993. Promising Technology – The Dynamics of Expectations in Technological Developments, Enschede, NL.

VIDAL, J. 2000. GM foods company decides to turn over a new leaf, *The Age*, 2 December.

McCarthy, J. and Wright, P. 2004. *Technology as Experience*, MIT Press, Mass.

Waldby, C. 2006. Umbilical Cord Blood: from Social Gift to Venture Capital, *Biosocieties*, 1, 1

Technological Transformation of Society in Children's Books

DOMINIQUE GILLEBEERT

A lot of investigation on children's literature is done in the field of pedagogy and literary studies. However, almost no attention is paid to ethical and moral themes in children's books. Nevertheless, it is interesting to study those themes as especially children's books are not neutral: they always impart and transport directly or indirectly values, norms, moral ideas, opinions, argumentation, etc. While reading and analysing lots of children's books – within the scope of my PHD project[1] – I noticed that some explicit as well as implicit norms and values deal with or react to technological transformation in society. That is what this paper will be about[2].

In general I consider children's literature, as Dixon (Stephens 1999: 20) and Lehnert (Lehnert 1994: 31), to be part of a complex cultural system which can be analysed – beyond its esthetical and pedagogical evaluations – to tell us something about the social, economical and political world in which the book was written and its author lives. However, a book always first tells us a story. This story can, of course, relate to the real life of the described persons or phenomena, which exist in reality. Nevertheless, it cannot be taken as an exact description of this reality. Literature simplifies, symbolises, fantasises, condenses, etc. and creates in this way its own reality. Precisely that is why literature is so special, important and also interesting to investigate. Children's books thus present a special context for the operation of values and norms, because,

1 www.ifs.tu-darmstadt.de/fileadmin/gradkoll/Personen/GillebeertD.html

2 As my own research has not that much to do with technological transformation of society in children's books, I incorporate my own findings in the investigations of Schilcher (Schilcher 2001), Ewers (Ewers 1995, 1999), Daubert (Daubert 1999) and Schweikart (Schweikart 1995).

citing Stephens (Stephens 1992: 126f.): “narrative texts are highly organised and structured discourses whose conventions may either be used to express deliberate advocacy of social practices or may encode social practices implicitly”.

The concept “norm”, however, does not only refer to the semantic complex of normativity which postulates an ethical discourse that refers to qualitative values. The concept “norm” also refers to normality, which is about a gradual category. For the following discussion on normality I refer to the studies of Link (Link 1997)[3]. A moral and ethical investigation of children’s literature can directly be joined to his ideas.

In children’s books, as in lots of other mass media, pictures and ideas of what is normal and what is not are established. A specific image of normality, which is subject to changing processes, is generated (i.e. the images of children, the extent of autonomy which is conceded to children). Because many people strive for normality, these representations of normality can manipulate reality via a feedback strategy. According to Link, the basic anxiety of modern society exists in having fear of being abnormal. Hence people create imaginary boundaries of normality and compare their own position to the center.

However, not all subjects try to be as near as possible to the normal center zone. Limits are tried out. Just as in modern literature border characters, marginalisations, irreversible denormalisations, as well as crossing over and back across the borders of normality are of interest. Such border characters can also be found in the children’s literature of the past 20 years and are, in general, very typical in children’s literature (cp. Schilcher 2001). In this way children’s literature expands the zone of normality and the borders become flexible. Children’s literature integrates particular societal discussions about normality and experiments with its borders.

The following Discussion deals with the question, to which extend the participation in the normality discourse and the implementation of ethical norms in regard to the technological transformation in society exist in parallel as well as to which extend their confrontation is brought up in children’s books.

Nowadays an important part of children’s literature can be described as realistic (Budeus-Budde 1997). Lots of authors occasionally try to make the social life a topic of discussion and they make an effort to grasp the latest manifestations of the world of children and their way of living. Daubert (Daubert 1999: 92) states: “[…] it catches one’s eye, that Literature and Reality closely correspond here and that some authors

3 See also Link’s Paper in this volume.

trace seismographically the changes of childhood and translate them literarily. The realistic children's and youth's literature become a medium of temporal diagnostics". In this paper I will speak mostly about this realistic children's literature (because of limited space there are no possibilities to consider all the different categories of children's literature).

With reference to the technological transformation of society and the environment of children, the following topics are discussed in the scientific literature about children's literature: changes in experience of family life and diversity and the destabilisation of gender casting – both of which are described realistically and accurately in children's books, yet sometimes quite idealistically. In addition economical childhood, education, media consumption and technological childhood are elaborated – topics which are hardly ever described or discussed in children's Literature. By means of example, I would like to elaborate shortly on each of these topics.

Changes in experience of family life

Family is a very important value in children's books. Family, however, doesn't assure stability and reliability for children. It cannot counteract processes of modernisations, as "Verinselung", Isolation, etc. and is affected by individualisation trends (see Richter 1995: 32; Schilcher 2001). *Bolles toller Trick* can exemplify this. Bolle has moved from what he calls a podunk town in the forest to the big city of Stockholm because:

> Mama Berit wanted to make Carrier and Papa Stellan... What he had in mind, Bolle didn't know that well (9).
> Berit worked at a company, whose name was EFFEKTA (Making the working life more affective), she earned a lot of money by traveling from one company to the other giving lectures [...] Her major task seemed to be the cultivation of big companies. Bolle didn't know exactly what that meant, just that lots of people lost their job and owners of these companies became even richer (18-19)[4].

In this book it is made very clear: processes of modernisations have a direct influence on Bolle's life.

Bolle knows that family life can be different, as it was different before, and at his friend Ellen's place things are totally different. Even

4 All citations which are originally in German are translated by the author of this article.

though Ellen says that her parents can't think about anything else other than the baby and diapers, Bolle experiences Ellen's happy family life, where someone is always home. But individualisation and compliance of luck for the individual, here Bolle's mother, are unquestioned values, which also count within the frame of the family. As in lots of other children's books, these values override traditional norms, i.e. differences in power and status between parents and children, obedience of children, the abandonment of fixed gender casting. Parents in children's books try to reconcile antagonistic values which one another. Some families, as Bolle's family, thereby lose stability: Bolle's parents decide to split up (36).

Thus, as also Annita Schilcher has made clear in her study of children's books, in children's books, limits of flexible norms are searched and tried out (Schilcher 2001: 198). In doing so, it amounts to the oscillation of still okay and not okay any more with regard to the well being and the positive development of the child, two values which are unquestioned and of importance in children's books. Children's books also show that children are at the mercy of their parent's decisions. They must comply with the new situation brought about by the parents:

> "Please Bolle,", she said, "try to see all that a little bit more adult". As she wanted to pet him, he avoided her. They have to be adults after all, he thought. Eventually, I am the child" (43).

Nevertheless, children's books also show that relationships in families can ideally gain intimacy and qualitative intensity from processes of modernisation. Communication and autonomy become very important values, which corresponds to the educational values of the higher society. Bolle, for example, acts independently and tries to gain more time with his mother by changing the appointments and names in his mother's time manager. Yet more important, he tries to change something (76). A lot of things happen; Mama Berit almost even loses her job. The problem is that Bolle can't tell that he is responsible for Mama's trouble and that he had a good reason for doing so (103). He can't express himself until he brings Mama and Papa together in the pizza place with one last trick and, in doing so, finally succeeds in creating an opportunity for them to speak with one another. The end of the book suggests that the little family of Bolle will find a way to combine Mama's career, Papa's wish to become an inventor and Bolle's need for parental care, as they are now able to speak with each other. Communication is considered an important value. It is the premise for a prosperous family life because it is an adequate instrument to solve problems and to express oneself, one's feelings and wishes. In children's books it is then tried out to what extent the norm of parental care can be delimited by the autonomy and independence of children and their communication skills (cp. Richter 1995: 33).

Processes of modernisation are criticised by children's literature when parent's jobs prevent them from spending the necessary time with their children. As long as children have the feeling they can handle the tasks they are given and can fall back on the love and care of the parents whenever they need to, autonomy is portrayed as a positive value that must be supported by parents.

Because of the social modernisation processes, the individual is detached from social and class specific commitments and ties and, therefore, gains more freedom. The family structure is affected by those individualisation tendencies too: in children's books conventional family structures and ways of living are often rejected and the modification of the concept of family and its functions is described (see Daubert 1999, Ewers 1995, Schilcher 2001, Ewers and Wild 1999).

Recapitulating, one can say that modernisation processes in families are rated positively if

– normativity in upbringing and education is reduced as much as possible so that children obtain more freedom and, in the process, adults can live a modern life and pursue their goals

– Simultaneously, children need psychological and emotional support to be able to develop a positive self perception. Only then does a child's autonomy makes sense.

Only if those criteria are accomplished, do children in children's books become happy children (cp. Schilcher 2001: 199).

Diversity and destabilisation of gender casting

Children's books favour the diversity of gender casting and, thus, stand for the democratisation of gender relation. This democratisation is very advanced in the lives of the children, whereas parents and grandparents still remain in more traditional roles. In children's books, gender relation is questioned. The values presented correspond to the values of the higher society. (Schilcher 2001) found out that the fixed, traditional norms about how boys and girls, men and women, have to behave are dissolving and are being replaced by new norms, which result from the process of normalisation. If one follows the logic of children's literature, for girls, being normal means being self confident and ballsy, whereas for boys it means being sensitive and imaginative.

The presentation of Nella in *Nella Propella* is prototypical for the presentation of girls in children's books. With an active attitude towards her environment, she doesn't feel the victim of her situation but, rather, tries to fit the situation with her own needs and ideas. She is self confident, articulates herself very well and doesn't look like a traditional girl. Futhermore, she is cute and smart and surprises the reader once in a while with her sharp witted and original ideas. Dirk in *Oma und ich* is an example of a positive presentation of a boy. He is not characterised by strength and courage. He is not a hero. Dirk is sensitive, different and responsible. He is called Dicki, as not many children like him. However, Jutta discovers that Dirk is really nice. She likes him precisely because of his otherness and the fact that he is a good listener. She feels comfortable talking to him. They share a whole afternoon in Dirk's secret place where Jutta tells Dirk about her worries about her sick grandmother. Dirk comforts her and makes her smile. Elias' mother in *Elias und die Oma aus dem Ei* represents a typical female image. Elias' mother is pretty (she looks like a princess), young and dynamic. She is employed (writes articles about castles) and wants to realize herself. She is independent, strong and tries to bring Elias up with modern values and norms. In the beginning of the story she doesn't have much time for Elias, but little by little she starts to understand that Elias needs his parents. At the end of the story she then also incorporates traditional female values as empathy, readiness to take on responsibility and emotional spontaneity. It is the mixture of these modern and traditional values which make her a modern woman. In *5 Minuten vor dem Abendessen,* Babete's stepfather stands for a picture of a positive male. He is sensitive and empathetic. Not only has he helped Babete's mother, who is pregnant and was left by her boyfriend, but he has also a lot of time for his little family. He tells Babete stories and plays with her. He is funny and imaginative and incorporates an ethical attitude.

As Schilcher (Schilcher 2001) says, modern children's literature also reflects, social processes of progress and uses strategies of normalisation to introduce normss of what is normal. Variations on these norms are accepted only to a certain degree. To be too normal means being boring, overly-adult and adapted to society; one doesn't respond to the social norm of being individual. To be too different from normal is to be abnormal. Moreover, children's literature shows that conformance with gender roles makes people sad and unsatisfied. Individuality and individual structuring of life are promoted as a central value. However, this individuality must suit the norm of normality. Thereby individuality doesn't mean egocentrism, but incorporates relationship, friendship and interaction.

In summary, it can be ascertained that the democratisation of gender casting in children's books is beyond the democratisation of gender casting in reality.

Education

In children's books education as a value is taken for granted; its meaning and sense don't have to be explained. In books, almost no children do have problems in school. Being good at school is simply normal.

Jakob in *Spürnase Jakob Nachbarkind* is a good example to illustrate this. Jakob has always had good marks. Because of the quest for money and his job in the old house, which his neighbours and friends Tete, Pups und Wuzi have inherited, he doesn't have a lot of time to do his homework and study. Still, he knows that it is very important to be good at school and to learn. At the end of the story, when the money is found, he makes it clear that school has to come first again. Once he has finished, he can do his job. Also in the following quotation:

"I was studying for hours", he lied. Very affected Frau Pamperl shaked her head.
"But Jacobchen", she said. "you have good marks anyway. Don't rack yourself like that! I'm also happy, if you get a 2 or a 3. Honestly!" (157)

Even when children in books have problems at school, they don't doubt the norm of education. One can take Lena in *Lena wünscht sich auch ein Handy* as example.

This afternoon at Katrin's place they don't practice mathematics. Normally Lena always asks Katrin, to help her with her homework, as Lena finds that maths are difficult and Katrin thinks it is very easy. But today they don't have to practice. Today Lena got good marks for maths (7).

Incidentally, in *Lena wünscht sich auch ein Handy* the first two chapters are dedicated to the school achievement.

Why are children in books presented as dutiful children who do their homework with pleasure and without being nagged? This image doesn't fit with the image of independent children who take their lives into their hands as I wrote before. To exaggerate, one could say that in order not to be at mercy of the mechanism of the job market later on and to be able to live an "alternative" life, integration in the system of school is required beforehand. However, this integration and the outer forces that come with it contradict the values and norms which are put forth in the texts. Therefore, this area of conflict must stay hidden as far as possible and this is managed most easily if the protagonist is inherently intelligent (cp. Schilcher 2001).

Economical Childhood

I would like to start this section with a quote of Ewers (Ewers 1995: 47): "it is rarely aimed at children as experienced consumers, as adapts of the range of toys, computer games and the clothing industry". Nevertheless, children in books often own a lot of goods. This is not reflected at all. Katrin in *Lena wünscht sich auch ein Handy* can be taken here as an example.

"Computer?" she asks. Although she cannot imagine, that Katrin would be so happy about that.
"We have one in the basement, dummy!", Katrin says. "No, much cooler! Guess one more time!"
Lena furrows her brows. They have DVD in the living room and Video too (8-9).

Katrin's parents have more money and so Katrin can have a mobile phone and take riding lessons, whereas Lena cannot. However, more is not said on this topic. Another example can be found in *Spürnase Jakob Nachbarkind*

Jakob's soul was deeply moved by pity. Getting nothing else as building material for Christmas, that was really a hard fortune. He felt downright guilty, because he could expect a ski suit, a computer and a remote controlled sail boot – under guarantee – and 20 other things as surprise (161).

However, Jakob is one of the few children who are described as a consumer. It is striking that he doesn't buy something for himself, but for his friends: because they don't get presents for Christmas, he buys them presents them with the money he earns in his job. Moreover, it is important to notice that an environment-conscious life-style is part of the self-evident background of most children in children's books. In children's books, child and nature stand close together. Not controlling nature by using technology is important, but nature for itself becomes a high value. Here too, the norm environmental awareness and ecologically healthy life style is an incorporated norm which is taken for granted. For example Lena:

In her trouser pocket she feels the money she got back after she bought the healthy chewing gum yesterday. (34)

Media consumption and technological childhood

When reading children's books, one gets the impression that the environment of children doesn't have anything to do with media consumption. The life of children is described as an area free of any kind of technology (Ewers 1995: 47, Richter 1995: 26). In the last 5 years this situation has changed slightly. Some stories are built up around the topics of computers und mobile phones and new technologies are introduced in children's books. Still, most books simply don't refer to it at all.

Only very few books, for example, *Ich ganz Cool* or Christine Nöstlinger's *Fernsehgeschichten vom Franz* und *TV-Karl,* talk about television consumption and how to deal with it (see Schweikart 1995). Some other books tell about the handling of media consumption casually, whereas restrictive media consumption (for example, television consumption) is unquestioned and is not discussed. Watching television kills the imagination. It seduces one to useless television consumption, makes children stupid and get them away from more meaningful things such as social contact and reading books. Low consumption of television is the norm, even if children in children's books protest against it. The books show that children only protest when they don't have anything better to do. Children are presented mostly not as passive receivers, but as active, critical and emancipated media users (Schilcher 2001: 247-248).

Just as television, the use of computers is not often discussed in children's books. Mostly, children have a computer but somehow seem not to use it. When children use the computer, it is to do useful things, not to play games.

However, in some books the computer becomes the central element around which the story is built. An example is the book *Die cityflitzer jagen den Computerhacker. Die cityflitzer* is a series which tells about the adventures of a group of children who ride through the city on in-line skates. In *Die cityflitzer jagen den Computerhacker,* they hunt a computer hacker virtually online and physically with in-line skates. The cityflitzer are presented as computer detectives. One often gets the impression that the computer is introduced to attract the attention of children and to try to make them read the book.

All kinds of possibilities to use the computer are introduced and described: chatting, surfing the internet, writing emails, doing homework, playing games, etc. These books, where using the computer forms the base of the story, integrate factual information. They also show a lot of words which are correlated to computer language, i.e. in *Oskars ganz persönliche Glücksdatei*

My heart is like a big hard disk and there are still at least 3 Gigabyte free for you (73)
I explain to him, that his old computer is hopelessly outdated. My notebook has a modern operating system, is a lot faster, has more RAM and a much larger hard drive (92).

The computer and internet and the possibilities which they offer are described enthusiastically and experienced by children in children's books as exciting and great. One could say the computer is almost celebrated.

It seems that these books communicate a positive attitude towards computers; it is a new technology, of which the use is cool and fun. Computers are not viewed as negatively as television. No book refers to any negative aspects or possible dangers. Computers do not make one stupid but, rather, offer children fun and an adventurous and exciting life. No book warns about exaggerated use and no restrictions are given for the use of internet. New ways of communication are tried out. In *E-mails und Geheimnisse*

That's sheer madness, such a chat room. A meeting point, which is no more than a computer address. But with one click on the address, one can talk to people from all over the world. By using the keyboard of course (p.12).
I think we can leave out the question where you know each other from! You can write: "from the Internet". That is definitively correct!" proposes Oliver. (119)

Still, also here children are active, critical and emancipated computer users.

Both friends clap like blazes. "Chatting is somehow really nice!" murmurs Oliver to his friend. "But in real life everything is even much nicer!" (121)

What all books make clear in the end is this: The most important thing in life is to gain experience and not to waste too much time on the computer. That is what the children in children's books themselves express.

Summary

As we could see, literature is not a reflection of real life, rather, it reflects prejudices ideas about the reality of society (Kaulen 1999: 126) and the progressive, liberal utopia of the authors (Daubert 1999: 103).

On the one hand, realistic children's literature is situated in a field of tension between ideas and stereotypes and, on the other hand, in the factual knowledge about life's circumstances. As Ewers (Ewers 1995: 36) says: "The worlds drafted in children's books can be ideological constructs or social critical reality reconnaissance or, and this may be the case most of the time, they can be a mixture of both, difficult to see through".

It is very interesting to observe that the consequences of technological transformations in society are discussed in children's books, whereas what I called media consumption and a technological childhood itself are hardly handled.

Modern literature is a moral literature which has a clear attitude regarding certain ethical principles which regulate the social life and interaction. At the same time, they participate in the discourse about processes of normalisation and modernization. Processes of normalisation are only positively evaluated when they do not conflict with the implicit ethical values (Schilcher 2001: 249).

The modern children's literature presents an image of a modern and post-modern childhood. This childhood contains modern family constellations, modern gender roles, modern forms of interaction and contributes to an expansion of the borders of normativity by presenting ways of life and attitudes which are not yet likely to be accepted by the majority in society. Dissenters and children who lag behind tend to experience a negative presentation, whereas those who are modern tend to experience a positive image. The consciousness of upper-class normativity shapes children's literature and, in this way, can contribute to the enlargement of the borders of the normativity of the readers of children's books (cp. Schilcher 2001: 41, Mattenklott 1997: 22).

List of References

Arold, Marliese. 2004. *Die cityflitzer jagen den Computerhacker*. Wien: Bertelsmann Verlag.

Arnold, Marliese. 2001. *Oskars ganz persönliche Glücksdatei*. Frankfurt/M.: Fischer Verlag.

Boehme, Julia. 2001. *E-mails und Geheimnisse*. Bindlach: Loewe Verlag.

Boie, Kirsten. 1994. *Nella Propella*. Hamburg: Oetinger Verlag.

Boie, Kirsten. 1992. *Ich ganz Cool*. Hamburg: Oetinger Verlag.

Boie, Kirsten. 2005. *Lena wünscht sich auch ein Handy*. Hamburg: Oetinger Verlag.

Bröger, Achim. 2003. *Oma und ich*. Berlin: Rowohlt Verlag.

Daubert, Hannelore. 1999. "'Es verändert sich die Wirklichkeit'. Themen und Tendenzen im realistischen Kinder- und Jugendroman der 90er Jahre." In Renate Raecke, ed. *Kinder- und Jugendliteratur in Deutschland*. München: Arbeitskreis für Jugendliteratur e.V., pp. 89-105.

Ewers, Hans-Heino. 1995. "Veränderte kindliche Lebenswelten im Spiegel der Kinderliteratur der Gegenwart." In Hans-Heino Ewers and Hannelore Daubert, ed. *Veränderte Kindheit in der aktuellen Kinderliteratur*. Braunschweig: Westermann Schulbuchverlag GmbH, pp. 35-48.

Jaensson, Hakan. 2002. *Bolles Toller Trick*. Frankfurt/M.: Fischer Verlag.

Nöstlinger, Christine. 1992. *Spürnase Jakob Nachbarkind*. Hamburg: Oetinger Verlag.

Kaulen, Heinrich. 1999. "Vom bürgerlichen Elternhaus zur Patchwork-Familie. Familienbilder im Adoleszenzroman der Jahrhundertwende und der Gegenwart." In Hans-Heino Ewers and Inge Wild, ed. *Familienszenen*. Weinheim/München: Juventa Verlag, pp. 115-127.

Lehnert, Gertrud. 1994. "Phantasie und Geschlechterdifferenz, Plädoyer für eine feministisch-komparatistische Mädchenliteraturforschung." In Hans-Heino Ewers, Gertrud Lehnert and O'Sullivan, ed. *Kinderliteratur im interkulturellen Prozess*. Stuttgart/Weimar: Metzler, pp. 27-44.

Link, Jürgen. 1997. *Versuch über den Normalismus. Wie Normalität produziert wird*. Opladen: Westdt. Verl.

Mattenklott, Gundel. 1997. "Die Kinder- und Jugendliteratur als moralische Anstalt betrachtet." *Tausend und ein Buch* 3: 20-29.

Nöstlinger, Christine. 1994. *Fernsehgeschichten vom Franz*. Hamburg: Oetinger Verlag.

Nöstlinger, Christine. 1995. *Der TV-Karl*. Weinheim: Beltz & Gelberg.

Prochazkova, Iva. 1992. *Fünf Minuten vor dem Abendessen*. Stuttgart: Thieneman Verlag.

Prochazkova, Iva. 2003. *Elias und die Oma aus dem Ei*. Düsseldorf: Sauerländer Verlag.

Richter, Karin. 1995. "Kindheitsbilder in der modernen Kinder- und Jugendliteratur." In Erich Renner, ed. *Kinderwelten. Pädagogische, ethnologische und literaturwissenschaftliche Annährungen*. Weinheim: Deutscher Studien Verlag, pp. 26-37.

Schilcher, Anita. 2001. *Geschlechtsrollen, Familie, Freundschaft und Liebe in der Kinderliteratur der 90er Jahre*. Frankfurt/M.: Peter Lang GmbH.

Schweikart, Ralf. 1995. "Medienkindheit. Dargestellt in Kinderbüchern von kirsten Boie." In Hans-Heino Ewers and Hannelore Daubert, ed. *Veränderte kindliche Lebenswelten im Spiegel der Kinderliteratur der Gegenwart.* Braunschweig: Westermann Schulbuchverlag GmbH, pp. 109-126.

Stephens, John. 1992. *Language and ideology in children's fiction.* London: Longman.

Stephens, John. 1999. "Ideologie und narrativer Diskurs in Kinderbüchern." In Hans-Heino Ewers, ed. *Kinder- und Jugendliteraturforschung 1997/98.* Stuttgart/Weimar: Metzler, pp. 19-31.

Aesthetics of Technology as Forms of Life[1]

WOLFGANG KROHN

Outline: impossible definitions

The terms aesthetics, technology, and life forms comprise a variety of heterogeneous phenomena. Any attempt to define their proper meanings necessarily comes up short in one of two respects. Either it unduly excludes relevant features or dissipates into too many marginal aspects. Nevertheless, in order to offer the reader some guidelines as to what I plan to talk about, I will attempt to determine my use of the concepts.[2]

Aesthetics: In the context of technology, relevant facets of aesthetics are, among others:

- Vision, construction, design
- Perception, use, experience
- Pleasure, admiration, fun, excitement.

The first facet refers to the production of technologies. It is the knowledge dimension of planning, realizing, and shaping technologies. It also includes terrifying visions, lousy constructions and bad design – in short, un-aesthetic technology. The second applies to the functional in-

1 Note on terminology: Throughout the paper I use the term technology as covering all aspects of technology, techniques, and technological knowledge.

2 There is no space to relate the paper to the relevant literature. My use of the concept of technology owes much to the works of Albert Borgman (1984) and Don Ihde (1990). Both are phenomenologists and have a Heideggerian background. At variance with this paper, they restrict technology to material devices. My concept of aesthetics borrows from Gernot Böhme (2001), despite the fact that his favourite point of reference is nature, not technology. The concept of technological life style has no specific reference in the literature.

terfaces between artifacts and users and includes unusable, dangerous, and disturbing technologies. The third relates to the feelings of pleasure and displeasure generated in observers and users. These aspects are fairly independent in the sense that we can dislike functional technologies and like dysfunctional ones. Likewise, we can admire the design of a piece of technology precisely because we do not even notice its existence in using it, e.g. contact lenses or artificial hips. However, usually they are connected and reinforce each other, as examples from everyday life illustrate. Riding a bicycle comprises simultaneously the bicycle as a material artifact embodying a host of technological knowledge, the interaction between the experienced rider and the bicycle, and finally the feeling of the ride. Further examples follow below. However, the question here is what is especially aesthetic in these dimensions. The intuitive answer supposedly is: if everything fits nicely together – the functioning of the bicycle, the operations of the user and the enjoyment of the ride – we tend to say, "That's great." The better these fit, the higher the aesthetic value or delight. Accordingly, imbalance amongst the dimensions generates negative values or the feeling of uneasiness.

Technology: Picking up the example just mentioned, I did not bring to the foreground the technology of the bicycle, but the technology of riding a bicycle. The technologies I am talking about are practices, not artifacts. At first glance there is only a minor difference between, on the one hand, conceptually fixating technologies to material artifacts and keeping the individual and social aspects of use apart, and on the other hand, conceptually fixating technologies to practices of interaction between artifacts and bodies. However, this small difference is a conceptual bifurcation with considerable implications for the philosophy and sociology of technology. I shall invest some time to lay the ground for an understanding of the aesthetics of technology as the aesthetics of technological practices.

Forms of life: Technologies constitute cultural identities. We are used to interpreting technology in terms of means-to-end relationships: technologies are good for something. However, *use* in the sense of practicing something is quite different from the *utility* of something. The seemingly 'natural' view of technology as means, instruments, devices to ends, purposes, or effects obstructs the perception of its cultural meaning. Culturally, technologies are specific forms of social existence in which – previous to all utility functions – self-formation and self-definition are relevant. Technologies form identities, both individual and social. For example, calling oneself a cyclist means something for oneself as *being* a cyclist. Individually it implies having built the bodily capacity to ride the bicycle; socially it means to be perceived as member of a special kind of road user or to be a member of the fraternity of cross country riders, etc. *Being a cyclist does not mean using some piece of technology, but rather existing in a certain form of technology.*

Of course, technologies allow for multiple identities. Cyclists can switch to the car or to the subcultures of SUV drivers or hikers and adopt attitudes quite antagonistic to cyclists. Another source of fragmentation and variety is founded on the fact that technological life forms are scale dependent. There is the broad community of bicycle riding traffic participants as well as that of the youngsters' subculture of artistic cyclists in the half pipes or the adventurous mountain bikers. As is well known, subcultures tend to distinguish themselves. Distinction is the essential sociological operation in the self-forming processes of cultural groups. Certain formative artifacts and skills amended and refined by equipments, signs, and codes contribute to the feeling of being different. Life forms can develop a dynamic toward life styles.

From these examples it is obvious that it is possible to describe the identity formation function of technology without any special reference to its utility in a means-end relationship. My rather arbitrarily chosen example of bicycle riding should be taken as a paradigmatic example of a broad variety of technologies in which artifacts, skills and institutional codes are combined.

Having given these preliminary definitions, I will attempt to formulate a comprehensive view. The term aesthetics of technology refers to the experience of practicing a technology in an institutional setting. This integrative perspective can, of course, be compartmentalized into the aesthetic components of material design, functional handling and enjoyment. It is possible to find somebody who enjoys riding a rotten bycicle in heavy traffic, or somebody else who owns a superb high-tech machine that he is afraid to use. But by and large it may be assumed that aesthetics refers to the practice that emerges from, and consists in, the interplay of the components.

I take the term practice in its original Aristotelean meaning which distances practicing something from producing something (*poiesis*). It is true, though, that Aristotle related technology – or its Greek relative 'techne' – primarily to the production mode. However, he would not deny that every product is the result of a process in which the technician acts as practitioner. Aristoteles' paradigm for practicing without producing an independent product was playing the guitar (or "guitaring", "kitharizein" in Greek, as he would say). The primary purpose of playing the guitar, he said, is guitar playing. Plausibly enough, in all societies with division of labor and markets, understanding technologies in terms of their end products superimposes their understanding in terms of practices. Music seems to be rather an exception. Previous to Edison's invention of listening to recorded music, you could not have the product without listening to the practitioner. I propose that not only some, but all technologies are practices (though not all practices are technologies). The internal aesthetic dimensions of technology can only be brought out if we do not look at technologies as objects to be observed, but as forms of our existence. We shape ourselves with shaping technologies.

It may be objected that this is a romantic view of technologies of an older era. Many technologies we live with are not activated by our practices. Instead, they exist by themselves, operate more or less automatically, and constitute our environmental infrastructures. The less we act upon them the better they function. Admittedly, the production chains of artifacts have become extremely long and manifold. Most things we employ in technological work are prefabricated. In this sense, most of the aesthetics associated with technology is also prefabricated. On the other hand, this is new only in degree, not in principle. Almost no technology produces something from scratch. Focussing on technological practice does not mean ignoring division of labor, mechanization, and infrastructure. Still, in all our operations – partly based on artifacts, partly realizing our technological existence, partly aiming at producing something – there are interfaces between artifacts, action and institutions. From a sociological point of view the analysis of these interfaces is at the heart of the sociology of technology. From an aesthetic point of view the shaping of interfaces deserves the highest attention.

Technologies as practices

If one describes technologies in terms of technological practices one usually needs to refer to the three constitutive elements that I have already mentioned in passing: *artifacts, conventions, competences.* Technology as practice – and in a broader sense, as a form of life – is constituted by joining together components relating to these three dimensions whereby the "joints" are provided by interfaces.

Artifacts: I take the concept of artifact to extend over a broad variety of objects, comprising not only instruments and machines, but also houses, streets, clothes, paintings, electric light and even songs, algorithms and papers. Other authors distinguish between material and symbolic artifacts, however, to keep the classification simple here, symbols will be thought of as constituting a subclass of material artifacts.

Conventions: Conventions cover all aspects of technology related to the social processes of regulation and control. Implicit conventions are customs, rules of behaviour, or dress codes. In former days implicit rules have been more important than codified rules. But technological modernization goes along with an ever increasing amount of explicit regulations. ISO-norms, legal standards, prescriptions in the domains of medication, nutrition, packaging, typification of materials, conditions of warrants, certificates of purity and quality and many more regulations contribute to the reliability, manageability, compatibility and – last not least – acceptability of technologies. Equally important are social orders of technological behaviour. A quick look at car driving (instead of bicycling) makes it obvious how important the institutional framework for the functioning of the traffic is. The relevance of conventions as components of all technologies can hardly be exaggerated. It may also be called the "law-and-order" aspect of technology.

Competences are the third kind of fundamental components of technologies. To be competent means to act qualified and adequate. Competences comprise all learned, trained and specialized abilities for technological action. Instruction and training form cognitive and bodily skills and routines. Obtaining technological competences is bound to learning by doing, even if it goes along with textbook knowledge and rules.

In coordinating these dimensions I assert that a technology is the technological mastering of technological devices in a technological social context. These contexts can be specialized domains with high access barriers, the best examples of which are occupational fields. Societal differentiation as well as cultural distinction is predominantly based on technological differentiation. Alternately, the barriers can be low as in the example of licensed traffic participation. Finally, the domains can be open to everybody as in most consumer technologies, in monetary transactions, or media based communication. From a historical perspective the overwhelming impression is the never ending increase of technicalization of virtually all action fields. On closer inspection this assessment needs to be distinguished from the incorrect assumption that previous to the process of technicalization an action field exists in a pre-technical state. Technicalization usually means to transform a technological life form by means of a new technology. Again, it was Aristotle who proposed to understand humans as being in themsleves a technological being. He took as a paradigmatic example the *handling of our hands*. He interpreted the hands as tools ("organs" in Greek), capable to being handled as a set of tools (for pushing, lifting, scooping, pointing, embracing, etc.) which again can be used for fabricating external tools. He makes the point that we *are* these bodily tools. We cannot put them aside:

For the most intelligent of animals is the one who would put the most organs to use; and the hand is not to be looked on as one organ but as many: for it is, as it were, an instrument for further instruments. (Aristotle, De partibus animalium 687 a 19-24)

Aristotle's second example supporting the thesis that the creator in his wisdom construed people as technological beings is language. If the hand is the work tool of all work tools, the communicative mind (logos) is the thinking tool of all think tools. Language is a technological fundament, appropriate for instituting all other social structures of society. Is language a technology? At least essential parts have technological features. Symbols and grammar are throughout conventional and the skills to speak are cognitively learned and bodily trained over years. Here is not the appropriate place for a thorough interpretation of the Aristotelean philosophy of technology. Nevertheless, from this brief glimpse we can take away the idea that technology is deeply rooted in our biological, anthropological and societal constitution. Technology is not a cultural or economic add-on. The cultural add-ons are in the differentiation and extension of what we are – a technological being. In the final analysis, technological aesthetics is rooted in the way we operate and communicate with our bodily organs – in our art of walking, talking and working.

The general proposition that technologies consist of artifacts, conventions and competences and that through these technological practices are constituted and forms of life are shaped needs some qualifications. There are examples in which one or thc other component is not present or is only weakly present. For example, memory technologies can be put to work without much assistance from external artifacts; fully automated technologies such as the watch can operate almost independently of user skills; and the machinery of bureaucratic technologies is based purely on legal and organizational rules. However, instead of generating a long list of variant cases, it may be easier to say that the relative share of the components can differ and can be reduced to zero in certain cases.

If this is agreed, we can raise the question of what makes the components fit together. The above mentioned idea that the aesthetics of technology consists in the interplay of components at interfaces can now be formulated more precisely. The smooth fitting together of competences and devices in adequate institutional settings yields the aesthetic functioning of a technology. Presumably it is not possible to narrow down the meaning of 'smooth fitting' in any general sense save for one aspect: All three components are technical in the sense that they are essentially determined by regularities. The most formal definition of regularity denotes a fixed – or at least highly probable – coupling between the input and the output of a system. Beneath this formal definition the types of regularities differ essentially: Artifacts are determined by causal rules, conventions by socially accepted and obeyed rules, competences by rules of self-control.

Causal rules: Artifacts function according to rules embedded in the states and operations of the materials. They are a subset of causal laws. Even in the case of calculation and control technologies, the formal rules of logic are executed by materialized causality. Causal and logical rules regulate the functioning of artifacts deterministically. This determinism is the source of the material reliability of technology. I take the technology of car driving as an example. Among the relevant components determining driving are the car's reaction to the driver's operating impulses with respect to direction, speed, lights and everything else. Ralf Nader's classic criticism of an American car manufacturer under the slogan "unsafe at any speed" (1965) was a bestseller precisely because he showed trust in reliability to be illusory. Some of the most high-tech cars of today encounter unexpected problems when electronics seem to give the cars some sort of indeterminist free will, though this is not exactly what these cars are made for. We also rely on the logic of traffic lights and on the proper reaction of the road surface with respect to the car's weight and friction. By and large the general basic value of artifacts is that their functions are causally regulated. It is this determinism that makes artifacts different from their environment. The more complex technologies are, the more difficult it is to achieve causal reliability from the very beginning. Failures in installations, expensive recalls in the automotive industry, the recently established "banana-principle" in software systems development (deliver early and let it ripen with the costumer), rapid prototyping as a methodology which integrates users into the testing phases of components, innovation networks between developers and users in high tech industries all show that the process of getting the functionalities of new technologies right and reliable is not an easy goal to achieve.

Conventional regulations are not based on causality but on agreement and compliance. Nobody is causally determined to stop in front of red traffic lights. Regulations need not be stated as generalized formal rules, though they usually are in modern society. Threats and sanctions are needed to stabilize the institutional legitimacy of rules. In a very broad sense, one can consider every explicitly stated and controlled rule a social technology. This was the view of Max Weber: "From a purely technical point of view, a bureaucracy is capable of attaining the highest degree of efficiency, and is in this sense formally the most rational known means of exercising authority over human beings" (Weber

1921/1968: 223). Independent of Weber's assessment of the bureaucratic perspectives of societal development, we can assume that trust in technology is to a considerable part trust in expert personnel and a readiness to observe rules. Anthony Giddens (1990) has described to what degree modern society depends on the proper action of experts systems. Even in cases of failure we have to rely on organized services. While causal regulations exist, even if they are not used and they do not change if they are ignored, conventional regulations exist only when they are observed and they vanish or change with their institutional framework.

Skills and Routines: Regulations referring to ourselves and constituting our bodily and cognitive competences are again different. In many cases the acquisition of skills starts with formal rules of the conventional type, but fairly soon it is extended to regulations of self control which the user can no longer formally describe. Basic social skills – as in walking upright – cannot possibly be completely described, and the best robots walking on two legs are still fairly simple when compared to our art of walking. Cycling is another example of skills that cannot easily be described. These embodied rules make us technical experts in fields of every day activities. Other rules make us specialists. There is a long lasting controversy as to whether the "know how" skills of experts are beyond rules or are rather based on the ability to manage the interference of many rules.

Regulation I consider to be the fundamental value of all technology and it is important to see that three types of regularity operations are involved: material/causal, conventional/institutional, and cognitive/bodily rules. They cannot be reduced to each other because they are generically different. They shape the natural, social and individual ingredients of any technology. The aesthetics of technology refers to their coordination. If the coordination is well arranged it forms a piece of self-contained individual, social, and material life. Technological aesthetics, therefore, is essentially a completely artificial aesthetics. To be sure, all sets of rules are enclosed in broader environments. Material artifacts remain to be embedded in nature, institutional conventions in society, and skills in individual life. But in performing something technical, we count the specific operations enabled by the rules as relevant and valid within the scope of a technology.

Practices are the smallest social units of technology. Such units are made visible in historical pictures. They present technical practices as stereotypes and emphasize the internal relationship between body motion and tools. In one instance we have the blacksmith's powerful body, in his hand the huge hammer welding the metal, in another the poor little tailor with his needle and thread. Pictures are snapshots of practices. Usually occupational features and signatures are added in order to let in the practitioners' institutional framework. Why do we feel these pictures show the aesthetics of these practices? My suggestion is: Since it is known to be difficult to make these components match, it is impressive to observe technology as an easy going process in which heterogeneous operations are integrated. If occupational work, especially craftsmanship, is the paradigm for this aesthetical observation, it is by no means the only field. The other examples I have mentioned apply as well: sports, music and arts, riding bicycles, driving cars and infinitely more. The tension between background knowledge about the difficulties in making everything fit together and the observation of smoothness of operations playing at the interfaces between instruments, bodies, and institutional set-ups triggers the aesthetic moment of technology. In many cases this tension is not consciously present because our spontaneous perception refers to the unity of a technology. Only at second thought do we analyze the complex coordination of hetereogenous operations. When we practice a technology, the feeling of the technological unity of which we are a part is predominant as long as everything and everybody works fine. In riding a bicycle down a windy road you normally cannot tell what you are doing or what the bicycle is doing. The technological action we perform constitutes the aesthetic experiences we have.

From technological practices to life forms

Technological practices are the smallest units of a sociological analysis of technology. They are circumscribed by a specific set of regulations which control artifacts, skills and institutional expectations. But in singling them out, one almost unavoidably refers to broader units which are networks of practices. Alluding to Marvin Minsky's "Society of Mind" one can call single practices 'simple agents' which constitute higher technological units. Large amounts of practices are interconnected and support each other. They establish a comprehensive field of action and set boundaries to other fields. These fields I call technological forms of life. In referring to someone as being a sportsperson, a craftsman, and the like, we connote and bundle a broad variety of different practices.

The aesthetics of life forms are not visible by means of direct observation, but by stories (or movies). No practitioner can immediately experience a life form but learns to live it through the more general format of social ascriptions and reflexive self-descriptions. The primacy of practice over action, of performance over result, of exercise over application becomes even more obvious on this integrated level. Even if it were possible to define a technological life form by listing certain sets and chains of practices, it would not bring to the fore what it means to live some sort of technological life. The higher level of technological aesthetics refers to the integration of practices in forms of live.

Again adopting a view developed by Aristotle, one can call a good form of life – individual and social – an aesthetically convincing form. As he said in a slightly different context: Doing something well (eupraxia) is a goal in itself (Politik 1325 b 21). And the moral stance would be: We should strive for technological life forms which coalesce good practices in well ordered patterns. Occupational worlds are the historical prototypes for this integrated mode of technological aesthetics – especially as they present themselves and are presented in images. They constitute special social identities. Similar examples can be obtained from the arts, music and sport. However, there are other areas where social identity relates to quite different social categories, for example settlements and especially cities. Virtually all components of city life are based on material artifacts. Formal regulations are manifold and dense. Taking an active part in city life presupposes a fairly high degree of individual self control. Descriptions of cities in terms of their (technological) specialties are usual in guides and encyclopedias. Especially suitable for the concept of the technological life form is the view that cities have their goal in themselves. With few exceptions it is not meaningful to ask what a certain city is *for*. The goal is rather the self organization of social life for which every city has developed a distinctive solution. The fit of the heterogeneous components can take rather different forms in different cities. There is the rigid charm of industrial centres, the appeal of towns still exhibiting their medieval origins, the pride of great capitals, the passion related to Las Vegas-like cities, or the peculiar irritations of modern mega cities.

Admittedly, there are many steps between being an expert in one single technological practice and being a participant in city life. It may stand here as a hypothesis that the composition of practices to complex life forms as well as their decomposition into practices can be performed.

Throughout the paper I have avoided bringing into play technological purposes and products. Instead of asking what technologies are good for, I have emphasized the alternative question: In what sense does society form and realize itself – in its individual, social, and material aspects – as a technological society? Or from a more subjective perspective: How do we feel, interpret and reflect upon ourselves as being components of technologies? I admit, of course, that we are many things apart from our technological practices and our lives are richer than technological forms. The form of living with printed books does not predetermine the content to be written and read. That is not the point. The focal point of an aesthetics of technology, it seems to me, is the coordination of regulatory interfaces. I am inclined to propose (a somewhat inverted) Kantian technological imperative: Never consider any technology as being soley a means for an end, but also as an end in itself. But I resist doing so, because it pays too much tribute to the means-end interpretation. Technological life forms are not good for something, but are media of a social existence. Sceptical literature abounds which repeats the rhetorical question of whether people really need some new invention. Instead, we should ask why people are willing – or forced – to transform their technological live style. The point I wish to stress from this discussion is the following: The step from a means-end utility view of technology toward the aesthetics of a technological lifestyle needs to not be taken into the direction of making things more pleasurable and enjoyable, but rather towards consciously shaping ourselves, our communities and our material realities to practices and life forms through which we would like to live.

List of References

Aristotle. 1965. *De partibus animalium*. The Works. Edited by J.A. Smith and W.D. Ross. Oxford: Clarendon.

Bijker, Wibe. 1995. *Of Bicycles, Bakelites, and Bulbs, Toward a Theory of Sociotechnical Change*. Cambridge, MA: The MIT Press.

Böhme, Gernot. 2001. *Aisthetik. Vorlesungen über Ästhetik als allgemeine Wahrnehmungslehre*. München: Wilhelm Fink.

Borgmann, Albert. 1984. *Technology and the Character of Contemporary Life*. Chicago: University of Chicago Press.

Giddens, Anthony. 1990. *Consequences of Modernity*. Stanford: Stanford UP.

Ihde, Don. 1990: *Technology and the Lifeworld. From Garden to Earth*. Bloomington: Indiana UP.

Minsky, Marvin. 1986. *The Society of Mind*. New York: Simon and Schuster.

Nader, Ralph. 1965. *Unsafe at any Speed. The Designed-In Dangers of the American Automobile*. New York: Grossmann Pubs.

Weber, Max. 1921/1968. *Max Weber on Law in Economy and Society*. Max Rheinstein (ed.). Translated by Edward Shils and Max Rheinstein. New York: Simon and Schuster.

On the Symbolic Dimension of Infographics and its Impact on Normalization (With Examples Drawn from Demography)

JÜRGEN LINK

Before entering into a discussion on infographics including mainly the role they play in the context of normalization, let me start by proposing some theoretical concepts and tools concerning the terminological field of the "normal". The first problem is due to the fact that the notion of "normalization" is all but clear even on the simple linguistic level. I must briefly mention some linguistic differences between the terms for "normal" in English, German, and French. Whereas in German there is only one noun for the concept (Normalität), there are two in English: "normalcy" and "normality". It seems to me that the first one refers mainly to collective social or political conditions, and the second one mainly to psychological states of individuals. The french "normalisation" is in 90 % of the cases the equivalent of the English "standardization" – a fact that in my view has led to some misreadings of Foucault, especially in Germany where "Normalisierung" means 'rendering normal' or 'returning to normal'. If we now were to have a look at the actual use of "normal", "normality" and "normalization" and its equivalents in other languages all over the different discursive fields, from sociology and psychology to the media, politics and everyday language, we would be faced with utterly vague and contradictory results. Is there at all a concrete meaning in the declaration of a politician that a certain measure should be regarded as "ein Stück Normalität" or as "Normalisierung"?

This vagueness can be seen as a kind of challenge to us as theorists of discourse: Should we dispense with normalcy as a concept of criticism because of its very vagueness or can it be defined and developed in an operational way? While I am prepared to honor the first choice, I myself have opted for developing a theory of what I call "normalism", including the entire semantic field of "normal" – the main reason for my choice lying in the fact that "normalcy" undoubtedly plays a major role within the object discourse of media and politics as well as in several specialized discourses.

In order to approach a workable definition of normality and normalcy, I have suggested six restricting inequalities (Ungleichungen):

1. normality ≠ normativity
2. normality ≠ everydayness
3. normality ≠ bio-homeostasis
4. normality ≠ cybernetics and technocracy in a general understanding
5. normality ≠ aesthetical banality
6. normality ≠ reality socially constructed in a generally epistemological understanding

As you will know, all these inequalities are regarded by important scholars of diverse approaches in diverse fields of research as equations. I will discuss here in some detail only the first two inequalities. By the third one, I am claiming that normality is socially and historically constructed and not a naturally given entity (a thesis which may be relevant for the demographic problematic). By the sixth one however, I mean that not the entirety of socially constructed reality should be called "normal", but only a particular sector of modernity characterized by particular structures that can be described more or less exactly. Inequalities four and five would need a detailed discussion which I cannot enter into here. So let me come to inequalities one and two.

As to inequality one: Here I would like to stress that – regarding normality and normativity – we are dealing with, in the former, an ultramodern and, in the latter, an ancient and probably (in a literal sense) antediluvian phenomenon. According to concurrent interpretations of ethnology, anthropology, and sociology, all human societies possess and have possessed "norms" and "normativity". These are explicit and implicit regulatives, which are reinforced through sanctions that pre-scribe a specific action to materially or formally determined groups of people. "Norms" are, therefore, always pre-existent to (social) action: they are already known to at least a few professionals of the norm before such action. It is this quality which can explain the more or less legal overtone of all norms in the realm of "normativity".

On the other hand, according to my thesis, "normality" (including normalcy) is a historically specific "achievement" of modern Western societies, which has never before existed, and even today, in numerous societies or cultures does not exist, or exists only in its beginning stages. Furthermore, in my thesis "normality" presumes – quite fundamentally – statistical dispositives and is defined in relation to averages and other statistical sizes. If one takes this defining criterion seriously, there are (now formulated differently) "normalities" only in data-processing societies (auf Verdatung basiert): only in cultures that continuously, routinely, comprehensively, and institutionally make themselves statistically transparent. This kind of statistical transparency, which Foucault in many ways also had in view, is surely related to panoptic transparency, but is not identical to it: they can be as different as the secret police (e.g. the Stasi) and public opinion polls. If a "normal" action is now statistically constituted as "average" (or, is situated on a distribution curve within a "normal" distance from the average), then "normality", in contrast to "normativity", is – ideal-typically observed – essentially post-existent to action (instead of pre-existent). If an action is to be valid as "normative" (i.e., corresponding to a "norm" in the normative sense), it is, as previously stated, already known beforehand – if it were "normal", on the other hand, it is certainly capable of being first established retrospectively through its positioning on the concrete-empirical statistical distribution curve. This difference is absolutely fundamental for the functioning mode of contemporary Western societies (that is, those that I have suggested, within this corresponding parameter, naming "normalistic".) In such societies, there is in effect, namely, a final functional dominance of normality over normativity (which, of course, does not foreclose conflicts, but rather presumes them outright: one thinks of topics relevant for demography such as abortion, birth control, DINK (double income no kid) marriages, gay marriages, AIDS and so on).

Now as to inequality two: Admittedly, the operational praxis of all the specialized discourses (i.e. institutionalized sciences in academia) relevant to normalism (particularly, medicine, psychiatry, psychology, sociology, economics, and last but not least, demography) corresponds to my postulated regime of data processing and the determination of "normalities" through statistical distribution. On the other hand, however, there is the custom of an a-historical, pan-chronological concept of "normality" in ethnological sociology. According to this custom, "normality" is similar to "everydayness" in a historically all-encompassing sense, which affects all ages and cultures. (The equation of "normality" with "banality", which has progressed since a decade or so mainly in French – "banalisé" in the sense of the German "normalisiert" – , but to a certain extent also in English, can be regarded as an aesthetically connotated divergence of the everyday thesis.) One can thus talk about both "normality" in shamanistic society and "normality" in antiquity and the

Middle Ages. I have already explained why I suggest restricting this concept to data-processing societies. "Everydays" that are data processed are entirely new emergences: through them, we adapt ourselves to the average speeds of massive traffic flows; we respect critical values or not; we work according to normal workdays in normal working relationships, or, when we are unemployed, we live with the help of unemployment insurance, which is calculated on the basis of mathematical statistics and other factors; we attempt to adjust our weight (i.e., our "figure" – "Linie") according to data about "normal" and "ideal" weight, and, even when we get divorced or plan a late first birth, we orient ourselves (at least sub-dominantly) to the relevant statistical curve-landscape.

So let me try to put forward now a definition of the concept of normalism, which I (with reference to my book (see Link 2006a)[1]) will condense here: "Normalism" can be understood to be the entirety of all discursive as well as practical-interventionist procedures, dispositives, instances and institutions through which "normalities" are produced and reproduced in modern societies. Constitutive are especially the dispositives of data processing on a massive scale, i.e. statistical dispositives in the broadest sense. This includes questionnaires on the level of data collection, it includes mathematical-statistical theories of distribution on the level of valuation, and includes all social dispositives for re-distribution (Um-Verteilung) on the level of practical intervention. The produced and reproduced normalities are essentially characterized through "average" distributions, and ideally approach a "symbolically Gaussian distribution." An example of this "average" distribution might be a bell curve that has a wide middle area in which the majority of the data fall, a "normal range", and two symmetrical, "abnormal" extreme zones that contain little of the data. These dispositives of regular, systematic and ground-covering data processing represent, according to my estimate, the historical *a priori* of normalism and thus would have originated not earlier than the 18th century.

To put it in a somewhat simpler way: the vagueness of statements about the normal in modern societies can be controlled up to a certain point by statistics so that a "normal" society would be characterized by a more or less bell-shaped distribution of wealth, of knowledge, and of political preferences in voting, around the symbolic "center" measurable by statistics. Likewise, loss of normalcy (de-normalization) would be characterized by an askew shape of the distribution.

1 In English see four essays on normalism by Jürgen Link (with an introduction by Rembert Hüser) in: Cultural Critique, Number 57 (on "Normalization"), Minneapolis, Spring 2004.

Before entering the subject of my second part, i.e. the subject of infographics, I must briefly address the concept of "normal" development over time. The normal distribution and the entirety of symbolically related distribution curves, under the aspect of time, all are situated in a moment of synchrony. What then about the diachronic dimension of normality? As I have discussed in more detail in my study, the basic diachronic curve of normalism is constituted by the S-shaped (or logistic) curve or, rather, by an endless series of such curves which I call "the endless growing snake of normal progress". This fundamental curve results from the bending backwards (or curbing) of the symbolically "exponential" curve of modern growth. Here I am speaking of another important constituent factor of the fundamentally "modern" historicity of normalism: normalization and normalism must be seen as the response (in the sense given to this notion by Arnold Toynbee) to the challenge of modernity – in the view of the symbolically exponential (positive and negative) growth curves since the take-off of industrialization in the eighteenth century: that is, the growth of population, knowledge, work productivity, capital, transportation, and so on. In this fundamental historical context, normalism proves to be, so to speak, the "braking-insurance" (bremsende Ver-Sicherung) that stops the fear of denormalization resulting from the symbolically exponential trends. Or, now formulated from an entirely symbolic viewpoint: normalism is the necessary brake for the engine of modernity, which often seems on the verge of exploding.

The entirety of the statistical knowledge which generates what I call the normalistic "curve-landscape" or "curvescape", belongs to a more or less professional or even scientific type of knowledge elaborated by experts. All this normalistic knowledge claims a more or less "objective", empirically controllable, character. Now, if my guess should be right that normalism constitutes a fundamental orientation device in modern cultures, the question arises as to how the objectivity of data can be transformed into the subjectivity of acting people. And here exactly is where infographics come in. The main idea of my second part will be that infographics add a symbolic dimension to the statistical one and can thus make an important contribution to the "subjectivation" of allegedly "objective" data constellations. By performing this transformation, i.e. by "subjectivizing" the normalistic curvescape – or put somewhat more pathetically: by introjecting the normalistic curvescape into the souls of normal people, infographics can, in my view, contribute a lot to generate normalization processes which are not only forced upon people by authoritarian disciplining, but are "voluntarily" and willingly performed by a majority, i.e. by a "normal range" of otherwise allegedly autonomous individuals that only adjust themselves along so-called "corridors" of statistically underpinned normal behavior.

When we regard the symbolic dimension of infographics more closely, we will find that there are different degrees of such a dimension. The first one can be viewed as a kind of implicit symbolicity connotated by the very shape of a certain curve itself. The perhaps most classical example up to now was put forward by Thomas Malthus right at the beginning of normalism in 1798, when he claimed a "geometrical" curve for demographic growth and only an "arithmetical" one for the growth of the means of subsistence. By this formula, he established two fundamental symbolic connotations for two basic normalistic curves or combination of curves. The first one is the curve of symbolically "exponential" growth as we call it today with its ambivalent connotations: In its first segment, this curve provides positive connotations of progress, upswing, strong recovery and so on, whereas the following segment, with its ever steeper increase, sends the symbolic message of overheating and of danger of explosion – of "de-normalization" in the terms of my study. The second emblematic curve symbolism established by Malthus lies in the combination of both the "geometrical" and the "arithmetical" curve which together form the famous "opening scissors" formation. This formation also clearly signals a major risk of de-normalization. But as we will learn in a moment, such de-normalization messages of statistical data configurations should be seen within normalism merely as a so-called "early warning signal" which simply calls for normalizing interventions such as curbing the exponential growth curve to the S-shaped logistic curve in order to normalize it. (See figures 1 and 2 as examples of infographics in the Malthusian variety.)

If normalizing the exponential growth means curbing the trend down to the S-shaped logistic curve, there are at least three further major problems for such a normalizing procedure: The first problem would turn around the question of whether the curbing should be regarded as provisional (i.e. leading to further exponential upswings in the future) or as definitive (i.e. leading to stable zero growth). In the case of demography, there has been a large consensus in favor of the second option for more than a century. (I have discussed in the third edition of my study the case of the American physician and demograph Raymond Pearl who was already opposed to demographic alarmism during the 1920s and predicted the normalization of the tremendous growth trend in the future[2].) The second problem is about the average growth over time which depends on

2 See Pearl 1925. On Pearl and subsequent theoreticians of demographic dynamics cf. Höhler 2007.

the width of the time window foreseen for the normalization of the exponential curve segment. But here is where the overall average growth has to be split into the concrete averages of continents and countries which differ strongly. Pearl was already convinced that the fact of differential population growth rates was fundamentally due to a difference in cultural development, so that he expected every population to follow the same normalizing logistic curve one after another, only with different chronologies. Be that as it may, the differential birth rates obviously do constitute a fact of major symbolic impact since Darwin's cousin Francis Galton, perhaps the most important normalistic thinker of the Victorian age, rang the alarm bell over the opening scissors' curves of the upper vs. the lower classes in England. Today, we all know these scissors' constellation from the differential birth rates between the First and the Third World. Here we are faced with infographics bearing evidently strong symbolic connotations of de-normalization and triggering what I call de-normalization fear (see illustrations 1 and 2).

...UND MEHRET EUCH

Entwicklung der Erdbevölkerung

um sich der Mann aus Rom ausgerechnet um das Leben vor der Geburt so sehr sorgt. 300 von 1000 bolivianischen Kindern sterben vor ihrem sechsten Geburtstag.

Statt staatliche oder private Aufklärungskampagnen zu unterstützen, torpedieren kirchliche Würdenträger im Auftrag des Papstes jeden auch noch so zagen Ansatz der Regierungen, die Geburtenrate zu drosseln. So boykottierten die Bischöfe von Peru (95 Prozent Katholiken) jahrelang ein Familienprojekt der Regierung des Präsidenten Alan García.

Nun hat dessen Nachfolger Alberto Fujimori den Kampf mit den Moraltheologen aufgenommen. Der Katholik läßt kostenlos Verhütungsmittel verteilen, die mit Uno-Geldern finanziert werden. Den lautstark protestierenden Kirchenoberen schleuderte der Staatschef entgegen, er wolle verhindern, daß immer mehr „hungernde Kinder Abfälle essen, auf der Straße leben und sich prostituieren". Den Widerstand aus Rom gegen Geburtenkontrolle nannte er „mittelalterlich und verstockt".

Auch in Chile holte sich der Papst mit seiner Fruchtbarkeitsideologie eine Abfuhr bei den Gläubigen. Vor 90 000 Jugendlichen predigte er im April 1987 im Nationalstadion von Santiago, wo die Häscher des Diktators Pinochet nach dem Putsch 1973 rund 15 000 Chilenen eingesperrt und Hunderte zu Tode gefoltert oder erschossen hatten.

Während der langen Ansprache verurteilte der Papst aus Polen die Diktatur mit keinem Wort. Als er die rhetorische Frage stellte: „Nicht wahr, ihr weist das Idol des Sex zurück?" brüllten ihm die Massen entgegen: „Nein!"

Mit besonderer Vorliebe missioniert der Heilige Vater in Afrika. Bereits 1978, bei seiner Amtsübernahme, hatte Johannes Paul II. den Schwarzen Erdteil, der unter seinen 600 Millionen Menschen nur knapp 15 Prozent Katholiken zählt, „reif für die Ernte" genannt. Anders als in Europa, wo die Gläubigen der Kirche in Scharen davonlaufen, breitet sich der Katholizismus in Afrika weiter aus.

Aber auch dort wehte ihm auf seinen bislang sieben Reisen in 29 Länder ein zunehmend rauherer Wind entgegen. „Was soll ich einem völlig verarmten und unwissenden Ehepaar raten, das immer wieder gespenstisch anzuschauende Kinder bekommt, denen der Hungertod sicher ist?" fragte vergangenen September Ruandas Regierungssprecher Christophe Mfizi im überfüllten Stadion der Hauptstadt Kigali den versteinert dreinblickenden Papst.

Mfizi hatte noch mehr Fragen auf dem Herzen. „Was soll ich einem jungen Mann raten, der zu ewiger Arbeitslosigkeit verdammt ist und den die Gesetze der Gesellschaft und der Religion dazu verurteilen, keusch zu bleiben, der aber keine Kontrolle über seine Sexualität hat, während Aids die Menschen umbringt?"

Der mutige Katholik nannte in seiner Rede das kirchliche Gebot der Ehelosigkeit für Priester „schwer zu verstehen und noch schwerer zu leben". Der Amtskirche warf er vor, eine Bastion weißer Vorherrschaft zu sein und die nichtwestlichen Werte Afrikas zu ignorieren. Typisch: Bei seiner nachfolgenden Ansprache ging der Papst mit keinem Wort auf die Fragen Mfizis ein – weil dazu die Zeit gefehlt habe, hieß es später.

Nicht ohne Grund war es gerade das bettelarme Ruanda, wo den Papst die triste Wirklichkeit eines Kontinents einholte, der sich selbst nicht ernähren kann und den wachsende Verschuldung sowie ein dramatisches Bevölkerungswachstum (3,3 Prozent) immer tiefer ins Elend sinken lassen. Im dichtbesiedelten Ruanda (sieben Millionen Einwohner) liegt die durchschnittliche Zahl der Lebendgeburten einer Frau bei 8,5 Kindern; jedes vierte Kind stirbt in den ersten fünf Lebensjahren.

Ob in Burkina Faso, im Tschad, in Togo, Kenia, Burundi, Sambia oder Zaire – stets redete der Papst den an sexuelle Freizügigkeit gewöhnten Afrikanern energisch ins Gewissen. Im Uhuru-Park von Nairobi (Kenia) geißelte er künstliche Verhütungsmethoden als „gegen das Leben ge-

Familienplanungs-Plakat in Ghana: „Seid fruchtbar ...

DER SPIEGEL 52/1990 125

Fig. 1: Spiegel 24.12.1990.

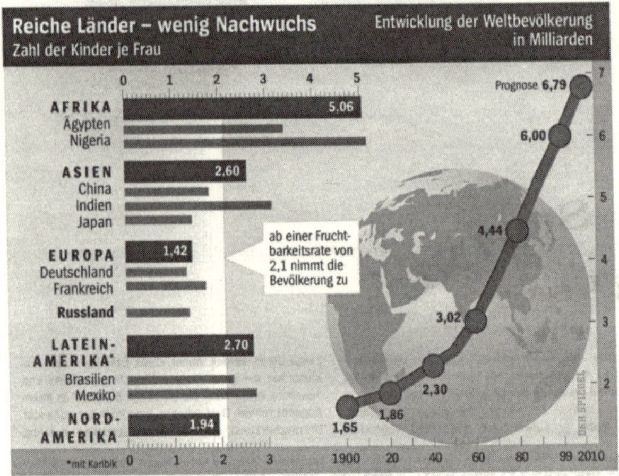

Fig. 2: Spiegel 30.08.1999

The third and most crucial problem about demographic growth is, of course, the question of intervening procedures for normalization (birth control or its interdiction, abortion or its interdiction, several types of sexual relations and their promotion or deterrent, family allowances and, last but not least, immigration or its interdiction). There are infographics for all these factors whose connotations as to de-normalization or normalization are mostly ambivalent. What then is a normal population growth anyway? The only certain statistical mark seems to be the average 2.1 child rate per woman which allows for a stable reproduction, i.e. the zero growth at the end of the logistic curve. But this is true only for the overall development and cannot insure the de-normalization fear of differential rates – on the contrary, it fuels this fear all the more when the rate of one's own country falls short of the magic threshold of normalcy.

Up to now I was dealing with connotations inherent in the very shapes of curves or bundles of curves as such. There is a special kind of these connotations represented by symbolic "letter" configurations. Some of these "letters" are bearing connotations of normalcy, others of de-normalization (see figure 3).

We now must discuss a new type of infographics which represents a higher degree of symbolicity and of subjectivation of the underlying data. These infographics are the most common ones in the mass media and are characterized by a combination of graphs or curves with human figures apt for identification or, in the terms of Michel Pêcheux (Pêcheux 1984), "counter-identification". There are photos as well as sketches of individual persons as well as of crowds (see illustrations 4 and 5).

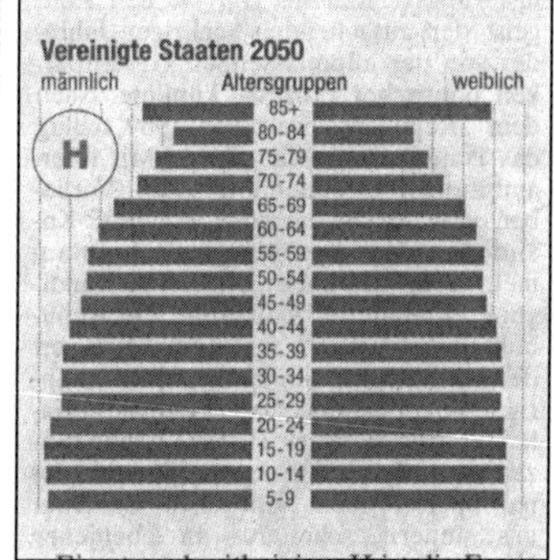

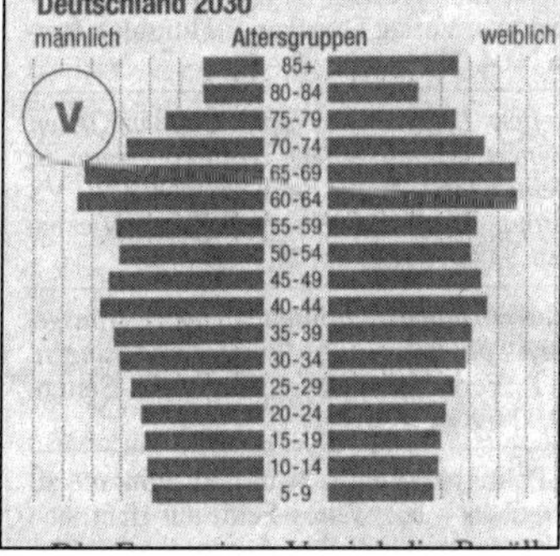

Fig. 3: FAZ 15.03.2005

Besides human figures, there are numerous other images of symbolic objects. The examples show further typical elements of what I call collective symbolism. Under this, I would like to understand the totality of all forms of collectively shared "figurality" (Bildlichkeit) of a culture – hence, not only symbols in the narrow sense of Goethe and Freud, but also allegories, emblems, illustrative models, comparisons, analogies, metaphors, metonyms, and synecdoches. In addition to allegories, all other symbolic personifications – including typical friend and foe images – also belong to collective symbolism. To mention some of the most usual collectively shared symbols of Western cultures which are currently integrated into infographics: the healthy body vs. cancer, AIDS, SARS or other diseases; the human body vs. rats or other ugly animals; the family house vs. the danger of fire; the dam vs. the flood;

football, boxing or other sports; railway, car, aircraft or other vehicles etc. (Link 1989, 2005, 2006b)

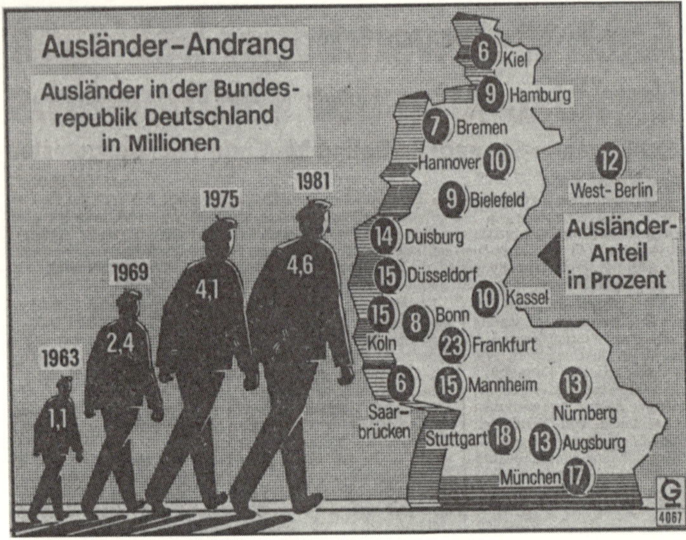

Fig. 4: WAZ 20.11.1981

By integrating collective symbolism, the normalistic curvescape represented by infographics attains an even higher degree of symbolicity. On the one hand, the curves are completed through symbolic figures, that is, subjects with positive and negative appeals for identification. On the other hand, the curves themselves send out positive and negative appeals to the subject in which they mark and offer for "application" zones of supernormal chance, subnormal risks (hence, positive and negative abnormalities) and insuring normality. More than other issues, the demographic curve-landscapes still continue to radiate an appeal of power for subjective application today – above all those of identification and "counter-identification", especially in the field of mass migrations.

If we take into account the totality not only of the infographs in the narrow sense, but also include all media images that by connotation can function as part of the curvescape, we finally reach the last and highest degree of symbolicity. Demography presents a very 'strong' case of this structural integration. Let me discuss the example of what I would like to call images of the race configuration (see illustration 6).

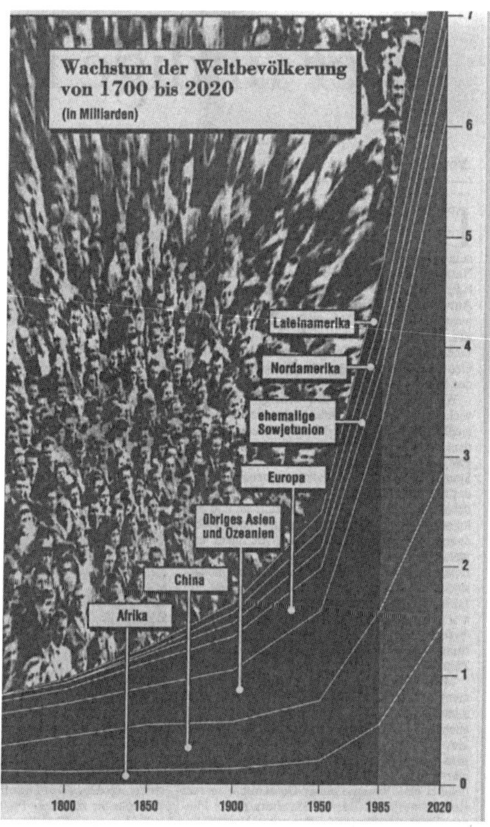

Fig. 5: Zeit 23.10.1992

My guess is that "normal" readers or spectators (i.e. a more or less vast majority of readers or spectators) will project onto these images the curvescape knowledge of the differential birth rate and so invest the white, yellow, brown and black figures with a kind of symbolically distinctive procreativity or potency. I can cite in this context the topical event of the September 2005 general elections in Germany which bore a certain connotation of loss of normalcy already by the non-regularity of their date which was then strongly increased by their results not foreseen by normalistic public opinion polls and, finally, by the formation of a "Große Koalition" usually linked to a situation of great risks and deep crisis. There certainly was some smoke of de-normalization in the air. Part of this atmosphere was the idea of a so-called "Jamaica coalition" propagated by several important media sources. Interesting enough, the purely symbolic color-constellation of black, yellow and green did also evoke a race constellation in the media (see illustration 7) which was strongly opposed by some readers because it was "utterly lacking in taste". Here, we are witnessing the possible overdetermination of topical events by mythical narratives based on collective symbolism in connection with data-fuelled normalistic narratives.

Fig. 6: Erziehung und Wissenschaft Nr. 16/1999

Fig. 7: WAZ 20.09.2005

The "Jamaica coalition" debate in Germany stands for connotations of what I suggest to call "flexible normalism" as opposed to a more rigid kind of normalism which can be called "protonormalism". To close my argument, let me add some reflexions on this "strategic discursive bifurcation" of normalism.

As a result of an analysis of the history of normalism during the last two centuries, it can be stated that the subjective character appeal of the normalistic curve-landscape can function within the framework of two ideal-typically discursive strategies in polar opposition. This appeal can first be imperatively directed toward narrow normal values: that is, "authoritarian" (Adorno/Horkheimer), "repressive" (Marcuse), "other-directed" (Riesman), and "disciplining-training" (Foucault). Through this discursive strategy, the normal range is established as narrowly as possible and is protected against the risks of de-normalization through both symbolically and pragmatically robust limits of normality. I am speaking in this case of "protonormalism". Thereby, the symbolic weighting of the limits of normality takes place through the articulation of pre-normalistic ideologies like those of "natural law". The pragmatic emphasis on limits of normality takes place above all through the link with juridical normativism (excluding specific varieties of "abnormality" as "criminal" and confinement behind the walls of a prison as real existing limits of normality) – or with medical indication in the widest sense (excluding other varieties of "abnormality" as "intellectually or spiritually deviating" behind the walls of institutions). Hence, the more restricted the normal range becomes, the "wider" the respective field of "abnormalities" must appear.

From a purely theoretical perspective, the principle of constancy and continuity of the normal curve allows an exactly opposite discursive strategy: If the transition between normality and abnormality is continuous, constant and fluid, then the limits of normality could also be placed "outside" of the "middle" as widely as possible, and thereby the normal range would be spread out to the maximum. In this case, large parts of the protonormalistic "abnormal" allow themselves to be completely integrated, while other parts can still be symbolically included into wide transition zones. A flexible structure for appealing to the subject corresponds to the strategy I call flexible-normalistic: The curve-landscape now serves not as an imperative, but rather as orienteering help for self-normalization. Hence, the subject is not fixed on pre-given normal values, but rather it can attempt to profit from the game of statistical compensation. The clash between both normalistic discursive strategies represents an important key to understanding the history of normalism, which I cannot address here. In rough summary, the flexible strategy in the leading Western countries after World War II could have reached cultural hegemony, even though protonormalism continues to play a role (since about 2000, protonormalism appears to be playing a larger role again). Michel Foucault's concept of "normalization", which is indeed closely connected to "disciplining" and "training" (dressage), corresponds primarily to protonormalism.

In our infographs, both strategies reveal their difference through the opposition of either narrow or wide corridors for trends of normal growth, as well as between either distinct or blurred markings of the limits of normality. The first tactic is exemplified by linking threatening exponential curves with enemy pictures as they usually appear in the immigration statistics of "rich" countries, for example. Specific types of "zero solutions" and "stop demands" (like "zero tolerance of criminality", "zero budget deficit", "zero diet", and "immigration stop", for example) are in general symptomatic for protonormalism, which theoretically remains a constant option for normalism in the future. Frequently, the symbolic opposite pair *hard* vs. *soft* signals the opposition of both the normalistic main strategies of protonormalism and of flexible normalism: typically protonormalism does not want to acknowledge discussions of *soft* drugs.

Without doubt, one can no longer imagine the media without infographic images of curve-landscapes that already belong to its normality by the extent of their large number. Less banal, my thesis is that these images must be characterized more precisely as "normalistic" insofar as we are dealing with socio-culturally necessary "nourishment for the modern subject". Through this "nourishment", these subjects are produced and reproduced foremost normalistically so that they can culturally orient themselves and can pursue their own normalization (as it were) in a new variety of "inner direction" that Riesman did not anticipate. Those normal-fields that are currently leading in the curvescape are of high relevance for the synchronic analysis of contemporary culture – just as the question of where (in each sector) the limits of normality are placed, how they are disputed, and which of the two normalistic strategies is thereby favored.

Coming back to demography, the career of this important normal field throughout history has known both peaks and lows. The denormalization fear triggered by the differential birth rate of a century ago has fuelled racism and given rise to eugenics which was invented by Francis Galton himself. In Germany, the first symptoms of a decrease of the then still exponential procreation rate gave way at around 1900 to an utterly hysterical debate over degeneration of the race at the horizon. So it is mainly the differential birth rate which in the normal field of demography tends to encourage protonormalistic discursive tactics. Nevertheless, it could not have been overseen even then that all kinds of those protonormalistic tactics generally prove ineffective in the field of demography because of its being linked intimately to sex. So, if we are witnessing today, a century later, a sort of *déjà-vu* of the fertility alarm (cf. illustration 8), it is clear from the beginning that the protonormalistic offensive must now confront a much stronger position of flexible normalism than a hundred years ago. Let me quote from the newspaper review of a recently published book on Germany's lack of kids problem which may be seen as a more than molecular discursive event. The author of the book is Supreme Court judge Udo di Fabio, its title being "Die Kultur der Freiheit" (The Culture of Freedom). Before praising di Fabio's book as a kind of relieving way out of what the reviewer calls a "dead end of hysteria", it is said:

> The nearer the election day, the more feverish the public debate around the child. It now enters into the grotesque. As soon as someone says something in favor of a particular lifestyle, be it that of the married housewife or of the double career couple, that of the gay marriage or of life without marriage at all, we immediately can hear outcries of indignation by others who feel neglected. We now are trapped in a situation of total deadlock of communication. It has become almost impossible to exchange arguments in favor of or opposed to the child because the issue is in such a degree personalized that every clear argu-

ment in one direction or the other is felt as a personal insult. (Frankfurter Allgemeine Zeitung, September 12, 2005, 41)

What the reviewer sees as the unsolvable problem is the fact that today protonormalistic measures are seen by many people as de-normalizing threats to normalities created in the meantime by flexible normalism, such as equal rights and equal opportunities for women, especially the opportunity of female careers in both business and academia. Look at the infographics which show the correlation of high female academic grades and low differential birth rates (see fig. 9):

Fig. 8: FAZ 18.01.2005

What about their implicit connotations? It is impossible to add images of deterrent and ugly feminists to the statistics like the images of "floods" of black or colored immigrants in front of "our doors". But do the naked figures as such with their symbolic connotations not "send a clear enough message"? Here the two main normalistic strategies seem to run into a dead-end – unless cloning will one day resolve the problem.

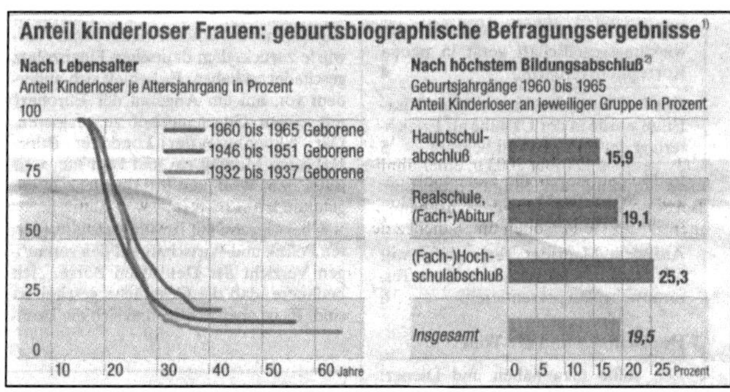

Fig. 9: FAZ 08.03.2005

List of References

Höhler, Sabine. 2007. "The Law of Growth: How Ecology Accounted for World Population in the 20th Century." In *Distinktion. Scandinavian Journal of Social Theory* No. 14., special edition "Bioeconomics", edited by Thomas Lemke and Lars Thorup Larsen (coming May 2007).

Link, Jürgen. 1989. "On Collective Symbolism in Political Discourse and its Share in Underlying Totalitarian Trends", in: Reiner Schürmann (ed.), *The Public Realm. Essays on Discursive Types in Political Philosophy*, Albany N.Y. (State University of N.Y, Press), 225-238.

Link, Jürgen. 2004. Essays on normalism, in: *Cultural Critique*, Number 57 (on "Normalization"), Minneapolis, Spring .

Link, Jürgen. 2005. "On the Specificity of Modern Collective Symbolism: The Techno-Vehicle Body and Normalization", in: *Symbolism*, Vol. 4 (2005), 207-227.

Link, Jürgen. 2006a. *Versuch über den Normalismus. Wie Normalität produziert wird*, third, revised and enlarged edition Göttingen: Vandenhoeck & Ruprecht, first edition 1996.

Link, Jürgen. 2006b. "German Mass Media Facing the Crisis: Limits to Normalization?" in: William Urricchio/Susanne Kennebrock (ed.), *Media Cultures*, Publications of the Bavarian American Academy, Vol. 5, Heidelberg (Winter), 125-142.

Pearl, Raymond. 1925. *The Biology of Population Growth*, New York.

Pêcheux, Michel. 1984. "Zu rebellieren und zu denken wagen: Ideologien, Widerstände, Klassenkampf", in: *kulturrevolution. zeitschrift für angewandte diskurstheorie*, number 5 (1984), 61-65, number 6 (1984), 63-66.

The Mental and Practical Impact of Pre-Bacteriological Quality Criteria for Water in the 1870s

MARCUS STIPPAK

In the late 19th and early 20th century many city governments in, for example, Great Britain, France, the United States and in Germany decided to build up a centralized and mechanical water supply system. At the beginning there usually stood a critical consideration of the contemporary facilities to supply cities with water. In the end it was often said that wells and spring water pipes were inadequate to deliver water in sufficient quantities and in good, i.e. hygienic harmless, quality.

In this context I will analyze in the first instance the validity of pre-bacteriological quality criteria for water. For this reason I will discuss especially the Hygienic movement's efforts to find general valid, scientific criteria in the 1870s and early 1880s. As will be seen, these efforts eventually turned out to be fruitless.

Nonetheless, local politicians, citizens, engineers and physicians appropriated these criteria as a scientific point of reference. Relying themselves on these criteria, city dwellers demanded to replace wells and spring water pipes with a modern water works. Thus, I point at the meaning these quality criteria had and were given, respectively.

With reference to two case studies, the German cities Darmstadt and Dessau, I also intend to show the decisive role hygienic standards played despite their limits in introducing a modern water supply system. By analyzing the local controversies, I will finally work out two circumstances: the symbolic character that technology was given by society and the practical consequences the mentioned appropriation process initiated.

Defining hygienic harmless water: German attempts between the 1860s and early 1880s

The question whether water out of a certain well, river or spring is harmless or not is as old as humanity. What was new in the 19th century was the attempt to provide a scientific definition of harmless and risky water. In the early 1860s the water supply commission for Vienna made a first step towards this direction. In its 1864 published report (Veitmeyer 1871), the commission made the statement that water for drinking and household purposes had to be cool and fresh, also pure and colorless. Furthermore, water should be free from "mechanische Beimengungen" (mechanical admixtures), organic matter and from substances resulting from decayed or fermented organic matter, respectively. Good water, it was said, should not change its outward appearance, e.g. water kept in a glass over a certain time should not get muddy, precipitate and start to smell foul. These "soft" criteria were already known, but for the first time also "hard" criteria, i.e. threshold values for the amount of "Abdampfrückstand" (residue of solid matter after vaporizing a certain quantity of water), lime, nitric acid, chlorine, sulfuric acid and organic matter were supplemented. In 1871, Veitmeyer, a well-known engineer, called the report from Vienna a thorough and most comprehensive piece of work. Simultaneously, he criticized and scrutinized, respectively, the whole efforts to identify limits for substances that allegedly imposed dangers for the individual health. According to his own work (Veitmeyer 1871), it was neither sure which substances contained in water could cause epidemic diseases nor which amount of a certain, presumed hazardous substance was acceptable from a scientifically-hygienically point of view. Veitmeyer had not been the only critic at this time. However, the research provided by the Viennese commission remained the only relevant scientific guide for analyzing water and assessing its quality, at least in the German-speaking area till the end of the 1870s. According to historian Shahrooz Mohajeri, there had not been a better alternative (Mohajeri 2005).

Around 1880, the "Deutsche Verein für öffentliche Gesundheitspflege" (German Association for Public Health Care, founded 1873) tried to establish a catalogue of methods to analyze and assess water. In the German-speaking area, this association was probably the central proponent of the idea of a public hygiene based on and executed by political institutions (Münch 1993; Witzler 1995). For numerous engineers, architects, social reformers, physicians, economists, urban planners, civil servants and wealthy or at least influential members of the bourgeoisie, it formed a common public platform to discuss and take influence on the urban development. On the occasion of its meeting in Düsseldorf in 1876, Ernst Grahn, an engineer and prominent member, gave a well known speech on the demands that physicians and engineers should make on the urban water supply systems (Grahn 1877). According to Grahn, the state of knowledge concerning water, its tinges and their properties, was characterized by uncertainty. On the one hand, it was generally accepted that consumed water could affect the individual health in a negative manner. On the other hand, it had turned out to be extremely difficult to furnish scientific proof of this assumed interaction, because the methodically heterogeneous disciplines chemistry and microscopic observations provided ambiguous results. In the end the assembly passed a resolution that the association should constitute a commission and instruct it to develop standardized methods of analysis, to catalogue a universally valid list of substances contained in water and to set threshold values for these substances.

Seven years later, in 1883, the belief to come to a general applicable canon of methods and widely accepted limits was gone. In his talk to the association's general meeting in Berlin, Gustav Wolffhügel, senior executive officer in Berlin and member of the 1876 established "Kaiserliches Gesundheitsamt" (Imperial Public Health Authority), summarized the results of the efforts which had been made since 1876 (Wolffhügel 1883). The effect was quite disillusioning. From the etiological point of view, water remained an unsolved riddle over a longer period. Additionally, the chemical composition of water changed with the seasons, and water quality also varied within geographically small areas. Both parameters gave Wolffhügels' colleagues and listeners a hard time. Unlike the English "Society of public analysts" that had reached an agreement in 1881, the methodical pluralism in the German-speaking area prevailed. Lacking a homogenous analytical ground, the German analysts had been unable to bring about a useful compilation containing results of water analyses from different places. Against the background of this inability, Wolffhügel postponed the idea of generally accepted threshold values indefinitely and took a step backwards. Instead of waiting for a central institution or association to create an analytical and methodical framework, local authorities and institutions, respectively, were called upon to take the matter into their own hands. He asked local acting analysts to find comparative values within their region of research. Furthermore, he appealed for a flexible assessment of local water sources. He argued that, in case of doubt, it is more comfortable to have water of lower quality in sufficient quantity than water of higher quality in insufficient quantity. If local acting analysts announced their chosen method of analysis, this procedure, as Wolffhügel said, would finally help to replace the existing, unsuitable quality criteria for water. Until then, the analytical and methodical heterogeneity would and did indeed predominate.

Appropriation of pre-bacteriological quality criteria for water on the local level: Darmstadt and Dessau in the 1870s

As I explained above, there had not been any generally accepted quality criteria for water in the German-speaking area during the 1870s. Nonetheless, till the end of this decade already 195 local authorities of German communities with at least 2000 inhabitants decided to implement a central water supply system (Grahn 1904). Obviously, the threshold values from Vienna could not play an important role when, for example, Hamburg in 1848 or Berlin in 1856 introduced their new water supply systems. Similarly obvious, several local governments did not wait for Wolffhügel and his colleagues from the German Association for Public Health Care. Above all driven by the processes of industrialization, urbanization, population growth or the ambition to represent themselves as modern communities, local authorities took their matters into their own hands.

Unfortunately, at this point at least, many German historians only describe but do not explain or scrutinize, respectively, the role the disputed limits played in this time. Usually they singly reproduce analysts' statements about the quality of a certain surface, well or spring water. Here I am trying to refer to the – in fact controversial – scientific basis of these expert's opinions and outline briefly the processes how these had been used to influence political decisions that led to a desired technological development.[1]

Around 1870 in Darmstadt (the medium-sized main capital of the grand duchy Hesse and seat of its monarch and his ministries) entrepreneurs, especially railroad companies and manufacturers, the chamber of commerce, publicists, local politicians, military personnel, inhabitants, and the influential local newspaper "Darmstädter Tagblatt" judged the local water supply situation as critical. Already between 1820 and 1860, the increasing population of the expanding city had to learn what water

1 The following expositions in this paragraph go back to my PhD thesis which exemplifies the historical development of urban water supply and sewer systems in the 19th and 20th century based on the case studies Darmstadt and Dessau. Therefore I forego here from detailed bibliographical references. I am planning to finish my PhD thesis in 2007.

shortage meant several times. The sixteen public and an unknown number of private spring water pipes as well as the 61 public and an unknown quantity of private wells (status from October 1872) failed to satisfy the water demand. A thoroughly compiled memorandum concerning the water supply of Darmstadt, published in 1871, authored by a high-ranking artillery officer and land surveyor, initiated a process of discussions, controversies, trials and errors lasting for almost ten years until finally leading to a successful outcome. At first, as the officer put it, the problem was primarily a quantitative one. Figuratively spoken, Darmstadt's closer natural surrounding was no longer efficient enough to keep pace with the city's development. Relying on his experience and observations, the officer did not see any reasons why the water available from sources out of town could not be used for drinking or household purposes. The last assessment was rejected after a chemical analysis of the wells and spring water was carried out in 1872. According to these findings, neither the water from the wells within the city, nor the spring water from outside could be called hygienically harmless. The analyst did not name the source for his judgment explicitly. But after a look at the considered substances and their threshold values it becomes clear that he was referring to the water supply commission of Vienna. It is interesting that one year later, in 1873, a local physician announced the sanitary congress of Brussels from 1852 as the scientific point of reference. I therefore suppose that the sanitary congress of Brussels laid the foundations for the work of the Viennese commission. Also of interest: the very same limits from Brussels and Vienna, respectively, disputed by the experts served as a prominent key argument for a totally new arrangement of Darmstadt's water supply system. The dominating local newspaper "Darmstädter Tagblatt" incorrectly called the threshold values universally valid principles which the Hygienic movement had set up. Astonishingly, nobody objected to this interpretation. In a first step, the use of the actually controversial quality criteria led to a break with the traditional water pipe and wells system. In a second step, these criteria were used by the municipal authorities to give the impression they were not only acting on a scientific, but also on a solid basis. And again almost nobody objected. The use of the criteria legitimized and reinforced the authorities' turn to the at that time unpopular, because on ground water based water supply system. The unpopularity derived from the circumstance that expensive machines were necessary. Furthermore, the ground water had to be pumped upwards, i.e., in an "unnatural" manner. Only the successful development after the new waterworks' opening in 1880 let the people forget how thin the scientific-methodical ice was the municipal authorities were moving on.

The municipal authorities of Dessau made this painful experience. Dessau was likewise a main capital but here of the duchy Dessau-Anhalt and seat of the duke and his government. Similar to Darmstadt, but smaller in size and extent, Dessau ran through a demographic, economic, spatial and town planning change. In January 1874 the here dominating local newspaper "Anhaltischer Staatsanzeiger" described the new water supply system as an almost realized project. Originally, the government placed – among other things – an obligation on the municipal authorities in 1872 to build a new sewage system within three years. Very probably, it intended to spruce up the main capital of the young, only since 1863 existing duchy with the implementation of a modern sewage system. Furthermore, a new water supply system did not appear as an issue because of the city's geographical location. In the north, Dessau is situated to the Elbe, in the east to the Elbe's tributary, the Mulde river. The inhabitants got their water for their daily purposes either from the Mulde river or from the around 700 private and public wells within the city (status from 1872/73). Although there was neither any serious water shortage nor any form of public pressure worth mentioning, the municipal authorities changed their mind. They asked the government in spring 1874 to replace the sewage by a mechanical and centralized water supply system. Which reasons did they have? The main argument constituted the results of two chemical analyses carried out in 1872 and in 1873. As the local newspaper reported in January 1874, the water from the wells was not only varying in quality, but also partly unhealthy. According to an expert's opinion that was ordered mainly by the municipal authorities and provided essentially by a local pharmacist, the scientific inquiry included not only the threshold values set up by the sanitary congress of Brussels. It also included criteria the Viennese commission had formulated considering the guidelines from Brussels. Wrongly, the pharmacist described his working basis in line with scientific standards generally accepted by all scientists who were working in this field of research. Moreover, the analyst conceded that the physical condition of Dessau's inhabitants could be judged as a good one. But despite the fact that Dessau has the right to be called a healthy city, the pharmacist urged the local government to initiate prophylactic measures in order to prevent a future marked by stench, feces, illnesses and death. Similar to the events in Darmstadt, there had been two forms of reaction at first. On the one hand the chemical analyses or their interpretation were called a good piece of research. On the other hand, well's water consumers heavily criticized the municipal authorities' intention to pump ground water into the buildings. Dissimilar to Darmstadt, these critics also aimed at the conclusions about the wells' water quality within the city which were drawn on the ground of the chemical analyses. Regardless what the chemical analyses would tell, they argued that the well's water was and will always be better than tap water, which would stay in the pipes indefinitely. Notwithstanding, the analyses produced the same consequence as in Darmstadt. Well and river water was brought into disrepute, and in May 1876, after the government gave the go-ahead, the city's first centralized and mechanical waterworks was put into operation. But already six months later all delight with the tap water vanished. Now, the analyst declared, although the tap water was totally impeccable, compared with the criteria the Viennese commission had set up in 1864, he could not recommend it as drinking water. Containing iron, the tap water – now disrespectfully called "Eau de Dessau" – too often took on a sometimes brown, sometimes yellow color, tasted and smelled marshy, and created a brownish mud right in the moment it came into contact with oxygen. The case study of Dessau makes it clear that the implementation of a mechanical and centralized water supply system solely founded on the chemical threshold values which the sanitary congress of Brussels respectively the Viennese water supply commission had formulated were not a guarantee for success.

The mental impact of pre-bacteriological quality criteria for water and its practical consequences

Why did the municipal authorities of Darmstadt and Dessau appropriate the threshold values despite their scientific limits? The described events took place in a time characterized by an enormous social esteem for the new natural and exact sciences, respectively. Although the young discipline of hygiene was not counted as an exact science, many politicians, architects, civil servants, physicians and so on, saw her as a practicable instrument to change the urban status quo (Rodenstein 1988). The threshold values in question offered a range of options to alter this status quo. Appropriating the limits also gave the municipal authorities the opportunity to represent themselves as modern communities. In cities like Darmstadt, where the water supply situation obviously left much to be desired, the values fell on fertile ground. They were widely accepted quite quickly and without any stronger objection. If a powerful opposition resists the authorities' political position it seems that it lacked a comparable scientific basis. The opponents in Dessau, for example, had nothing more to refer to than to their (individual) experience with well water and their common aversion to tap water. These emotional arguments were not at all insignificant but in the short and medium term they proved to be too weak. In the long run the argument of prevention, the consumer and comfort experiences made in other cities and the idea of a technical grounded urban modernization pushed by political authorities' with reference to scientifically legitimized values were stronger.

Soon after the quality criteria had been introduced into the discussion, a self-constructed and self-imposed practical necessity followed: the municipal authorities convinced themselves, the government, and at least parts of the local public that something had to be done. Pointing to the events taking place elsewhere, inhabitants too made it their local authorities' duty to break new ground. This mixture of motifs, ideas and interactions between various local actors led in the long run to the implementation of a "large technological system" (Hughes 1989). But already during the debates a new attitude towards technology became audible, reflecting desires and expectations. Proponents of a centralized water supply system emphasized the increasing conveniences within buildings and apartments as well as the improved firefighting conditions. Furthermore, as Darmstadt's mayor in 1877 emphasized for example, the new system itself was seen as the city's central technological net which would allow a controlled urban development (Ohly 1877). Thus it would help to prevent anarchy and chaos not only upon the drawing board, but also in the peoples' daily life and with it the municipal authorities' daily routine. With the introduction and popularization of tap water, the inhabitants' relationship towards nature also changed basically (e.g. van Laak 2001). Physical strains were no longer necessary to bring water from a well up the stairs. At last the new technical system symbolized a chance to get and keep a deficient and capricious nature – at least so it seemed – under control. Unlike wells or spring water pipes, the water works were interpreted as a guarantee for a never ending influx of water in sufficient quantities and quality. From now on, additionally reinforced by bacteriology since the 1880s and 1890s and later by microbiology, water was pumped up, optimized in quality and distributed by machine. Only the water consumption remained an experience a tenant or house owner was allowed to make. Today we know that this taming and exploitation of nature, initiated by various actors and perpetuated by (local) politicians and engineers, does not weaken the cities' and society's, respectively, dependence from nature at all. But starting from the 1870s, despite failures like that in Dessau 1876, the idea of hygiene transformed into technology characterized the way society was dealing with nature till at least the mid of the 20th century.

List of References

Grahn, Ernst. 1877. "Die berechtigten Ansprüche an städtische Wasserversorgungen vom hygienischen und technischen Standpunkte aus." *Deutsche Vierteljahrsschrift für öffentliche Gesundheitspflege*, vol. 9: pp. 80-120.

Grahn, Ernst. 1904. "Die städtischen Wasserwerke." Robert Wuttke, ed. *Die deutschen Städte. Geschildert nach den Ergebnissen der ersten deutschen Städteausstellung zu Dresden 1903*. Leipzig 1904: Friedrich Brandstetter, vol. 1, pp. 301-344.

Hughes, Thomas P. 1989. "The Evolution of Large Technological Systems." In Wiebe E. Bijker, Thomas P. Hughes, and Trevor P. Pinch, eds. *The Social Construction of Technological Systems. New Directions in the Sociology and History of Technology*. Cambridge. Mass./London, MIT Press, pp. 51-82.

Laak, van Dirk. 2001: "Infra-Strukturgeschichte." *Geschichte und Gesellschaft*, vol. 27: pp. 367-393.

Mohajeri, Shahrooz. 2005. *100 Jahre Berliner Wasserversorgung und Abwasserentsorgung 1840-1940*. Alzey: Franz Steiner Verlag.

Münch, Peter. 1993. *Stadthygiene im 19. und 20. Jahrhundert. Die Wasserversorgung, Abwasser- und Abfallbeseitigung unter besonderer Berücksichtigung Münchens*. Göttingen: Vandenhoeck & Ruprecht.

Ohly, Albrecht. 1877. *Vortrag des Großherzoglichen Bürgermeisters Ohly an die Stadtverordneten-Versammlung betreffend die Wasserversorgung von Darmstadt (20.5.1877)*. Darmstadt.

Rodenstein, Marianne. 1988. "Mehr Licht, mehr Luft" – Gesundheitskonzepte im Städtebau seit 1750. Frankfurt/M. / New York: Campus.

Veitmeyer, Ludwig Alexander. 1871. *Vorarbeiten zu einer zukünftigen Wasser-Versorgung der Stadt Berlin. Im Auftrages des Magistrats und der Stadtverwaltung zu Berlin ausgeführt in den Jahren 1868 und 1869*. Berlin: Reimer Verlag.

Witzler, Beate. 1995. *Großstadt und Hygiene. Kommunale Gesundheitspolitik in der Epoche der Urbanisierung*. Stuttgart: Steiner Verlag.

Wolffhügel, Gustav. 1883: "Ueber die hygienische Beurtheilung der Beschaffenheit des Trink- und Nutzwassers." *Deutsche Vierteljahrsschrift für öffentliche Gesundheitspflege*, vol. 15: pp. 552-574.

War and the Beautiful: On the Aestheticising of the First World War in Film – Yesterday and Today

SILKE FENGLER/STEFAN KREBS

This paper is an enquiry into the medial construction and aestheticizing of the war in present day TV documentaries on the First World War. The analysis refers mainly to the exemplary 90-minute documentary "Der Moderne Krieg", which the German-French station, ARTE, broadcast in summer 2004.[1]

Two temporal phases converge here: at the first phase it can be seen how film, as a new medium during the First World War, lent a hitherto unknown aesthetic dimension to the industrialized war events, which oriented itself at the same time toward traditional image forms and motifs. In the second, present-day phase, this film material is respliced and loaded with additional meaning. Both temporal phases are inseparably intertwined – both construe the modern myth of the clean war: each in its own manner, each according to its own era.

In general, the official pictorial propaganda put forth an effort to blot out the horror of the industrialized waging of war. Instead, it perpetuated the scenario of a pre-modern, romantic war (e.g. Hüppauf 1997: 887f.). One outstanding motif of German film reporting, as well as of war photography, was the front, i.e. pictures from life at the front line. On the other hand, a contemporary photographic iconography and aesthetic of destruction were also developed. Thus, the improved photographic technology and faster shutter speed made possible a new type of presentation of the dynamics of war, e.g. in the shots of soldiers using their equip-

1 On the occasion of the ninetieth anniversary of the outbreak of World War One ARTE broadcast the contribution "1914-1918. Der Moderne Krieg" by Heinrich Billstein and Matthias Haentjes (Germany 2004, first broadcast July 30, 2004, at 10:15 p.m. on ARTE).

ment. Film technology was used for the first time for the construction of the reality of war as visual battleground. Thus, soldiers themselves participated in combat exhibitions that were staged especially for the camera (e.g. Paul 2004: 106).

Overall, wartime film material remained indebted to popular aesthetic conventions and delivered with its manifold staged scenes, which were filmed without sound, hardly a realistic view of battle events. The fictional portrayal of war in cinema films, starting in 1916, which wrote played-out scenarios into real war experiences, further blurred the difference between fiction and reality, turning war into entertainment.

In 1992, William Mitchell attempted to rehabilitate thinking in and about images with his concept of the pictorial turn. Gottfried Boehm (2005: 40f.), who first coined the term *iconic turn,* recovered the picture as an autonomic authority to the core of hermeneutics and philosophy with his conceptual recommendation. The demand went so far as to analyze, as much as possible, all contemporary visual fields – pictures of the mass media, from natural sciences, plastic arts, etc. – using a logic of images, to be developed gradually.

But how can an assumed logic of images be decoded? The focal point here is in the analysis of film material with respect to its specific formal language, and the contents and messages connected to it. We shall concentrate on the moment of the aestheticizing of the war, asking how it is mediated over the composition of pictorial contents and edition, but also over the later combination with audiovisual elements.

Firstly, contents and themes of the film images are to be determined and grouped according to selected aesthetic categories. These are foremost the categories *beauty* and *sublimity*. In the second phase of the enquiry, they are to be complemented with the categories of the *comical* and *tragic*.

Wartime reporting took up traditional iconographical stylistic methods and pictorial aesthetic conventions. The category beauty played a central role. Beauty in the aesthetic sense showed itself in perfection, purity, the right measure, clarity, order and symmetry of the individual pictorial motifs.

In propaganda film, but also in still life photographs of World War One beauty is represented in the classical subcategories of the *idyllic*, of the *picturesque*, but also *cleanliness* and *order*. The idyll of war coheres with its supposed picnic character.[2] A frequently recurring scene shows soldiers preparing or distributing food. In these images, the impressive

2 In the Crimean War, the photographer Roger Fenton took numerous photographs for the British Royal House which idealised war events as a picnic (e.g. Becker 1998: 73).

abundance and plenitude of the food available is meant to show the wartime viewers the excellent efficiency of the supply lines at the front. The happy faces of the soldiers reflect the enjoyment of a simple meal in undisturbed nature. Just as the picture of a soldier calmly smoking a pipe, these images emphasize the impression of the picturesque. They are reminiscent of the genre painting of the 19th century.

The picturesque as an aesthetic category rests upon the variety- and contrast rich properties of a mostly untouched, yet bizarre natural landscape. This conventional model was in part reproduced with the help of new film technology, in part, though, an aesthetic of destruction all of its own was developed. In the aerial shots as in the panorama of destroyed forest landscapes the impression of harmony and calm is preserved. Thus, nature loses nothing of its beauty, in spite of the most brutal destruction.

Contrary to the factual chaos of war, the film images suggest a constant, perfectly preserved order that permeates both the everyday life of the soldiers and the waging of battle, i.e. the actual combat. Shells stacked horizontally or in many graded rows for storage show neat, straight lines in the pictorial composition.

The traditional commander's eye view of the battlefield, i.e. the view from an elevated point of observation of a field that disappears into the horizon, a field on which soldiers are at once no more than tiny figures, was also reproduced in wartime films. In an exemplary scene, one sees a chain of artillery projectiles stretching vertically into the picture and, on the horizon, row upon row of riders riding toward the front line.

In contrast to the aesthetic category of the beautiful, the sublime is not so much associated with feelings of joy or pleasure, as with those of terror, of fear or of disturbance. This emotional reaction is called forth by the *grandiose*, the *splendid*, that which *transcends* that which can be sensually experienced, or which is *majestic*. In order for the observer to experience the sublime, some distance from the image is necessary. Especially the sublime in its negative variation – the *threatening*, the *fearsome*, *death*, and *power* – incites wonder and fear at once in the observer; the aesthetic enjoyment necessarily requires that the observer be not immediately threatened by the experience. For only the distance, as a rule physical distance, enables him to gain some overview of the entire matter as well as the quality of individual aspects, while at the same time experiencing the sensual-emotional effects of the observed image (e.g. Sontag 2002).

As already in the case of the presentation of the beautiful, the images of battle equipment and the battlefield contribute little to the portrayal of the reality of the industrialised war. Artillery guns were as a rule portrayed as imposing military machinery. Whether the cannons seemed threatening and intimidating or offensive to the observer depended upon the camera perspective chosen. Sometimes the observer gazed quasi from the front into the barrel – wherewith the threatening aspect of the weapon was immediately given. Usually, though, wartime propaganda filmed cannons either from the side or diagonally from behind, so that the camera gazed along the weapons barrel, or seemed to follow the line of fire.

The expressive images of exploding shells follow a recurring aesthetic of the immensely threatening: earth splattering up, and in its midst, a rising column of dark smoke. In the portrayal of artillery, the new medium film could still outdo photography: the quick loading of cannons by the artillery teams, the turbulent dynamics of the discharge and the impact of the shells develop their destruction aesthetic only in moving pictures. It is conspicuous in the differing film images that the artillery shells almost always impact on a field completely deserted by people; the effect of the shells on a human body thus remains hidden. This is firstly a hint as to the artificial character of the pictures. Secondly, the observer is thus not confronted with the actual reality of injury, death and suffering, but can yield himself over completely to the clean aesthetic enjoyment of the scene.

The proverbial firepower was also illustrated by pictures of the employment of the flamethrower. This new weapon seems especially threatening when the stream of flames moves toward the camera, and thus toward the observer. The viewer is thus given a perspective that is usually exclusively reserved for the mortally threatened victim of the attack (e.g. Hickethier 2001: 65). But instead of being exposed to the real terror of the column of fire spray accelerating out of the picture, the distanced observer experiences a queer thrill at the fascination of the power of arms.

Death on the battlefield was unwelcome in propaganda, and is therefore rarely shown in wartime films (e.g. Paul 2004: 126, 129). The images of dead soldiers all have in common that the camera remains mostly distant, and that the scenes appear rather peaceful. This effect was achieved by the use of extremely slow camera movement, or the resting of the camera in one position. Nor do any mangled corpses appear in these scenes, but rather dead soldiers who, corporally unscathed, appear to be sleeping.

Much less ambiguous are the photographs of horribly disfigured war victims, such as those published by Ernst Friedrich (2004) in his book "Krieg dem Kriege" in 1924. The pictures have their origin in the documentation of injuries for didactic and research purposes in medicine.

The feeling of displeasure that befalls the viewer of these blunt pictures is transformed into a softened horror.[3] The spectators' emotions sway between pity for the victims and the apathy of the voyeur: the horrible becomes a spectacle (e.g. Sontag 2002).

The TV authors rely in their portrayal of the First World War largely on historical film material. The portion of original footage, i.e. films and still photos recorded later in film, amounts to nearly three quarters of the entire broadcast duration. The compilation film[4] derives its claim to have portrayed the past as authentically as possible from the extensive use of historical footage. This effect is also intended by the reduction of the stylistic devices employed in the new type of contemporary historical documentary successfully introduced in the 1990's by the German TV historian Guido Knopp. These rely mainly on witness testimonies and historical film footage to tell history and to enable the TV viewers to gain a direct emotional access to historical reality. Other classical stylistic devices of the compilation film, such as expert interviews and the audiovisual citation of historical documents, sink into the background (e.g. Lappe 2003: 97).

If Siegfried Kracauer's paradigm (1993) is followed, the documentary film draws its authenticity from the depiction of supposedly non-staged, real actions. In its efforts to show past events the way they really were, modern historical documentaries aspire to objectivity. The supposedly unchanged reproduction of the original material is intended to lend credibility to the historical feature.

At first sight, the film footage of World War One seems then very well to open a window[5] to the war, to document its hidden reality. As the analysis of motifs and scenes above shows, however, instead of an authentic depiction of the reality of war, rather a veiling of it with an aestheticizing and glorification of war events occurs. This can hardly be surprising considering the origin of the film footage in war propaganda departments of the participant armies. Before this background, it appears secondary whether the reality of war can even be portrayed, as the propaganda film's depiction of war is bent on achieving exactly the opposite (e.g. Paul 2003: 60). This brings us in the following section, in which we turn our attention to the contemporary phase of documentation, to the question of how TV authors treat film resources.

3 The concept of the softened horror is from Edmund Burke.

4 In the production of compilation films, historical cuts, which have been extracted from their original context, are recombined on the editing table and thus placed into a new context (e.g. Hattendorf 1999: 203).

5 This is in the sense of Leon Battista Alberti's window metaphor.

Whereas the first part of the paper deals with the analysis of individual pictorial motifs, the television viewer is confronted with the film resource not as a single picture, but as a sequence. The process of pictorial editing creates narrative and aesthetic structures which themselves produce the narrative continuity of the documentary.

Editing is already a first interpretation of the raw material, and structures the viewer's reception in advance. The authors use only a very small variety of techniques: mostly the individual adjustments are simply connected with clean cuts. Doing without elaborate editing techniques, such as fading in and out or time lapse, reinforces the impression of fundamental authenticity. Only the cut frequency is adapted to the viewing habits and aesthetic expectations of today's viewers. A dynamic is lent to the wartime film material which is alien to its original form, especially by making sequences in which attacks are portrayed, to conform both to the fragmentary visual aesthetic of the younger audience and to contemporary ideas of the modern war.[6]

Whereas especially those scenes of military actions are made more dynamic through this splicing technique, TV authors resort to longer cuts and slow pans for their narrations of destruction on the battlefield and of the death or wounding of soldiers. Editing serves the fundamental rhythm of a documentary. The frequency thus conforms to the dramaturgy of the narrative.

Only through editing can the film reality be generated for the viewer of a documentary. So the TV authors can rely on firm conventions of combinations and standardized meanings. The viewer bridges two consecutively shown positions independently with his own imagination. Thus, the firing of a shell and the hit belong together in the context of a film, even if the two shots are completely independent of one another. The association is then the real narrative process, or the core of film narration. The viewer is given the impression of a closed, uninterrupted action, a continuous motion. This all fortifies the impression of authenticity, and gives the viewer an apparently immediate insight into the reality of war (e.g. Hickethier 2001: 145).

A prerequisite of this associative play on conventionalized images is that the television viewer is acquainted with a certain set of picture icons of World War One. Only someone who is already familiar with the pic-

6 In an especially succinct sequence the firing of a railroad cannon and a vertical shot of a gigantic explosion from above are spliced together. This scene borrows from the pictorial aesthetic of reports on the first Iraq war. There shots were mounted so that the discharge of sea-based cruise missiles and the target videos of American fighter planes were taken out of their original contexts and combined (e.g. Frohne et al. 2005).

torial connection between soldiers in gasmasks and blinded gas victims can follow the visual narrative. In the case of iconographic scenes, their recombination – as briefly touched upon – is only possible through the iconological bridging on the part of the viewer, independent of their original context. So the TV authors rely on the observer producing causal connections between images which are not present in the sources. This yields a certain haphazardness in the use of historical pictorial material. It can most easily be seen in the reuse of certain images at various places in the film narrative.

Instead of longer shots, the TV authors generally use a multiplicity of images from a single group for the visual production of their narration. Aside from catering to the aesthetic expectations of the television audience, a well-known rhetorical figure is also hidden within: accumulation (e.g. Borstnar 2002: 45). Through the juxtaposition of thematically similar consecutive shots the pictorial argumentation is fortified at the same time with the claim to a faithful portrayal of reality.

A soundless event seems incomplete, unreal. To the television viewer it seems an aesthetic inadequacy. Because the film footage was originally silent, due to the technological limits of its time, television authors employ standardized sound icons for the auditory illustration of their narrations, such as hoof beats, cannon sounds, marching boots and explosions. Especially the so-called synchronous noises which the viewer can localize in the picture serve to fortify the impression of reality (e.g. Hickethier 2001: 96).

But non-synchronized sound also often finds its place in the examined documentaries, being easy for the viewer to associate with the context of war events – as in the case of the shots and explosions often heard. The non-synchronous sounds often are of symbolic character, and support the narratives of the eyewitnesses.

Due to their ability to endow continuity, non-synchronous noises add an additional cohesion for otherwise disparate images. Thus one can speak of atmospheric sound, i.e. the military sound icons emphasize the realistic impression left on the viewer by the historic images of World War One.

The video aesthetic restraint exercised by the TV authors leaves the original aesthetic much leeway: the film images speak for themselves, and the glorification and romanticizing of war events which they convey remains mostly unbroken. In this way the aesthetic of war engraved in the wartime images is conserved – as is seen in the unbroken fascination with technology exuded by the extravagant and dramatic presentation of the war machinery.[7] Thus, the supposed reality of war – war as a game, contest or outing – also coagulates to historical reality (e.g. Mikos 2003: 147).

Wartime film material is presented in the documentaries we examined as an authentic source, and localized neither temporally, nor spatially, nor by situation in the voice-over commentaries. Although these commentaries often provide information on the economic or social dimension of the First World War, something which extends beyond the images themselves, the fact that the historical images shown are mainly propaganda is not mentioned.[8]

The depictions of the everyday war by veterans of World War One, just as the original pictures, are intended to convey credibility. In addition, historical witnesses offer the observer the opportunity for identification and contribute to the overcoming of the temporal distance from the events. The statements of eyewitnesses are embedded in a comparatively reductionistically composed narrative whole. Each sequence is opened with a panorama of a present day landscape or building that gives us a point of reference as the setting of a war event. The landscapes in question are filmed in color, just as are the eyewitness interviews. They zoom the observer into the event. The interviews alternate in each sequence with a series of consecutive original pictures. Sometimes the voice of the witness takes over the commentary of the subsequent images. By this means the historical footage often gains an additional credibility.

As communications science has shown, the visual impression in television documentaries always takes precedence over the commentary (e.g. Prase 1997: 62). Where these stand in contradiction to one another, the viewer will be much better able to remember the visual information in the long term. On the other hand, a commentary which is close to the image enables the viewer to remember the pictorial information better (e.g. Bock 1990: 78).

At various points in the documentaries there is a flagrant contradiction between the auditory and visual narration. This shall be briefly explained here with the use of one example: The statements of witnesses

7 It is conspicuous how often the TV authors place shots of railway cannons – the most immense artillery weapons of World War One – into scenes. The images are, however, empty of content, being localisable neither temporally nor spatially for the viewer.

8 Corresponding information is found neither in the opening credits or introductory commentary nor in the course of the documentary. Furthermore, an indirect critical assessment of sources is dispensed with entirely, e.g. such that wartime film material were presented as film in the film, or its origin identified in subtitles.

and commentaries on the insufficient state of provisions are combined with pictures of soldiers at the distribution of food and heaps of rations, waiting to be passed out. The narrations are in a way denied, falsified by the pictures. If we should choose to speak in aesthetic categories, then the tragic aspect of hunger and want is thrown into the comical by the incongruity of sound and image, and contributes thus to the playing down of the war. These sequences make it clear how the voluntary compulsion of the TV authors, who wish to strew their narration with historical film footage through and through, hinders the portrayal of the horror of war.

In summary, the TV authors allow themselves to fall prey to wartime propaganda through the extensive use of historical film footage. The propaganda film's presentation of the war's trade corresponds entirely to the viewers' expectations of the reality of war. So the individual motif groups, the front, sickbay and artillery, their aesthetic production and the myth of the clean war which lies therein find multiple counterparts in contemporary war reports. Even the aesthetic of destruction first developed during the First World War – its fascination with arms technology and its destructive power – differs little from today's war images.

On the other hand, some of the motifs shown in the documentaries are so strongly embedded in their time that they have become somewhat strange to the aesthetic perception of today's viewer. The scene of a soldier, like little Red Riding Hood, passing a picnic basket neatly packed with a clean white cloth into a trench to his colleague no longer seems idyllic and serene, but rather almost comical.

Regardless of the genre, there is in war reports – often due to a lack of visual alternatives – a hardly broken continuity of resorting to official propaganda images or images approved by military censors (e.g. Paul 2004: 365-403). By relying ever more, under the influence of Guido Knopp's productions, on the employment of historical footage in their narratives, the authors of contemporary historical documentaries make it all the more difficult for themselves to argue against the discourses engraved in these images.[9] If this procedure is compared to the historical documentaries of the 1970's and 1980's, an unambiguous break in content and in form can be discovered. This is all the more so in that the authors voluntarily renounce alternative stylistic devices that might make it easier to question the visual dominance of the apparently authentic film material and to tell a different history of World War One. With regard to the employment of pictures in historical science, the question remains

9 How difficult it is thus to paint a fitting picture of World War One is something we have tried to show elsewhere (e.g. Fengler/Krebs 2005).

whether a history of the war really needs illustration via movie footage and photography at all. Or do these not rather hinder the critical treatment? Susan Sontag (2003: 104f.) comes in this spirit to the following conclusions: "The first idea is that public attention is steered by the attentions of the media – which means, most decisively, images. When there are photographs, a war becomes 'real'. [...] The second idea [...] is that in a world saturated, no, hyper-saturated with images, those that should matter have a diminishing effect: we become callous."

List of References

Becker, Frank. 1998. "Die Anfänge der deutschen Kriegsfotografie in der Ära der Reichseinigungskriege (1864-1871)." *20th Century Imaginarium* 2: 69-102.

Bock, Michael. 1990. "Medienwirkung aus psychologischer Sicht. Aufmerksamkeit und Interesse, Verstehen und Behalten, Emotionen und Einstellungen." In Dietrich Meutsch and Bärbel Freund, eds. *Fernsehjournalismus und die Wissenschaften*. Opladen: Westdeutscher Verlag, pp. 58-88.

Boehm, Gottfried. 2005. "Jenseits der Sprache? Anmerkungen zur Logik der Bilder." In Christa Maar and Hubert Burda, eds. *Iconic Turn – Die neue Macht der Bilder*. 3rd ed. Köln: DuMont-Literatur-und-Kunst-Verlag, pp. 28-43.

Borstnar, Nils et al. 2002. *Einführung in die Film- und Fernsehwissenschaft*. Konstanz: UVK Verlagsgesellschaft.

Fengler, Silke and Krebs, Stefan. 2005. "In Celluloidgewittern – Die mediale Konstruktion von Wissenschaft und Technik als Paradigma des Ersten Weltkrieges." *Technikgeschichte* 72 (3): 227-241.

Friedrich, Ernst. 2004. *Krieg dem Kriege*. Reprint. München: Deutsche Verlags-Anstalt.

Frohne, Ursula, Ludes, Peter and Wilhelm, Adalbert. 2005. "Militärische Routinen und kriegerische Inszenierungen" In Thomas Knieper and Marion G. Müller, eds. *War Visions. Bildkommunikation und Krieg*. Köln: Halem, pp. 120-152.

Hattendorf, Manfred. 1999. *Dokumentarfilm und Authentizität. Ästhetik und Pragmatik einer Gattung*. 2nd ed. Konstanz: UVK-Medien Ölschläger.

Hickethier, Knut. 2001. *Film- und Fernsehanalyse*. 3rd ed. Stuttgart, Weimar: Metzler.

Hüppauf, Bernd. 1997. "Kriegsfotografie." In Wolfgang Michalka, ed. *Der Erste Weltkrieg. Wirkung, Wahrnehmung, Analyse*. München: Piper, pp. 875-905.

Kracauer, Siegfried. 1993. *Theorie des Films. Die Errettung der äußeren Wirklichkeit*. 2nd ed. Frankfurt/M.: Suhrkamp.

Lappe, Christian. 2003. "Geschichte im Fernsehen." *Geschichte in Wissenschaft und Unterricht* 54: 96-103.

Mikos, Lothar. 2003. "Zur Rolle ästhetischer Strukturen in der Filmanalyse." In Yvonne Ehrenspeck and Burkhard Schäffer, eds. *Film- und Fotoanalyse in der Erziehungswissenschaft. Ein Handbuch*. Opladen: Leske und Budrich, pp. 135-150.

Paul, Gerhard. 2003. "Krieg und Film im 20. Jahrhundert." In Bernhard Chiari, Matthias Rogg and Wolfgang Schmidt, eds. *Krieg und Militär im Film des 20. Jahrhunderts*. München: Oldenbourg, pp. 3-76.

Paul, Gerhard. 2004. *Bilder des Krieges – Krieg der Bilder. Die Visualisierung des Krieges in der Moderne*. Paderborn et al.: Schöningh.

Prase, Tilo. 1997. *Das gebrauchte Bild. Bausteine einer Semiotik des Fernsehbildes*. Berlin: Vistas.

Sontag, Susan. 2003. *Regarding the Pain of Others*. New York: Farrar, Straus and Giroux.

Sontag, Susan. 2002. *On Photography*. Reprint. Harmondsworth: Penguin Books.

Soldiers on the Screen. Visual Orders and Modern Battlefields

STEFAN KAUFMANN

I.

In the modern era, the shaping of soldiers has always been, in a way, an aesthetic project, besides other considerations. To form soldiers meant to work on the appearance, on the visual design of the soldier. The outward appearance, where weaponry and posture, military technology and discipline join and merge, has always been seen as an essential medium for soldierly self-determination. The modern military has come up with two prominent projects to define this appearance of the soldier: the neat, well-drilled soldier of the 17th and 18th century, marking the age of mechanical machines, and the steel figure who emerged from the material battle of industrialised war. When one considers the military from this standpoint of a visual order, the two projects are seen to represent two opposite poles: one order which focuses on visibility, and another order in which the essential prerequisite is to disappear from sight beneath the visible surface.

In the 18th century, all battle functions were executed in geometrical structures and precisely determined, mechanical patterns of movement. The military corpus was thought of as a machine, composed of human bodies and guns, moving and firing to a mechanical rhythm. And in keeping with this idea, the formation of the soldier started with physical training as its basis. The soldier was to be formed into a well-fitted mechanical component of the larger machine. In this context the visual representation of the soldier, as typically found in the military handbooks of the 18th century, takes on the appearance of a an engineer's plan: The image becomes a kind of detailed construction manual. (Fig. 1)

Fig. 1: Drilled Soldier (in: Lottes 1998: 227)

The pictorial representation is here an essential element of the disciplinary constitution of the soldier. The dominating semantics in which the pictures are set also indicates that outer posture and inner discipline are mechanically coupled together. Consequently, discipline is entirely on the visible plane; it completely merges into what one sees as the appearance of the soldier. The soldier-state is manifest in a specific bodily presentation or habitus (cf. Bröckling 1997: 39-87; Lottes 1998).

Contrastingly, the steel figure is the project of an order of disappearance, an almost anaesthetic mode of battle which emerged in the wake of the trench warfare of World War I. The steel figure steps out against the backdrop of an empty battlefield. He comes as a response to the widening gulf between industrialised war machinery and the individual soldier, and to the disappearance of the visible order, which had led to the usual soldierly posture, the upright movements, becoming totally dysfunctional in the military operation. The energetic machine which the military semantics saw at work here was now being viewed rather as a more complex tactical combination of divided labour. And now soldierly obedience could no longer be thought of as a mechanical principle, as a reflex, an incorporated reaction to orders. This high-technology warfare was now demanding work on the inner attitude, on the mental qualities of the soldier. It is precisely this double requirement, to have the technological competence as well as a deep-seated mental awareness of soldierly identity, that the image of the steel figure was designed for. The figure is an amalgamation of man and machine. For, the warrior is created as both a technoid body and an anthropomorphic machine. He combines the qualities of the cool-headed technician of war and the archaic Ur-warrior (Fig. 2).

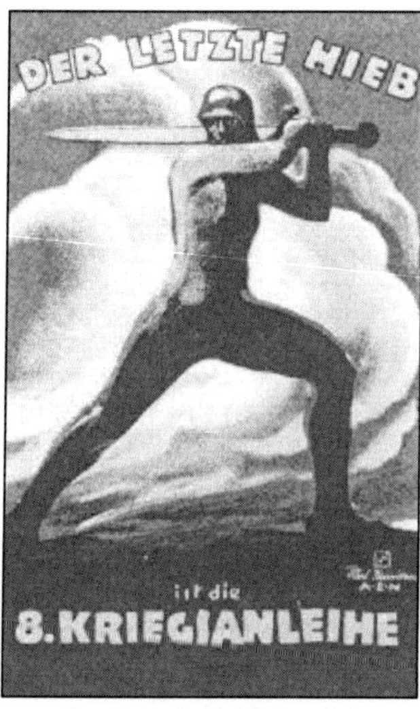

Fig. 2: The Steel-Figure

Technical skills are blended with the full spectrum of the warrior ethos: the eagerness to kill, the archaic esprit de corps of warriors and the instinct for sensing the key elements of the battle situation, which at the height of industrialised warfare translates to the terms of sense physiology, as the readiness to detect gas, to identify the sounds of shell trajectories, to size up the situation and the movements of the enemy (Hüppauf 1991; Horn 1998). In this steel figure we find on the one hand a greater focus on developing the cognitive skills of the senses and mental capacities, to be able to handle technical devices and machines even in the thick of battle, and on the other hand building up the inner attitude of the soldier, conditioned to duty and self-sacrifice.

Clearly, however, this image of the steel figure is not being worked out like a fixed set of construction plans, but is primarily on a symbolic level. It is hardly a thing that can be anchored in disciplinary regulations, rather it gains efficacy by its power for political mobilisation and as a cultural code for the soldier's self-esteem. But by the end of the 2nd world war, at least, this cultural sheen of the steel figure as a model for the dismounted soldier to identify with, began to fade. There seemed to be no place left for the simple foot-soldier on the battlefields of the end of the 20th century, with their threat of possible contamination by tactical atomic weapons. One could no longer present an aesthetic representation of the dismounted soldier of the cold war: the idea of "the lonely soldier" came into circulation (Bigler 1963).

II.

Work is underway on how to give the dismounted soldier a new design, the main impulse coming from the US army. Just as the steel figure was not only a figure of military disciplinary discourse, but one of literary fiction, so now we find that the new soldier can be understood as a project of a "military entertainment complex" (Lenoir 2000): the soldier design is inspired by the image aesthetics of Hollywood productions, and its technical realisation is based on technologies and skills which can be traced back to computer games.

As *Land Warrior,* the dismounted soldier is to be fitted out with wearable computer, permanent radio contact, global positioning system, heads-up-display and other optional items of kit, such as gun sights with night-viewing, infrared or video capabilities. The Land Warrior, whose development was planned in the mid 1990's and is soon to be ready for mass production, is evidently influenced by the image aesthetics of Hollywood. The figure of the soldier is familiar to us. We have met the design, with its amalgamation of the body and a weapons technology based on sense physiology, in the cinemas in 1992 as *Universal Soldier.* (Fig. 3) And as *Future Force Warrior*, the successor to Land Warrior equipped with nanotechnology, still only at the planning stage, he even shows clear similarities to the *Robocop* who appeared on the screens in 1987. Such illustrative comparisons are not infrequently found in official statements, in publications and internal expert reports on present and future developments (cf. Sterk 1997; NSD 2001). However, they are not presented in a way which might show up the possible ambivalences of cinematic semantics. Instead we find that promise of the aesthetic form is held out as a kind of advertising poster to gain support for the project of the new soldier.

The reference to *Robocop* and *Universal Soldier* shows Land Warrior not merely in the context of a progress in weapons technology, it suggests that the project can make a leap forward into a new technological age. Land Warrior is said to be an integral part of a Revolution in Military Affairs. Since the end of the Gulf War in 1991, the military has been using this term to describe its programmatic, technical und organisational work aimed – as they often say – at bringing the methods of warfare up to the level of the information age (cf. Sloan 2002, Kaufmann 2003). The new design allows the soldier to appear as a thoroughly scientific and technical project, with a conscious reference to the Cyborg topos. Originally deployed in the context of space travel research, the (bio-)technical equipment was intended to make the human body adapted for life in space (Gray 1989). Even if the actual project details focus only on the technical enhancement of the senses and do not, as in the original vision, strive at a reconfiguration within the body itself, the reference to Hollywood is still used in order to evoke the striking image of a basic anthropological equipment which can be radically transformed. Of course long before this, the networking together of technological and social features was extended to include the body, but only are such wide-ranging comments being heard from such a wide variety of people – from future-oriented computer and nano-scientists, to avant-garde contemporaries such as the performance artist Stelarc – saying that it is now the time "to newly shape the human being so as to make him better compatible with his machines" (Stelarc, cited in Rötzer 1998: 609). In the context of such visions as these, Vilem Flusser (1994) has spoken of a transition "from subject to project". The underlying idea is that the power of scientific analysis has now come to be so developed, that it is deconstructing every unit. This current trend is seen as leading to a radical loss in security and orientation, but at the same time opening up boundless combinatory possibilities of design. The world can be redrawn in all its facets, including even the physical nature of the body. With Land Warrior this vision is to be realised as a military programme.

Fig. 3: "Land Warrior" (US-Government, Public Domain)

By linking the figure to Hollywood productions, the promised model is given a boost in capacity. Equipped with information technology, the soldier is reinforced with new potencies, he is connected up to an information network spanning the entire battlespace and kept permanently online. The sense-expanding prostheses allow the soldier to enter fully into an augmented reality. The classical machine type of socio-technical combination, where the clothing, guns and other equipment are combined with the body and senses, is now with the integration of mediate transmission processes, transformed into a network-type of combination. The concrete expectations are that with this combination, the sense-physiological attributes which are needed to face the demands of machine warfare, can now take on an actual technical form. Features that the steel figure was compelled to compensate for in imagination, are now provided by automated context sensitivity: the sensing of gas has become effective in the form of sensors which are integrated in the clothing and which warn of atomic, biological or chemical contamination. The vague sensing of the presence of the opponent, of his concealed or nocturnal movements, is now replaced by precise technical visual presentation, for example by radar or heat ray observations, or by the feeding in of surveillance data that are circulating in the net. In this context the military speak of a "transparent battlefield" and envisage a kind of "god's eye view" circulating in the network. (Sterk 1997). "The network will enable you to see all that can be seen", runs the specific promise (General Gorman in NSD 2001, Composite Vision, p. 2). The Army Soldier System Command (SSCOM 1997: 42) proposes that: "If we are really good and we are, the soldier of 2025 will be as effective as the tank of 1995". And the ideal which army leader General Shinseki put forward in 1999 for the army is: "persuasive in peace, invincible in war". In this context the appearance of the soldier works in a way very similar to that of the steel figure: he is presented as part of an enthusiastic project, which admittedly sets itself an unattainable goal, and yet does effectively mobilise enormous forces of every kind.

It is not only because of its design that the *Land Warrior* profits from the suggestions of the entertainment industry: the forming of the soldier's perception also participates in those developments; in the field of computer games, for example. Infantry training is increasingly based on simulations of actual battle situations – like training pilots. It is no longer just a matter of practising technical skills, now what is emphasised is the experience of battle missions by visual experience, with the new kinds of simulation. This kind of visual experience was adopted for military training at the end of the 60's, when the flight and tank crews were equipped with simulators featuring head-mounted displays. The coupling of machine-guidance and the physiological-sense experience of the body-machine movement, which was already provided to some degree by the older flight simulators, has now been extended by the experience of moving through a computer-simulated 3-D space. The graphic revolution which was initiated by the military, making the display into a window, "through which the user looks at a virtual *conceptual* 3-D universe", (Lenoir 2000: 295) rapidly led to spin-offs in the entertainment sector. While the year 1983 saw the marketing of the first ego-shooter game *Battlezone*, in which a tank gunner shoots his way free through a meagrely presented three-dimensional region, by 1997 an official study could conclude that the computer games industry had taken over the military simulators, in their sceneries, character modelling and quality of graphics and sound. In the same year the US marines then took the ego-shooter game *Doom II*, available as shareware, and developed it further so that it could be used for the training of their squad leaders. And in 2002, commissioned by the US army, an ego-shooter game was released which was designed as a recruiting tool: *America's Army: Operations*. It can be downloaded free from the internet. Finally the Institute for Creative Technologies (ICT), which was set up with funding from the US army, developed *Full Spectrum Warrior*, a video game which was used both as a market product and a training medium for the infantry. The simulation of the battle experience is finding entry into the most varied levels of tactical training. Ego-shooters, such as *America's Army*, or again *Full Spectrum Warrior,* which is played from the perspective of an unarmed team leader, serve as screen simulations for the training of the private, the simple, dismounted soldier. Within the frame of the army's JFETS programme, missions from desert battles to house-to-house conflicts are modelled in installations and on large screens using various modules. Thus, for example, large screens fixed in windows or inserted in walls show the scenes of Arabian town buildings, with enemy fighters darting through them, their fire simulated by shots whipping through the air (cf. Korris 2004).

It is not only in the technical sense that the boundary line separating game and war-game is dissolving. Games like *Full Spectrum Warrior* are played equally in the barracks and in the living room or children's bedroom, and are even taken along to the battle regions for entertainment. Warrior-skills are gained at the game consoles. Some officers even see in present-day Iraq a new generation of Spartans who have grown up with ego-shooters and computer proficiency (Vargas 2006). The computer game is becoming the trainer of the soldier's new skills of perception which are needed for modern-day warfare. For it is necessary to switch on, perceive and interpret the data which are circulating in the network, be they in the form of pictures, iconographic symbols, or signals, text or other forms. And more importantly, the games emphasise not only the learning of technical skills, but how to deal with emotionally provocative battle situations. And the training focuses not on general rules but on exemplary situations: realistic stories are modelled – such as incidents during peacekeeping missions or guerrilla warfare in towns. In some models of events, for example the simulation of an accident occurring during a peace mission, an attempt is made to simulate socially plausible, emotionally expressive agents within the microregion (individual persons, rapidly gathering crowds). The models of events are derived from a sophisticated background of highly informed psychology covering social, cognitive, learning and behavioural aspects. Training on simulators is becoming part Hollywood, part theme-park. A hyper-realism of sense-physiology is made possible, which erases the dividing line between war and war-game not only for those at the planning table, but also in the battle itself (cf. Lenoir 2000; Morris 2002).

The online-connection of each individual soldier promises to release the infantryman from the sense of loneliness and being lost on the battlefield. The cohesion of the whole military body, which in earlier times was guaranteed by the mechanical military machine, is now to be ensured by the network. This cohesion is kept up by the "Common Operational Picture", to use the new military term, to which everyone is connected. This Common Operational Picture is the data repository from which the infantryman gets his information. It is the repository that lifts him into an augmented reality. And the Common Operational Picture always serves the added role of a control room. For every soldier also exists as a data copy in the net. So each soldier is acting as his own supervisor, watching his actions as represented on the screen. And he always acts under the supervision of others for whom the same rules also apply. Thus a permanent control room is set up in the net where everyone is under their own supervision as well as that of the others. The order of the visible is superseded by the order of the augmented reality, and the mechanical cooperation is replaced by one created medially. Consequently, augmented reality means that the integration into the military body is effected not only by external supervision, but also increasingly by self-supervision.

Land Warrior does, in fact, take up elements of the steel figure, which had its origins in trying to confront the difficult problem of how to achieve cohesion among the functionally differentiated fighting units. The raison d'être of the steel figure was that, given this situation of differentiated parts, the cohesion of the military body could only be ensured by strengthening the archaic warrior's sense of unity. Thus when we come to network-centric warfare, which takes shape in the figure of the Land Warrior, the developers feel the need to emphasise that participation within the Common Operational Picture requires a far more intensive type of cohesion. A joint effort of this nature calls for a strong feeling of mutual trust, based on a common interpretation of the situation. It requires deep-rooted doctrines and also totally congruent values. It is not surprising, then, that the army is developing a military creed where the soldier is defined as "Warrior". Each step that a soldier takes during his military training, at and on the screen, is accompanied by this warrior creed. In military service, equal importance is attached to the training of technical skills, and the nurturing of this revived warrior-culture. Thus we find the soldier's cultural self-conception expressed by images such as this: "The American Warrior is an adaptive brotherhood elevated by technology […] and decisive across the spectrum of conflict." (NSD 2001: panel one, p. 1) As in the case of the concept of the steel figure, here too the military community or brotherhood relies on the archaic warrior type. Behind the type of the individualised soldier, there lies hidden the mentally and emotionally related collective. In the concept of a network-centric warfare, archaic collectives and medial interconnection appear as congenial complements. Thus we find this statement in a programmatic paper: "Unlocking the full power of the network also involves our ability to affect the nature of the decisions that are inherently made by the network, or made collectively rather than being made by an individual." (Alberts/Garstka/Stein 1999: 105) Collective principles are no longer resisted, but incorporated. Perhaps one can discover the true face of the new warrior not in the individualised Cyborg, who serves as the advertising placard, but rather in the Borgs, who embody the anti-American principle of collectivism in the series Star Trek. Warfare has always heavily relied on the imitation of the enemy.

List of References

Alberts, David S., John J. Garstka and Frederick P. Stein. 1999. *Network Centric Warfare. Developing and Leveraging Information Superiority*. Washington D.C.: CCRP Publication Series.

Bigler, Rolf R. 1963. *Der einsame Soldat*. Frauenfeld: Huber.

Bröckling, Ulrich. 1997. *Disziplin. Soziologie und Geschichte militärischer Gehorsamsproduktion*. München: Fink.

Flusser, Vilém. 1994. *Vom Subjekt zum Projekt. Menschwerdung*. Bensheim: Bollmann.

Gray, Chris Hables. 1989. "The Cyborg Soldier. The US military and the post-modern warrior". In Les Levidov and Kevin Robins, eds. *Cyborg Worlds. The military information Society.* London: Free Association, pp. 43-72.

Horn, Eva. 1998. "Die Mobilmachung der Körper". *Transit* 16.1998/99: 92-106.

Hüppauf, Bernd. 1991. "Räume der Destruktion und Konstruktion von Raum. Landschaft, Sehen, Raum und der Erste Weltkrieg" *Krieg und Literatur* 3(5/6).1991: 105-123.

Kaufmann, Stefan. 2005. "*Network Centric Warfare*. Den Krieg netzwerk-technisch denken". In Daniel Gethmann and Markus Stauff, eds. *Politiken der Medien.* Zürich: Diaphanes, pp. 245-264.

Korris, J. H. 2004. *Full Spectrum Warrior: How the Institute for Creative Technologies built a cognitive Training Tool for the X-Box.* ICT, Marina del Rey (http://www.ict.usc.edu/publications/korris-fsw-asc.pdf)

Lenoir, Tim. 2000. "All but War Is Simulation: The Military-Entertainment Complex." *Configurations* 8, pp. 289-335.

Lottes, Günther. 1998. "Zähmung des Menschen durch Drill und Dressur." In van Dülmen 1999, pp. 221-239.

Morie, J. F. et al. 2002. *Emotionally Evocative Environments for Training.* Marina del Rey 2002 (http://www.ict.usc.edu/publications/ASC-Morie-2002.pdf)

NSD. 2001. *Objective Force Warrior. Another Look. The Art of the Possible. A Vision.* Prepared for the Deputy Assistant Secretary of the Army (Research and Technology) by the National Security Directorate Oak Ridge National Laboratory. December. (http://www.natick.army.mil/soldier/WSIT/OFW_Vision.pdf; 23.03.2005)

Rötzer, Florian, 1998. "Posthumanistische Begehrlichkeiten. Selbstbestimmung oder Selbstzerstörung." In van Dülmen 1999, pp. 609-632.

Sloan, Elinor. 2002. *The Revolution in Military Affairs. Implications for Canada and NATO.* Montreal: McGill-Queen's UP.

Sterk, Richard 1997. "Tailoring the Techno-Warrior." *Military Technology,* 21(4/5): 69-72.

SSCOM. 1997. *Yearbook. Maintaining the Edge.* (http://www.sscom.army.mil; 23.03.2005)

Van Dülmen (ed.). 1998. *Erfindung des Menschen. Schöpfungsträume und Körperbilder 1500-2000*, Wien: Böhlau.

Vargas, J. A. 2006. "Virtual Reality Prepares Soldiers for Real War." *Washington Post*, 14. Febr.

Military Bodies, Weapon Use and Gender in the German Armed Forces

CORDULA DITTMER

Introduction

The following shall consider the relationship between military, masculinity and weapons as a key to military ideology. Since the emergence of standing armies in the 18th century, warfare and bearing arms have been firmly established as primarily, if not solely, male concerns. Since then, women have been formally excluded from military service, and citizenship has been directly linked to the ability to use and carry weapons for defence purposes. Indeed, weapons are still considered the main symbols of nationality and masculinity in many cultures up to the present day (Schroeder et al. 2005). The rise of technology and its incorporation into military life, comradeship and the permanent struggle to become a "real" soldier are central to a military bodies' ability to function effectively. It has been shown by various researchers that technology, comradeship and the idea of a soldier as such are nevertheless historically loaded with male concepts and associations (see e.g. Apelt 2005; Theweleit 1977). However, there is a paucity of research investigating both the current relationship between and meaning of military, gender and weapons and the relationship between soldiers and weapons in the German Armed Forces in general. There are several central questions which therefore remain to be answered in light of modern circumstances. Firstly, how can a soldier come to identify with weapons and develop a willingness to kill while the armed conflict is taking place "out of area"? Secondly, how do technologies and body interact in military socialisation? From a gendered perspective I will analyse whether weapons are still the "bride of the soldier" and whether they still serve as a phallic symbol as formerly argued by psychoanalysts. Furthermore, how do women deal with weapons and their traditionally "male" connotations? I will argue that since the end of World War II a semantic shift has taken place in this regard. With some exceptions – weapons are no longer described in female terms but rather as a) incorporated into military bodies; b) symbols of comradeship and c) symbols indicating one's status as a "real" soldier, in particular for women. In the first section, I will analyse the process of identification with weapons and military socialisation as a general disciplinary technique. In the second section, I will focus on the question of gender and how female soldiers deal with the male connotations of weapons and with identification processes. Finally, I shall conclude with some remarks on the meaning of weapons for the soldier of the future. [1]

On becoming a "cyborg" – Military socialization as a disciplinary technique

The general aim of military socialization is to prepare soldiers for war and situations of conflict, to engage in violence, to kill and to be killed or injured (Apelt 2004). As Foucault (1976) has shown in his discourse-oriented study "Discipline and Punish", stable and obedient military bodies are the result of the disciplinary techniques of normalization, standardization and synchronization of military agency. Bodies (persons) and weapons are synchronized by approaching the body as an object which must be analysed, used, transformed and improved in accordance with general physical exercises. Soldierly bodies become "weaponized". They become cyborgs[2], man-machines who react as automatically as robots in situations of extreme danger, fading out all human feeling by simply fulfilling their tasks and missions. This exercise of power forges a new subjectivity, a new soldierly identity in which the subject identifies himself physically as well as psychically

1 The following considerations are based on the research project "Gender and Organization exemplified by the German Armed Forces" supported by the German Research Foundation. Beneath the author the project was carried out by Prof. Jens-Rainer Ahrens, Dr. Maja Apelt and Dipl. Soz. Anne Mangold at the Helmut-Schmidt-University, University of the Federal Armed Forces Hamburg (November 2002-April 2005). We carried out 65 problem-centred interviews with male and female soldiers of the medical services, the navy and the army from all ranks and 60 experts of military issues.

2 Here "cyborg" does not mean a mixture of human body and technical implants in the sense of really "going under the skin" but the incorporation of technical logics in a symbolic discoursive sense (see Spreen 2004).

with the weapons. The development of this cyborg-identity seems absolutely necessary to overcome the general socialized aversion to the use of weapons. In the wake of the experiences of the Second World War, the theoretical foundations of the German Armed Forces now lie in the concept of "Innere Führung" ("leadership and civic education") and the "citizen in uniform" which provides for ethical and moral values and norms which prevent soldiers from becoming unthinking combat machines (see e.g. Dörfler-Dierken 2005). One soldier said that he "was taught that it is not weapons killing humans but humans killing humans" (enlisted personnel, m, navy[3]). Integration in and subordination to the military power system is now designed to take place through understanding and not, as in previous times, as a result of compulsion. In the words of Bröckling (1997), it is a model of discipline which is characterized by introverted obedience.

Although ethical and moral values play an important role in military training, it is extremely interesting to see how the disciplinary techniques described above appear in basic training in particular. The process of identification, of becoming a "real soldier" using weapons, is broken down into different parts: It has both an intellectual, rational, technical dimension as well as an emotional and physical aspect which aim to incorporate weapons as a normal part of action. Only in military training does the first dimension play a significant role. As Steinert and Treiber (1974) conclude, the point is not whether you *want* to kill someone, but whether you are *able* to kill someone[4]. Foucault (1988) has shown that the exertion of power is not only destructive but also constructive: The formal exclusion of emotional issues in modern organizations generates them at the same time. This ambivalence can be shown in the following paragraph in which one soldier describes the first steps in basic training, a kind of formal technical introduction:

> You learn about the gun's firing-range, its calibre, how many parts it is assembled from and all that theoretical stuff. How it is dismantled, the weight of the weapon and how the projectile flies whether it flies in the arrow or so. Then there are also texts we have to read through. But it is just cooler to become acquainted with a weapon that way (basic training, m, armoured infantry).

---

3 In the following I describe the soldiers in rank/position, sex and service branch.

4 The study of Steinert and Treibel (1974) refers to the 1970er. It is astonishing how topical it is for today's military training, although there had been a lot of reforms and changes of paradigms. The heart of military training seems untouched.

We may observe from this that this soldier distinguishes between the technical dimension, that is, learning theoretically and the real physical and emotional connection with the weapon. But for him to really get to know the weapon means to feel it and handle it. He wants to get to *know* it better and forge closer links with it and he personalizes and emotionalizes his weapon like a human being.

One confidence-building strategy in basic training is continually disassembling and reassembling weapons. In many cases it is trivialised by games which can be conceptualized as a technique to separate the effects and functional aspects and the emotional, physical dimension of weapon use: "We incorporated little games into the disassembling and reassembling. For example, we would do the assembling and reassembling blindfolded" (basic training, f, medical service). The soldiers get to know their weapon as well as their own bodies and they learn to trust themselves and their arms blindly. They are trained to close their eyes to the effects of their action which results in decreasing awareness of the weapon's threatening potential. Nevertheless, all soldiers find their first actual experience with shooting as both an impressive and, at the same time, fearful moment: "Back then during basic training when I held a weapon in my hand I used to be shaking an awful lot. It was really hard to get over that just to shoot at the cardboard cut-out" (corporal, f, medical service in the navy). These physical experiences of touching, feeling blowbacks and trembling are controlled by detailed physical and disciplinary techniques:

> It was exactly the same as the first time using a weapon: the G36 puts a lot of pressure on you. And you didn't know how to handle it right, but you learned from your mistakes. You press it really firmly, that way you have it under control and then you shoot" (basic training, f, medical service). "But the breathing technique you learn for shooting ... it really is ingenious" (basic training, m, armoured infantry men).

In this way, the soldiers become totally at one with the weapon, even to the point of the most important and basic physical aspect of life, breathing. These physical techniques are lent the moral and emotional support of their superiors: "And then we started shooting really slowly. We had very kind superiors. They said, don't worry, we're behind you, if something happens you don't need to worry about it. That was o.k." (basic training, f, medical service). As a result of their training, the soldiers build up their trust in their superiors because in emergencies everyone is dependent on everyone else and subordinates in particular rely on orders from above.

In another stage of basic training the recruits must incorporate the weapon into their daily life. They must get used to and control their emotional as well as physical access to the weapon. As a result of this process their fear of the weapon declines and it becomes an intimate instrument losing its deadly character.

I myself can't imagine that you can kill someone with such a device. I can't imagine it because I haven't seen it happen. Because in the first two weeks we used it in our everyday routine. We had it close to our bodies constantly. We went to meals with it, we had to go to the toilet with it. Everything. So I just can't imagine that things can come out of it that would kill somebody else" (basic training, m, medical service).

This soldier was socialized in a civilian environment and had never been in contact with weapons before. He cannot imagine how a weapon could kill. As a result of disciplinary techniques the weapon is now a part of him. It has become ordinary because it "accompanied" him in his everyday life, when eating and even in private and intimate situations like being on the toilet. It functions like a comrade in whom you can trust, who will "stick with you through thick and thin", who will defend you if worst comes to worst and stay with you till the end. The success of these training methods becomes even more obvious in the following passage: The soldier talks about a situation in which he had to shoot and how he reacted. "Because you have so much to do out there, so much going on in your head and actually only achieve the right awareness when you think about it in peace and quiet afterwards" (staff sergeant, m, armoured infantry men). This soldier is perfectly socialized because he reacts automatically in a difficult situation without having to think about the possible effects of his actions. Instead he simply carries out his orders as most of the soldiers do. However, it seems that "Innere Führung" really is working in this case as the soldier can think about and reflect on the experience afterwards.

The level of differentiation between functional aspects and emotional effects seems to increase depending on how much use is made of new techniques. The ongoing process of transformation of the German Armed Forces is to a large degree a process of replacing the old weapons systems to more information-based technologies. It is interacting with new forms of warfare and new methods of confronting, or better yet, avoiding confrontation with a real human enemy. For example, the Navy relies on a lot of new technologies and the soldiers manning these complex technical weapons have neither emotional nor moral cause to deal with the effects of pushing a button. Military socialisation as described above is still working:

It's comforting to me that here (in the navy, C.D.) we don't have much to do with hand weapons, which means that the chance that I will end up in a situation where I have to take someone out with the gun is relatively small. Apart from that, the other advantage of the navy is that we aren't shooting at a target like in the armoured infantry where you can still look them in the eye afterwards. We shoot on a radar image, so the whole thing, this sounds horrible but it's all a little more impersonal, more anonymous and maybe easier for your conscience to digest. O.k., in the back of your head you know that a missile hit to a ship will mean that maybe two hundred sailors die, but primarily you just fought a target. You see a point and it disappears and that's good. That's the logic. (officer, m, navy).

This officer rejects the use of weapons and is happy that there are few opportunities for direct confrontation with the enemy. He mentions another advantage in the navy which is the use of highly technical weapons. These new weapons are complex machines in which human beings are transformed into targets on a computer screen. Emotional and/or physical issues no longer play a role. Pressing a button needs neither emotional detachment nor physical strength – anybody can do it. Now the potential enemy becomes the man-machine, in the words of Djuren (1996). "Technologies are the instruments for the dehumanisation ('Entlebendigung') of the real world" and making "clean wars" possible. The introduction of technology therefore means civilising the military body and, at the same time, it strengthens effectively as countermovement the discourse on the body. Focusing on bodies then becomes a sign of military humanity (see Djuren 1996).

Military-masculinity-arms: How women are placed in weapon use

To talk about a man-machine-complex means to talk about a male-machine-complex as "man" stands for humanity and a male person (see Frevert 1996). Djuren (1996) argues that nowadays women can take over soldierly positions because through the development of technology the soldier becomes disembodied and therefore genderless. Women thus serve in various branches of service in technical as well as in physically demanding positions. Considering the interaction between male and female soldiers, a recurrence of biological discourses could be found to be reviving the "old" gender dichotomy. With regard to their exposure to weapons, gender dichotomies arise when, in the first instance, female soldiers talk about the problems they have with weapons. In most cases they ascribe these problems to their sex, for example, carrying weapons depends on physical strength and, as women, they say that they do not have the necessary power to do so. However, the problem is not based on weapons as such but rather points to the ongoing discourse of "body", which in the military is very powerful (see also Cohn 2000). It treats the woman as too weak to fulfil difficult military tasks thus precluding women from being "real" soldiers.

A second instance of relevance to gender issues is when female soldiers would like to stand up to the men (or maybe us as interviewers) and appear particularly strong: They argue that there are many men and other women having problems with shooting (and they – as special women – do not):

> There were a few who said 'shooting's not for me' just like there were men who said 'I don't want to shoot, I don't touch weapons', one women told us. 'And, some of them just left it, and others we managed to persuade by arguing that in medical service it's only for self-defence or defending the patients'[5] (staff sergeant, f, medical service).

This woman has no problem with shooting but her male comrades do. In this interview the soldier constructs herself as being part of the military body – more than men.

It is immediately striking when one considers existing research on the relationship between weapons and gender that psychoanalytical concepts dominate. Weapons are seen either as the "bride of the soldier" (Vinnai 2001) or as phallic symbols (Pohl 2004). Both conceptions view weapons as loaded with sexual meaning. Warfare and the military are seen as the most important institutions in the construction of masculinity. The focus on fears linked to sexuality and femininity taking place in military training could be turned against themselves and therefore against the women. As Pohl (2004) further argues, femininity then becomes the enemy and warfare, a grand collective coitus aimed at reconstructing damaged masculinities. He concludes that through the symbolic connection of weapon and phallus men become almighty. Pohl's argumentation is almost completely based on 25 year old literature and he refers in many cases to US-army training methods which are completely different to those of the German Armed Forces. Above all he talks of a kind of "typical male fascination of weapons". By contrast,

5 It is interesting here to note how she reduces shooting by referring to her status as non-combatant.

many women in our interviews note that the chance to use weapons in the military had been one of their main reasons for enlisting (see also Kümmel and Werkner 2003) but that on the other hand many men hate handling weapons. However, looking at the reasoning of women concerning their weapons, it may be possible to adopt Pohl's argument to a certain extent: Female soldiers in the military have to deal with male organizational structures and must position themselves in relation to male-military-structure. They do not identify with the image of femininity which this military masculinity proposes – the weak and passive woman – and they adopt the sexualising and eroticising of weapons: "But this pistol was so light and small and sweet and you can move around quickly with it and then shoot and firing is actually pretty easy too" (staff sergeant, f, medical service). On a symbolic level it is possible to argue that by adopting the weapon as a phallic symbol, acquiring a "metaphorical penis" (Höpfl 2003) this woman becomes male while, as shown before on the physical biological level, remaining female.[6]

One might conclude that for women handling weapons is a means to show their ability to be real soldiers and to define themselves as soldiers in the military. Many women are enthusiastic about shooting and they talk about it more frequently than men. Their enthusiasm for weapons and the habit of mentioning men who have more problems with the weapons than they do could therefore be interpreted as an attempt to adapt to militarized masculinity and to approach the picture of the male soldier as combatant and comrade: You are a real soldier only if you carry a weapon and if can cope with the twin prospects of killing and dying.

Conclusion

At this point it is apt to mention here that although the above discussion examined some very impressive parts of the interviews, the question of killing and death are not themes which arise very often in the other parts of the interviews either when discussing the problems the soldiers face in general or in discussions about the particularities of being a soldier. I think this is due to the fact that most of them actually have very little contact with weapons apart from the basic training and military exercises. They work as technicians, in offices and in hospitals or are responsible for different materials. All soldiers emphasize that the rules of

6 Another effect of adapting to male military culture is the acceptation of the devaluation of femininity resulting above all in multiple conflicts between women (see e.g. Apelt and Dittmer 2006).

engagement which regulate weapon use in international missions are very restrictive for German soldiers and that they would use weapons only as a last resort for self defence and/or defence of comrades. Although many soldiers have fun when using weapons and admit to a slight fascination with them, it is true to say that all of the soldiers we interviewed hope they will never be in a situation where they would have to use their weapons (or they are glad that they did not have to use them in difficult situations during international missions or sentries).

However, with the increased popularity of long-range missions, the actual use of weapons has returned as an issue to military agency. All conceptualizations of new technologies, clean wars etc. negate the reality that "real" soldiers have to use "real" weapons in long-range missions maybe not to kill on purpose but to defend themselves and/or their comrades. They are instructed to carry the weapons at all times when abroad and in dangerous situations. For soldiers the use of weapons has become more and more casual. Thus, there is a real "retraditionalization" and redefinition of weapon use coupled with a growing importance of dealing directly with death and killing which at the same time results in a "remasculinization" of warfare for both men and women.

List of References

Apelt, Maja and Dittmer, Cordula. 2006. *Geschlechterkonkurrenzen – das Beispiel Militär: Zickenalarm und komplizenhafte Weiblichkeit.* Paper presented at "Geschlechterkonkurrenzen", 4th Conference AIM Gender/ AKHFG, 2.-4. February 2006. [Online]. Available from http://www.ruendal.de/aim/tagung06/programm.php3 [cited 23 April 2006].

Apelt, Maja. 2004. "Militärische Sozialisation". In Sven Bernhard Gareis and Paul Klein, eds. *Handbuch Militär und Sozialwissenschaften.* Wiesbaden: VS-Verlag, pp. 26-39.

Bröckling, Ulrich. 1997. *Disziplin. Soziologie und Geschichte militärischer Gehorsamsproduktion.* München: Fink.

Cohn, Carol. 2000. "'How Can She Claim Equal Rights When She Doesn't Have to Do as Many Push-Ups as I Do?': The Framing of Men's Opposition to Women's Equality in the Military." *Men and Masculinities* 3: 131-51.

Djuren, Jörg. 1996. "Die Entkörperung des soldatischen Mannes." *FAUST* 4.

Dörfler-Dierken, Angelika. 2005. *Ethische Fundamente der Inneren Führung: Baudissins Leitgedanken: Gewissensgeleitetes Individuum – Verantwortlicher Gehorsam – Konflikt- und friedensfähige Mitmenschlichkeit.* Strausberg: Sozialwissenschaftliches Institut der Bundeswehr.

Foucault, Michel. 1976. *Überwachen und Strafen*. Die Geburt des Gefängnisses. Frankfurt/M.: Suhrkamp.

Foucault, Michel. 1988. *Sexualität und Wahrheit. Der Wille zum Wissen.* Frankfurt/M.: Suhrkamp.

Höpfl, Heather H. 2003. "Becoming a (Virile) Member: Women and the Military Body." *Body & Society* 9 (4): 13-30.

Kümmel, Gerhard und Werkner, Ines Jaqueline. 2003. *Soldat, weiblich, Jahrgang 2003: sozialwissenschaftliche Begleituntersuchungen zur Integration von Frauen in die Bundeswehr – erste Befunde.* Strausberg: Sozialwissenschaftliches Institut der Bundeswehr.

Pohl, Ralf. 2004. *Feindbild Frau: Männliche Sexualität, Gewalt und Abwehr des Weiblichen.* Hannover: Offizin Verlag.

Schroeder, Emily and Farr, Vanessa and Schnabel, Albrecht. 2005. *Gender awareness in Research on Small Arms and Light Weapons. A Preliminary Report*. Bern: swisspeace.

Spreen, Dierk. 2004. "Menschliche Cyborgs und reflexive Moderne. Vom Jupiter zum Mars zur Erde – bis ins Innere des Körpers." In Ulrich Bröckling and Axel T. Paul and Stefan Kaufmann, eds. *Vernunft – Entwicklung – Leben*. München: Fink, pp. 317-346.

Steinert, Heinz and Treiber, Hubert. 1974. "Erziehungsziel Soldat." In Erhard Klöss and Heinz Grossmann, eds. *Unternehmen Bundeswehr. Zur Soziologie der Streitkräfte.* Frankfurt/M.: Fischer, pp. 103-122.

Theweleit, Klaus. 1977. *Männerphantasien*. Reinbek: Rowohlt.

Vinnai, Gerhard. 2001. *Sozialpsychologische Hintergründe von Krieg und Terrorismus.* [Online]. Available from http://www.vinnai.de/terror.html [cited 23 April 2006].

"A Horribly Beautiful Picture". Aesthetic Dimensions of Aerial Warfare

CHRISTIAN KEHRT

Flying as a key 20th century technology disclosed new dimensions of warfare and implied a radical shift of perspective on the battlefield. It made the mass killing of non-combatants possible and blurred the distinctions between soldiers and civilians, front and homeland. In contrast to the anonymous killing on the ground, images of military pilots helped to represent modern warfare as a sublime experience. Although it is cynical to talk about phenomena of destruction and devastation in terms of the beautiful and sublime, heroic narratives were quiet common in the Age of the Two World Wars [dt. *Zeitalter der Weltkriege*]. As David Nye pointed out, "the most powerful experiences of technology for many have long been encountered in warfare". (Nye 1994: xvi) Yet, the battlefield as a realm of the technological sublime has as yet been neglected and it is rather difficult to approach past aesthetic experiences of soldiers.

Kant's notion of the sublime will help to identify aesthetic dimensions of aerial warfare. I will ask in what situations pilots perceived war aesthetically and what role technology played as a medium of aesthetic perception. I will argue that only in situations when the pilot was not in actual danger that the sublime experience of warfare was possible. Thus the aesthetic experience of aerial warfare required the absence of actual bodily threat and a certain distance to the enemy made possible through technology. The examples I use are taken from published and unpublished war diaries from the First and Second World War. The medial construction of warfare is an integral part of heroic narration and thus images and photos will be as important as written texts in analysing the technological sublime in modern warfare.

Kant's notion of the sublime

For the historic understanding of flying, Kant's dynamic notion of the sublime is helpful.[1] He defined it as a negative kind of pleasure aroused by experiencing something that would be absolutely threatening and deadly under normal circumstances. The sublime [dt. *"Das Erhabene"*] literally means *to stand above something*. In relation to the sublime everything else seems small: "We call sublime that which is absolutely great." (Kant 2000: 131) – It is immeasurably big in scale, great and majestic.

The examples Kant refers to stem from the realm of all-mighty and powerful nature: "Bold overhanging, as it were threatening cliffs, thunder clouds towering up to the heavens, bringing with them flashes of lightning and crashes of thunder, volcanoes with their all-destroying violence, hurricanes with the devastation they leave behind, the boundless ocean set into a rage […]" (Kant 2000: 146)

Fig. 1: "Every day people enjoyed this overwhelming picture. Heini Dittmar turned immediately after his take off with the beautiful 'Condor II' in a dashing turn, in order to make his turns above the steep rugged wall of the 'Jungfrau'." [This picture is probably a montage.] (Brütting 1939: 160)

1 In difference to Kant's transcendental approach Edmund Burke deals with concrete phenomena of bodily pain arousing feelings of pleasure. I thank Stefan Kaufmann for his remark that Burke explained the sublime by a certain "physiological tension of the nerves". Burke's major example is death. Since Kant and Burke both address similar phenomena and explain it by a distanced relation to the dreadful object, the philosophical distinction between the empirical and transcendental, as well as the difference between the beautiful and sublime are not relevant for the historical focus of this paper.

In the interwar period flying and especially gliding promised to reconcile a dichotomy of myth and reality, nature and technology that seemed to be in a rigid opposition at that time: "Gliding is beautiful. While engine-powered flight is fun too, it will never bring you as close to nature and you will never feel as majestically immersed within the invisible forces of the sky. Engine-powered flight might be more needed and useful, but gliding is beautiful!" (Hirt 1933: preface)

In books like "Flieger sehen die Welt" ["Airmen See the World"] photos of modern cities like New York, ancient temples, arctic regions, alpine landscapes or oceans of clouds implied the sublime aesthetics of flight. In Ernst Udets autobiography "Mein Fliegerleben" (1935) flying was represented as an adventurous, romantic experience of the unknown. Together with Leni Riefenstahl he shot adventurous movies like "Die weiße Hölle von Piz Palü" (1929) "Stürme über dem Mont Blanc" (1930) and "S.O.S. Eisberg" (1933) directed by Arnold Fanck. In this case we have to ask whether the seemingly authentic and thrilling perspective offered by Udets spectacular flights were mere instances of mass-produced Kitsch and bad taste.[2]

Fig. 2: "Like giant fishes the slim bodies of the fighter planes are floating on their march to the enemy at 7000m in an ocean of clouds." (Orlovius 1940: 102)

2 Kant's criteria of authenticity is that everybody in the same position would experience the subjective but universal dimension of the sublime. Whereas Walter Benjamin criticized the role of mass media and their propagandistic illusion of authenticity in his classic "*Das Kunstwerk im Zeitalter seiner technischen Reproduzierbarkeit*". (Benjamin 1980)

Phenomena of destruction – chaos, lightning and thunder – are to be found in warfare as well. It is no surprise that Kant addresses the phenomenon of war and the warrior in his analysis of the dynamic sublime: „Even war, if it is conducted with order and reverence for the rights of civilians, has something sublime about it, and at the same time makes the mentality of the people who conduct it in this way all the more sublime, the more dangers it has been exposed to and before which it has been able to assert its courage." (Kant 2000: 146) The sublime, as Kant describes it, implies an ambiguous and repulsive relation of the subject facing a threatening, powerful and potentially destructive object. It can be defined as a feeling of negative pleasure, a repulsive fascination with the horrible. In Kant's understanding facing danger in this way has a moral quality.

However, such an appeased and virtual situation of "harmless danger" is not possible in situations of combat, where physical threats and bodily vulnerability prevail. Furthermore we have to take into account that the technological transformation of warfare leads to a distanced, technologically mediated relation to the enemy. For example mathematical equations of an artillery engineer or radar blips of an enemy air plane on the screen of a navigator take away the alleged heroic face-to-face encounter with the enemy. Therefore I will read moments of sublime experience in the context of modern warfare as idealistic constructions, propagandistic representations or detached, technological experiences of soldiers in situations that were either not heroic or not dangerous.

Men: The image of the "modern warrior"

People were fascinated by the first aviators. The Wright Brothers were portrayed as modern heroes and especially their facial features focus of heroic narration: "The face of the aviator was unshaken, chiseled into stone. You could feel that this man was hard as iron, and embodied energy, stamina and assurance." (Zeitschrift für Luftschiffahrt 1909: 377)

Fig. 3: Oswald Boelcke assuming the manly self-assured attitude of the pilot (Boelcke 1916: 80).

Boelcke symbolized the new ideal of the heroic war ace. In this picture of Boelcke, published in his "Letters from the Front", you see a self-confident, manly attitude that is assured of winning. The legs are spread and he is looking from a superior, mighty position. Boelcke represented an active, determined man of deeds, who has mastered flying and killing. His characteristic traits of calmness and control are opposed to nervousness and anxiety pilots were confronted with in the complex and deadly war arena. These manly features of the military pilot are comparable to Kant's definition of the warrior, characterized by his courage and "incoercibility of his mind by danger." (Kant 2000: 146)

To understand the mechanisms of visual representation, it is important to realize what is not explicitly shown. There is no sign of the chaotic, loud and anonymous war of the trenches. Boelcke, Immelmann and Richthofen appear almost soft and innocent, undamaged by the war. In contrast to the myth of Verdun (Hüppauf 1999) creating a new and harder image of the front soldier, the fighter pilot allowed a continuity and synthesis of traditional ideals of soldiering with high-tech warfare (Schilling 2002).

Fig. 4: Propaganda image of Göring as the last leader of the Richthofen squadron (Sommerfeld: 17).

Hermann Göring then took advantage of the myth of the successful World War ace at the beginning of the Third Reich. The constitutive elements of Göring's portrait suggest a close link between the airplane, representing the Fokker company and modern military technology, the traditions established by the myth of Manfred von Richthofen and the political leadership of Hermann Göring. Thus as the last leader of the Richthofen squadron, Göring alluded to a supposedly glorious past. He promised to overcome the Versailles treaty and its prohibition of military aviation, in order to realize a powerful future of an airminded nation (Fritzsche 1992).

In the Second World War the face of the pilot was still the focus of heroic narration. The attitude of crews flying bomber missions against Poland was expressed by highly idealized portraits.

Fig. 5: The face of the pilot (Eichelbaum 1939: 15).

This frontal portrait from slightly below expressed a manly, heroic attitude and a close and intimate relationship to technology. Obviously the dangers and contingencies of the battlefield did not enter the cockpit: „In the face of battle. Nothing in the faces of these two pilots betrays that they are on their first mission." (Eichelbaum 19390: 14) The pilot is firmly holding the steering wheel and seems to have everything under control. His glance is focused and alert, yet calm and relaxed. According to NS-ideology pilot and observer apparently worked closely together. These heroic images helped to personalize the experience of war and symbolized the soldiers readiness and willingness to fight.

Machines: military airplanes as symbols of power and modernity

Boelcke, Immelmann and Richthofen are closely associated with the Fokker legend. In their popular war accounts there are passages of affectionate love and a personal relationship to technology. Boelcke described the new Fokker as a Christmas gift. For him they were like his little children standing together in a tent, one with his tail under the wings of the other." (Boelcke 1916: 26) Immelmann felt "very comfortable and at ease in that little bird." (Immelmann 1916: 61) Indeed, the Fokker was easily controlled and reacted immediately. Nevertheless the failure of engine or machine guns were a real and permanent threat to the pilots. Reports from the front clearly show that German pilots complained about the bad performance of their airplanes (Kommandierender General 1917: 129). According to death statistics, more than half of the pilots in WW I died through technological defects and individual failures at home or outside the realm of battle (BA-MA 1918: Verlustliste). A strategy to downplay the risk of dying was to emphasize the will of the soldier controlling the "dead material" of the machine. In situations of enemy superiority the appeal to the traditional virtues of the warrior was a military strategy to motivate the pilots to sacrifice themselves for the nation.

A shift in aesthetics then took place in the thirties. The aerodynamic streamlined design suggested a modern aesthetics alluding to speed and power of military technology (Asendorf 1997: 89; Meikle 2005: 116-124; Schatzberg: 1998: 3, 154, 174).

Fig. 6: "The face of the airplane. Frontal view of one of our most successful fighters." (Orlovius 1940: 92)

The airplane was focus of heroic narration and an integral part of visual strategies to prove the superiority of German military technology. The "face" of Germany's most popular standard fighter in the Second World War, the Me 109, was depicted from below, like the pilot's face. The propeller and the engine suggested dynamics of speed and power.[3]

Aesthetic dimensions of aerial warfare

In comparison to giant airships strut-and-wire airplanes looked rather tiny and fragile. Yet, it was not their size, but the perspective they disclosed and their performances that were perceived as sublime. Pilots had to learn to see and recognize the battlefield from a third dimension and decipher those signs as enemy movements. In Kantian terms flying was a condition of possibility for a new, aesthetically structured perception of warfare: "The people were tiny, the houses like from a box of building blocks for children, everything so cute and small, in the background Cologne. The Dome was like a toy. It was a sublime feeling to float above all. Who could do me any harm now?" (Richthofen 1990: 36/37) For Richthofen a position of safety was a condition for this kind of aesthetic experience. Flying, he tells his audience, was like sitting in a com-

3 The sublime appearance of military technology resembled depictions of airplanes in other countries at that time, e.g. the advertisements in the British aviation journal "Aeroplane".

fortable chair, yet much more thrilling (Richthofen 1990: 38). When he flew over the Russian front lines he felt that it was a "horribly beautiful picture" to see burning villages and withdrawing Russians (Richthofen 1990: 41).

We also find this aesthetic perception of burning houses at the Eastern front in an unpublished war diary of the observer Ernst Eberstein: "It was getting dark already, when we departed, around the horizon burning villages, the sky blood red, in between bursting shells, an uninterrupted rolling thunder. Down on earth you are more under the immediate impressions of fighting than above in the air, where you have a better overview but feel more like a spectator." (DMTB Eberstein 1918: 59) The observer appreciated his distanced position and perceived the battle – its colours, the blood red sky and sounds, rolling thunder – aesthetically. Although at the Eastern front there was no serious aerial combat, Eberstein registered every single hit in his airplane and put some extra iron plates underneath his seat. Thus only in situations of relative safety the observer could enjoy the "beauty" of the battlefield, the winter landscape and sunsets.

Aesthetics of aerial combat

Boelcke downplayed the risk of aerial combat as a joyous and easy game. In his letters from the front he uses expressions like "joyous dance", a "cat-and-mouse game", a "comedy", a "funny shooting", a "joke" and a "big joy". The main metaphor to talk about aerial combat was hunting and the battlefield understood as hunting ground. Despite the rather short and sober style of Boelcke's field reports we can find situations of aesthetic experience. For example the explosions of the enemy aircraft were perceived as colourful burning torches. This aesthetic transfiguration of combat downplayed the high risk of dying and expressed an self-assured attitude of superiority and success: "I stayed about twenty to thirty feet behind him and was hammering [my machine guns] until [the airplane] exploded closely in front of me in bright shining flashes of flames and dissolved into atoms [...]." (Boelcke 1916: 119)

Can we find similar war accounts in the Second World War? In the unpublished war diary of Ernst Trautloft, leader of the fighter squadron JG 54, there are several passages that can be identified with Kant as situations of sublime experience. At the beginning of the campaign against the Sowjet Union he perceived war as a great historic event: „Everywhere I see the flashes of gunfire, indicating our advancing on earth. An exciting picture. To the North my view goes over the Memel to the south to the Rominter Heide. This powerful spectacle gives us the feeling that here we are opening the door to a new chapter in history, but also a new fate for us all." (Trautloft vol 2 1941: 143) Trautloft described fighting like Boelcke or Richthofen: "Like tails of a comet the Russians were going down. Everywhere I see fire in the sky. Death is having a big harvest" [...]. An eerie picture." (Ibd.: 166)

The sharp contrast between war and peace, front and homeland characterizes Trautlofts narration from the battlefield. There are situations of peace within the larger framework of war: "I am watching all the small animals of the Russian earth. That is like at home. There are flies, beetles, worms, snails, even a small mouse is watching the wheel of my airplane with interest. What a peaceful hustle and bustle." (Ibd: 178) Then he turns his attention to an aerial fight he watches from the ground: "A weird and wonderful sight. In a couple of minutes the Russian bombers are roaring down like burning torches. [...] It takes minutes until their remains are dancing like the falling leaves of autumn. Above all a wonderful sun, radiating peace. What a contrast!" (Ibd.) This aesthetic perspective from the cockpit is often described as an "eerie" and "strange picture" [dt. *ein unheimliches, ein merkwürdiges Bild*].

Using aesthetic images to express the irreconcilable difference between war and peace, Trautloft maintained a cult of the warrior. For him it was simultaneously exciting and horrible to realize, for example, that while he was in Munich watching a theatre play, a young lieutenant of his squadron died after an emergency landing at the Russian Front. (Trautloft Vol 5 1942: 38) With this episode he wanted to emphasize that war intensifies life and is preferable to an "ignorant" and "degenerate" life at home. Along similar lines Kant thought that the attitude of the warrior is preferable to the mere spirit of commerce in times of peace (Kant 2000: 146).

Aesthetic transfiguration does not mean that the pilots were unaware or ignorant of destruction or their immediate dependence upon technology. Yet, figurative perception helped to stabilize their actions, without reflecting further. While flying back from a mission against England Trautloft enjoyed the sunset and the spectacular colourful play of light. In such aesthetic moments after the battle, having a splendid overview, the pilot had time to meditate upon his experience without questioning it in principle: "A fighter pilot has to be lucky and believe in his star. Those who have lost that faith get unhappy and insecure." (Trautloft vol 1 1940: 74)

The propaganda of success

Fig. 7: "These Ju 87 dive bombers, that resemble giant belligerent hornets, were often of decisive help for our fighting troops on the ground" (Orlovius 1940: 142)

The view from the cockpit offered a perspective that suggested power and control over the enemy. From above the pilots could see vast landscapes of destruction while the airplane was flying untouched and undefeated over enemy territory. Aesthetically intriguing propaganda images of destroyed capital cities like Warsaw or London should prove the success of the German war effort: "The sun is setting in the western horizon. Her rays illuminate the smoke that is ascending from the former Polish capital. It seems as if the red glooming fires reach up to the sky." (Grabler 1940: 275) The war propaganda pointed out that only legitimate targets were chosen, while at the same time it dwelled on the devastating effects of its attacks. This aesthetics of destruction helped to show how powerful and successful the Luftwaffe was in comparison to its enemies.

Conclusion

In the period of the two world wars we find continuities of aesthetic perception in published and unpublished material that were an integral part of heroic narration. Lightning, thunder and colorful spectacles of light were to be found not only in nature but also on the battlefield. In Kantian terms the detached view from the cockpit was a condition of possibility for an aesthetic experience. Aesthetic perception indicates moments of security and allowed one to give sense to the war experience without questioning it fundamentally. On the level of propagandistic discourse the images of airplanes invading enemy territory, flying above the sky with a deadly bomb load, expressed the power and success of a victorious nation. This aesthetic transfiguration of warfare as a sublime experience maintained a cult of the warrior on the political level of the nation and gave sense to the individual actors on the battlefield.

List of References

Adler, Hermann. 1939. *Unsere Luftwaffe in Polen*, Berlin: Wilhelm Limpert Verlag.

BA-MA [Bundesarchiv-Militärarchiv] N 760 Nachlaß Hans Trautloft, persönliches Kriegstagebuch, Vol I-V.

Asendorf, Christof. 1997. *Super Konstellation. Flugzeuge und Raumrevolution. Die Wirkungen der Luftfahrt auf Kunst und Moderne*. Wien: Springer.

BA-MA PH 9XV/7 *Die Verluste der deutschen Fliegertruppen einschließlich der bayrischen Verbände, 1914-1918.*

Benjamin, Walter. 1980. *Das Kunstwerk im Zeitalter seiner technischen Reproduzierbarkeit*. In: Benjamin, W.: Gesammelte Schriften I, 2 (Werkausgabe Band 2), ed. by Tiedermann, Rolf and Schweppenhäuser, Hermann, Frankfurt/M.: Suhrkamp, pp. 471-508.

Boelcke, Oswald. 1916. *Hauptmann Boelckes Feldberichte. Mit einer Einleitung von der Hand des Vaters und zwanzig Bildern*. Gotha: Verlag Friedrich Andreas Perthes.

Brütting, Georg. 1939. *Segelflug erobert die Welt*. München: Knorr und Hirth.

Burke, Edmund.1958. *A Philosophical Enquiry into the Origin of our Ideas of the Sublime and Beautiful*. London: Routledge and Paul.

Der kommandierende General der Luftstreitkräfte (ed.). 1917. *Sammelheft, enthaltend die vor Erscheinung des Verordnungsblatts der Luftstreitkräfte erlassenen für die Fliegertruppe geltenden allgemeinen Verfügungen*, Mézières-Charleville.

Eichelbaum, Hans. 1939. *Schlag auf Schlag, Die deutsche Luftwaffe in Polen, Ein Tatsachenbericht in Bild und Wort*. Berlin: Adler Bücherei.

Fritzsche, Peter. 1992. *A Nation of Fliers. German Aviation and the popular Imagination*. Cambridge/Mass. and London: Harvard Univ. Press.

Grabler, Josef. 1940. *Mit Bomben und MGs über Polen:* PK-Kriegsberichte der Luftwaffe. Gütersloh: Bertelsmann.

Herf, Jeffrey. 1984. *Reactionary Modernism. Technology, Culture and Politics in Weimar and the Third Reich*. Cambridge and New York: Cambridge University Press.

Hirth, Wolf. 1933. *Die hohe Schule des Segelflugs. Anleitung zum thermischen Wolken- und Gewitterflug.* Mit Beiträgen von Robert Kronfeld, Hermann Meyer u.a., Berlin: Klasing.

Hüppauf, Bernd. 1999. "Schlachtenmythen und die Konstruktion des 'Neuen Menschen'". In: Gerhard Hirschfeld and Gerd Krumeich (ed.): *Kriegserfahrung. Studien zur Sozial- Mentalitätsgeschichte des Ersten Weltkrieges.* Essen: 1999, pp. 43-83.

Immelmann, Max. 1916. *Meine Kampfflüge. Selbsterlebt und selbsterzählt*, Berlin: Verlag August Scherl.

Kant, Immanuel. 2000. *Critique of Judgment,* edited by Paul Gyer. Cambridge: Univ. Press.

Meikle, Jeffrey L. 2005. *Design in the USA*. New York: Oxford University Press.

Nye, David. 1994. *American Technological Sublime*. New Baskerville: MIT Press.

Orlovius, Heinz [Hg.] *Schwert am Himmel. Fünf Jahre deutsche Luftwaffe.* Berlin: Scherl 1940, Vol 3/4 Adler Bücherei.

Richthofen, Manfred von. 1990. *Der rote Kampfflieger. Die persönlichen Aufzeichnungen des Roten Barons mit dem "Reglement für Kampfflieger".* Hamburg: Germania Press.

Schatzberg, Eric. 1998. *Wings of Wood, Wings of Metal. Culture and Technological Choice in American Airplane Materials, 1914-1945*. Princeton: Princeton University Press.

Schilling, Réné. 2002: *"Kriegshelden". Deutungsmuster heroischer Männlichkeit in Deutschland 1813-1945*. Paderborn: Schöningh.

Schüler-Springorum, Stefanie. 2002. "Vom Fliegen und Töten. Militärische Männlichkeit in der deutschen Fliegerliteratur, 1914-1939" In: Karen Hagemann and Stefanie Schüler-Springorum (ed.). *Heimat-Front. Militär und Geschlechterverhältnisse im Zeitalter der Weltkriege.* Frankfurt/M.: Campus, pp. 208-233.

Schütz, Erhard.1999. "*Condor, Adler und Insekten. Flugfaszination im 'Dritten Reich'*". In: *Jahrbuch Berliner Wissenschafliche Gesellschaft*, pp. 49-69.

Sommerfeld, Martin Henri. 1933. *Hermann Göring. Ein Lebensbild.* Berlin Mittler.

"Wilburg Wright auf dem Cento Celle in Rom." 1909. In: *Zeitschrift für Luftschiffahrt*, p. 377.

Postmodern Aesthetics: Manipulating War Images

RAPHAEL SASSOWER

I.

While some claim that modern aesthetics can be clearly "read" according to a set of criteria (however historically informed), there are those who claim that this kind of reading is too rigid, even hegemonic. Even art theorists, such as Ernst Gombrich (1960), who are more concerned with the psychological elements that inform our visual perspectives, define modern art in terms of "nonfigurative art" and associate it with twentieth-century art. There are the modernist-formalists, such as Roger Fry (1934 – on British art) and Clement Greenberg (1961 – on defining the Avant-Garde and Kitsch), whose concern is with the conventions that inform each stage of aesthetic production and appreciation. For them, the very definition of what is considered art is tenable, as well as the value and meaning of that art. All of them are gate-keepers of sorts who can tell us why this is art and what kind of art it is: they are the aesthetic rainmakers of the present. In some respects, each art critic has been able to determine the acceptability of pieces of art within the artistic community and with the public at large: criteria have been set, such as re-presentation, form, content, style, medium, context, social message, so as to set apart or bring together various artists and their productions into a "school" or "movement."

Postmodern aesthetics has been offered as an alternative to modern aesthetics, one that opens up to multiple interpretations according to multiple sets of criteria (see Hoesterey's collection of 1991). This openness has been lauded as both liberating and empowering, encouraging theoretical and practical participation from diverse groups whose own voices were historically marginalized. The postmodernists posture begins with a protest, a refusal to admit a single definition for art and likewise, a disruptive attempt to redefine accepted definitions, expand and revise, undermine and break-down that which the artistic establishment has claimed to be the case. It almost does not matter who says what about art, since whatever is being said is bound to be challenged as a matter of course so that the definitions of the modern and postmodern are less meaningful than the actual doing of art. The very attempt to define art once and for all becomes an occasion for defiance and for rethinking that which has been defined. This is not to say that there is no definition of or for art pieces, but rather that whatever is suggested remains open-ended, putative, and conjectural in the Popperian sense (1963).

It is in this spirit that any discussion of art and aesthetics remains an invitation and solicitation, a way to get more people involved in the debate over the value of art and the ways in which meanings are injected to it. In short, this is a recognition of the temporality of what we create and observe, the artifacts that surround us. Put differently, it is a way to engage the art world and ourselves, reexamining our prejudices and convictions, and expressing our willingness to participate in an adventure. Indeed, the postmodern condition in general and in art in particular has forced us to reconsider our perception of political images that subsume an ideological commitment regardless of a prefigured artistic intent. Whether we admit to it or not, we all subscribe to an ideology, a set of ideas and ideals that guide our life and the choices we make. Let us detail this reconsideration.

II.

The argument here is a three-step argument in regards to war images. First, there is the step that connects data, facts in the field, so to speak, with pictorial images. Second, there is the presentation of images in the popular media as means of explaining the war to the public. And third, there are the ways in which images are read by the public regardless of what they were meant to convey. Let me elaborate briefly on the three stages, and then explain why the postmodern option of reading images, though appealing in various ways, may turn out to be more problematic than the modern one.

Already in the 1920s Otto Neurath explained how powerful, useful, and socially progressive it was to use images instead of difficult and cumbersome arguments (1925 – International System Of Typographic Picture Education). Though there are problems of re-presentation no matter in what medium, whether words and reality in Wittgenstein's early (1922) and late (1958) senses or models of the marketplace and the economy, pictures and images have always been welcomed by politicians. Through images, politicians hope to convey a message, even an idea, on a good day, a set of ideologically refined principles. Seeing an image of a wholesome family, whether drawn by the American Norman Rockwell or propagated by the Nazis, immediately conveys a certain set of commitments: to one's family, to one's ancestry, to one's religious beliefs about Adam and Eve, all the way to heterosexual procreation and pretenses of harmony and happiness. Here we unsuspectingly gaze at a whole set of ideals that eventually get internalized and that might be questioned and undermined only in a more diverse and multi-cultural context, where heterosexuality and homogeneity, for example, are not taken for granted as the only standards according to which to establish and assess human relations. So, the first stage of the argument deals with the relationship between images and the events or facts they are supposed to represent. How does one represent the reality of wars? Is this about battle scenes or about personal devastation? Can one recreate the reality of pain, suffering, and heroism? Can an arrested image ever convey the fluidity of events and their context?

The second step of the argument deals with the specific context within which images are distributed and consumed, the particular media in which they are presented. For example, should a war image be presented only in the internal documents of government agencies or be leaked to the press? Is it more prestigious or credible for an image to appear in a newspaper rather than on the internet? Obviously it matters which newspaper is used, for there are some, like the National Inquirer in the USA, that are notorious for fabricating or manipulating images to such an extent that they are no longer seen as being "real" or "credible." But once the extremes of the media are excluded, such as hate-group Websites and the like, is there a hierarchy of credibility? Is there a way for all of us to know and to agree on a set of criteria according to which to judge which print-media is more or less credible than another? This is reminiscent of the debates over the authenticity of photographs and documentaries, black and white images that seem more detached and real than those characterized by color differentiation.

Some Americans are prone to believe that anything the government says is true and that whatever appears on television is "real," no matter what the circumstances. One can view the popularity of "reality shows" as an indicator of the perception that if it is shown on television, then it must be real and true. Similarly, the Paparazzi can always find a willing newspaper that will pay top dollars to see what real celebrities look like up close and what they are doing in their private lives! Of course the complexity of these assertions is compounded by the fact that some networks and media channels are public, in the sense of receiving government grants and having to abide by some government regulations and others are not. Should one give the same (public interest) status to a news-outlet that increases revenues through advertising and that invests in sensationalist broadcasting in order to increase viewership/readership to justify higher advertising rates?

It is within this context of public and private reporting that we confront the problem of how and where to present war images, that is, how gruesome or benign the images should be, so as to invoke patriotism and support for the troops. If the first step of the argument is about the content of the image and its relation to what it is supposed to represent, then the second step of the argument is about the format and context of representation in general, whether of war images or the economy, whether one chooses newsprint, television, or the Internet.

The third step of the argument is what was described elsewhere (Sassower and Cicotello 2005) as the predicament of having images read differently from the intent of the creator or sponsor. There we examined a variety of public displays, such as monuments and posters, from the Soviet Union, Italy, Germany, Mexico, and the United States in the period between the two World Wars. We discovered to our amazement and amusement that though diametrically-opposed ideologues thought they were putting forth a powerful message through posters, meant to invoke patriotism and support for all their respective regimes, they tended to use similar, and at times identical images. For example, one can find the same "hand with a hammer" in posters from the era between the two world wars in posters from the Soviet Union, the United States, fascist Italy, Nazi Germany, and Mexico. This means that in practice, once an image is put out in public, the public can read it however it wants, interpret it in a fashion that perhaps was not quite what was originally intended. There is, then, no linear or causal connection between an idea, its image, and its reading by someone else. Instead, a fascist or democratic reading can be gained from the same image: courage and heroism, diligence and commitment, national pride and camaraderie, family value and wholesome friendship. Who does not want them? Who would not endorse them? So, the third step of the argument is about the inherent ambiguity of images and the ways in which they lend themselves to multiple interpretations, none of which is the "right" one (except in the minds of the authorities who are responsible for their production).

III.

In what follows, I would like to suggest that however appealing postmodern aesthetics may turn out be, there is a high political and moral price to be paid when eschewing modern aesthetics. We all welcome the openness offered by the postmodern move as a way to read images however we choose, without the legitimating authority of politicians, in the case of wars, who justify military involvement or specific strategies to be used. It is perhaps this openness that ensures a critical dimension, a way to keep politicians honest and accountable to their decisions, suggesting that their decisions may end up in retrospect more problematic than they appeared at first. Our involvement in assessing images, of course, is already too late and inevitably compromised, since we have no access to the data that prompt the use of these or other images. Politicians and military leaders make decisions based on confidential information and then portray their decisions pictorially for public consumption. Here and there, photojournalists intervene, sometime on behalf of the authorities, sometimes in a subversive way so as to expose the weakness of decisions and their folly. As the Catholic tradition has taught us, Just War principles can sanction wars but also can illustrate the conditions under which one should not wage a war. In modern times wars have been waged for economic reasons, both domestic and colonial, to boost faltering political careers, and for sheer ignorance or arrogance. It is difficult at times to distinguish among these reasons, as they tend to merge and be self-serving along the way (especially when there are unintended consequences to the way in which ideas and images are eventually consumed).

The focus of the examples in this discussion will be war images, from a variety of combat-arenas: World War I, World War II, the Vietnam War, Iraq I, Iraq II, Afghanistan, and the Palestinian-Israeli conflict. These images can be crudely divided into those that can be read through the modernist prism, namely, as if they can be read in a straightforward manner: here are the facts, the reality that is captured by a camera with little ambiguity or confusion. Surely, one can add an interpretive layer to any reading, but let us refrain from this additional step at first and see what we come up with. Then there are those images that can be read through the postmodern prism, namely, as if there is no single reading that captures the meaning or the reality or the facts of the matter. Surely, one can agree on reading some elements in the image across cultural lines, but there is much room for interpretation.

The contention here is that it is precisely with war images (as represented in the media) that we come to the realization that interpretive manipulation is not merely a theoretical game (in Lyotard's sense 1984) but an enormously problematic and dangerous undertaking, akin to the worst fascist propaganda mechanisms of the past century. If all we know of the current war in Iraq, for example, is what is presented to us through official American channels (with strict military censorship), it may seem that it is a just war, a war for democracy, and a winnable war with little harm and suffering. Leni Riefenstahl would feel right at home within this propagandist context, when national pride and patriotic zeal overshadowed any and all critical disturbances. Should we then revert back to modern aesthetics in its classical guise, so as to expose manipulation and misleading images? Or should we provide intellectual frameworks that would alert audiences to the context within which images should be read?

IV.

It may be helpful to conclude here with a brief description of Just War Theory so as to explain some of the underlying issues that may be simmering below the political and the journalistic surface. Perhaps in doing so, we can more fully appreciate our own cognitive confusion and psychological discomfort when confronting war images. We will be following both Alexander Moseley's lead (2005) to illustrate the main arguments and principles associated with the justification waging wars and Robert Kolb (1977) to appreciate some of the misconceptions associated with the term. For example, though the Latin is used to distinguish between the causes and the conduct of war (*jus ad bellum* and *jus in bello*), these expressions were only coined at the time of the League of Nations and were rarely used in doctrine or practice until after the Second World War, in the late 1940s to be precise. What they all worry about is to provide a legal framework within which the problems of "might makes right" can be dealt with rationally, consistently, and fairly. The classical concerns of fighting honorably for an honorable cause have been historically transformed into a more concrete framework that acquired an international audience and legitimacy.

First, there is a theoretical framework that follows five major principles, all of which should hold in order to wage a just war. These are: having just cause, being declared by a proper authority, possessing right intention, having a reasonable chance of success, and the end being proportional to the means used. Possessing just cause is the first and arguably the most important condition of *jus ad bellum*. Most theorists hold that initiating acts of aggression is unjust and gives a group a just cause to defend itself. But, as the Encyclopedia reminds us, unless 'aggression' is defined, this proscription is rather open-ended. For example, just cause resulting from an act of aggression can ostensibly be a response to a physical injury (e.g., a violation of territory), an insult (an aggression against national honor), a trade embargo (an aggression against economic activity), or even to a neighbor's prosperity (a violation of social justice). The main argument regarding the initiation of war is self-defense, not unlike the Jewish, rather than the Christian interpretation of "thou shall not kill" from the Ten Commandments (which eventually finds its way into legal systems as degrees of culpability). Of course, the notion of "self-defense" can itself be stretched, as the United States has repeatedly done, so as to justify pre-emptive strikes, as in the war in Iraq. The threat, it was suggested by the Bush administration, was so grave with weapons of mass destruction, that it was a matter of "when" and not "if" Saddam Hussein would be attacking the western world.

The second condition that must be met in order to conduct "just war" has to do with the authority of the party declaring war. Should it be a government? What kind of a government qualifies? Must it be democratically elected? As the Encyclopedia suggests, some governments are more appropriately recognized as the ones with unchallenged authority. The Vichy puppet regime, set up in 1940 by the invading Germans, deserves no public support from French citizens who did not elect it. Here we come to questions of sovereignty as discussed historically by a range of political philosophers who insisted, ever since the trial of Socrates, that there must exist an underlying contract among all participants so that the government is merely a representative of the General Will (to use Rousseau's terminology). The people remain the sovereign power and have the ultimate authority in all state matters, domestic and international alike. The only condition for conferring any authority on the government is that when acting it indeed represents the will of the people and secondly, that when it does not, it can be removed without violence (through recall, impeachment, election, or referenda).

The third condition for waging a "just war" is that there should be a right or appropriate intention. The general thrust of the concept being that a nation waging a just war should be doing so for the cause of justice and not for reasons of self-interest or aggrandizement. Of course, moral theorists have some problems with this being a sufficient (though would allow for it to remain a necessary) condition for just war, for what happens if the right intention leads to many unfortunate and unintended consequences? For example, is it possible to draw a clear line between right intention and self-interest, especially when debating the vague notion of a national self-interest in the preservation of equal rights or freedom around the globe? Does the fact that according to foreign observers certain parts of a population seem to lack basic human freedoms justify an invasion, as was argued in the case of the Iraq war?

The fourth condition or another necessary condition for waging a "just war" is that of reasonable success. Perhaps one way of thinking about this condition is in term of a quixotic fight against evil monsters regardless of the outcome. Perhaps this sounds reasonable insofar as there are no costs or damages suffered in the process (except for a wounded ego or two.) Yet, in cases of domestic oppression by a police state, does it matter if one were to stand up to a bullying aggressor, knowing full-well that no benefits could emerge (except for gaining some self-esteem or national pride)? Should a sovereign state aid foreign people or declare war on moral grounds even if there is no conceivable chance of success? One could think of this condition, then, primarily as a deterrent to frivolous declaration of war when there is no chance to defeat another army so that the costs borne by the nation would not be disproportionate to what gains may be imagined. Many critics of the war in Iraq claim it has always been an un-winnable war, because a military offensive against civilian insurgents is bound to fail, as it did in the Vietnam War.

The final guide or condition of a "just war" is that the desired end should be proportional to the means used. This principle overlaps into the moral guidelines of how a war should be fought, namely the principles of *Jus in Bello*. This means-ends matrix is designed so as to ensure that a minimal provocation does not license another nation to overwhelm and destroy the instigator. This principle, used routinely in courts of law, ensures that the punishment fits the crime, so to speak. After the Japanese attack on Pearl Harbor during World War II, the United States felt justified in dropping atomic bombs on Hiroshima and Nagasaki, killing more than one hundred thousand civilians. Is the killing of more than fifty thousand Iraqi citizens (as reported in the UK) since the beginning of the second Iraq War the right moral "proportion" of civilian destruction?

Historically, questions related to just war theory have had to do not only with the reasons that justify or the conditions that ought to be met in order to declare war on another nation that can be retrospectively be deemed just, but also the guiding principles according to which a war should be conducted. Though Biblical references are inconsistent in permitting looting and even rape and enslavement, there is a sense already in that text that just causes are not sufficient to assess a war as being just, since the conduct of war might be inappropriate or even immoral. Has one fought honorably? Has one treated prisoners of war fairly, at least in the Kantian sense of treating them the way one would like to be treated? These questions were resolved much later, as in the Geneva Convention (1949), so that international standards could be appealed to. Moreover, as the two world wars of the twentieth century taught us, "crimes against humanity" is an actionable designation that can land a leader or a soldier in the International court in The Hague. In general, these principles, according to Moseley, can be divided into three.

The first moral principle relates to the appropriate targets of war, so that innocent bystanders or civilians should be excluded. Obviously, once in the throws of the battle, this kind of discrimination becomes difficult, especially, as the United States has done often, the targets are not directly seen because airplanes drop bombs from thousands of feet above ground. No matter how "smart" the bombs are, they may land on the wrong targets either because of faulty intelligence or because of operational errors. The second principle relates to how much force is appropriate to use in war (as already discussed). The third principle relates to responsibility, who should be held accountable for killing or for destroying the enemy? Is it the soldier who pulls the trigger of a machine-gun, the pilot who releases a bomb above a city, or the generals sitting at headquarters miles away but ordering this or that action? One can step even further and ask if it is not the politicians who decide launching a war that are ultimately responsible for war atrocities committed in the field. Elsewhere (Sassower 1997), it was made clear that scientists and engineers should be made responsible for the projects on which they work and for developing certain theories and devices that have the potential to be used against humans.

But here, too, there are moral problems that cannot be easily dismissed or glossed over. Every term is inherently problematic and raises new questions. What counts as a "threat" by someone? Must it be direct or indirect? What counts as "non-combatants"? Without support of munitions and gas and food, no battle could be waged, so even civilian mobilization for a war effort can be seen as direct engagement in the war effort. Are acts of "aggression" necessarily limited to force or occupation or can they be noticed in economic and financial terms? Hence the raising of tariffs or the nationalization of private industries alone could be deemed just cause for military action. Michael Walzer (1977), for example, has argued about the very concept of "war" as having been transformed once nuclear bombs were introduced into the arsenal of nation-states and even rogue states or fringe militant organizations). The ownership of nuclear weaponry alone can give cause for pre-emptive strike; the potential for having such weapons, as seen today in the case of Iran and North Korea, has brought about international intervention (sanctions, threats, talks, agreements).

It is with Just War Theory in mind, that one should examine images of war and terrorism. If the moral issues are related both to the causes and the proper conduct of war, and if these issues come to light in specific cases, then one photo or image can encapsulate everything that is right or wrong about a war. We can summarize hundreds of arguments and particular instances of a protracted war in a second, glancing at a pictorial representation. But this fact alone alerts us to the possibility that political or military leaders can censor or promote a specific image in order to justify a war or condemn the enemy, however construed. The potential for manipulation is so great that we are worried about the postmodern context and conduct under which images are produced, distributed, and consumed. So, however reluctantly, we may revert back to some modernist criteria of appropriate representation so as to ensure moral principles are not compromised or ignored. Perhaps the critical dimension of image producers, photojournalists, documentary movies directors, and other artists, will surface once again so that the Enlightenment ideals of modernity will be present.

List of References

Fry, Roger. 1934. *Reflections on British Painting*. London: Ayer Co, Publishers.

Gombrich, E. H. 1960. *Art and Illusion: A Study in the Psychology of Pictorial Representation*. Princeton, NJ: Princeton University Press.

Greenberg, Clement. 1961. *Art and Culture: Critical Essays*. Boston: Beacon Press.

Hoesterey, Ingeborg. 1991. (ed.), *Zeitgeist in Babel: The Postmodernist Controversy*. Indiana University Press.

Kolb, Robert. 1997. "Just War," *International Review of the Red Cross*, no 320, pp.553-562.

Lyotard, Jean-Francois. 1984. *The Postmodern Condition: A Report on Knowledge* [1979]. Translated by G. Bennington and B. Massumi. University of Minnesota Press.

Moseley, Alexander. 2005. "Just War," *The Internet Encyclopedia of Philosophy*.

Neurath, Otto. 1973. *Empiricism and Sociology.* Edited by Marie Neurath and Robert S. Cohen, Dodrecht and Boston: Reidel, (especially Ch. 7).

Paxton, Robert O. 2005. *The Anatomy of Fascism*. New York: Vintage.

Popper, Karl. 1963. *Conjectures and Refutations: The Growth of Scientific Knowledge*. New York: Harper & Row.

Sassower, Raphael. 1997. *Technoscientific Angst: Ethics and Responsibility*. Minneapolis and London: University of Minnesota Press.

Sassower, Raphael and Louis Cicotello. 2000. *The Golden Avant-Garde*. University of Virginia Press.

Walzer, Michael. 1977. *Just and Unjust Wars: A Moral Argument with Historical Illustrations*. New York: Basic Books.

Wittgenstein, Ludwig. 1922, 1958. *Tractatus Logico-Philosophicus*. London: Routledge & Kegan Paul.

Wittgenstein, Ludwig. 1958. *Philosophical Investigations*. Trans. G.E.M. Anscombe. New York: Macmillan.

Authors

Gernot Böhme, Professor emeritus for Philosophy, Technische Universität Darmstadt, Germany. Director of the Institut für praktische Philosophie e.V. (IPPh).

Thomas Borgard, Dr. habil., German Literature and Comparative Literature, Institute for German Studies, University of Berne, Switzerland.

Nik Brown, Ph.D., Sociologist, Deputy Director and Lecturer, Science and Technology Studies Unit, Department of Sociology, University of York, United Kingdom.

Cordula Dittmer, M.A., Sociologist, Ph.D. Candidate, Scholarship Holder of the German Foundation for Peace Research, Center for Conflict Studies, Philipps-University Marburg, Germany.

Silke Fengler, M.A., Dipl.-Vw., Chair for the History of Technology, RWTH Aachen University, Germany.

Heather Fielding, B.A., English, Ph.D. Candidate, Brown University, Providence, RI, USA.

Dominique Gillebeert, Dipl. In Ethics Studies, Associate of Post-Graduate College "Technology and Society", Officer in charge of Integration for Kreisstadt Neunkirchen, Germany.

Melanie Grundmann, Dipl. in Cultural Science, Cultural Scientist, Ph.D. Candidate, Europa Universität Viadrina, Frankfurt on the Oder, Germany.

Reinhard Heil, M.A., Philosopher, Ph.D. Candidate, Associate of Post-Graduate College "Technology and Society", Technische Universität Darmstadt, Germany.

Martina Heßler, Professor for History and History of Technology, University of Arts and Design, Offenbach, Germany.

Andreas Kaminski, M.A., Philosopher, Ph.D. Candidate, Associate of Post-Graduate College "Technology and Society", Technische Universität Darmstadt, Germany.

Nicole C. Karafyllis, Dr. habil., Philosopher, Assistant Professor at the Faculty of Social Studies, J. W. Goethe University Frankfurt on the Main, and the Institute of Philosophy, University of Stuttgart, Germany.

Stefan Kaufmann, Dr. habil., Historian and Sociologist, Vertretungsprofessur, Department of Sociology, Albert-Ludwigs-Universität Freiburg, Germany.

Christian Kehrt, M.A., Historian, Ph.D. Candidate, Associate of Post-Graduate College "Technology and Society", works now in the Research group Knowledge-production and Innovation at the Nanoscale, at the Deutsches Museum, Munich, Germany.

Stefan Krebs, M.A., Chair for the History of Technology, RWTH Aachen University, Germany.

Wolfgang Krohn, Professor for Philosophy and the Social Studies of Science and Technology, Department of Sociology, University of Bielefeld, Germany.

Jürgen Link, Professor emeritus for Literary Studies (and Discourse Theory), Department of Philology and Literature, University of Dortmund, Germany.

Laurent Mignonneau, Ph.D., Professor for Interface Culture, Department of Media, University of Art and Design, Linz, Austria.

Andrea zur Nieden, M.A., Sociologist, Ph.D. Candidate, Associate of Post-Graduate College "Technology and Society", Technische Universität Darmstadt, Germany.

Kathryn M. Olesko, Associate Professor, History of Science, Director of the Program in Science, Technology and International Affairs at Georgetown University in Washington, DC, USA.

Ingeborg Reichle, Ph.D., Art Historian, Research Fellow at the Interdisciplinary Research Group "The World as Image", Berlin-Brandenburg Academy of Sciences and Humanities, Germany.

Raphael Sassower, Professor of Philosophy, Department of Philosophy, University of Colorado at Colorado Springs, USA.

Thomas Sieverts, Professor emeritus for Architecture and Urban Design, at the University of the Arts, Berlin, Harvard-University and Technische Universität Darmstadt, Germany.

Christa Sommerer, Ph.D., Professor for Interface Culture, Department of Media, University of Art and Design, Linz, Austria.

Katja Stoetzer, Dipl. Päd., Educationalist, Ph.D. Canidate, Associate of Post-Graduate College "Technology and Society", Technische Universität Darmstadt, Germany.

Marcus Stippak, M.A., Historian, Ph.D. Candidate, Associate of Post-Graduate College "Technology and Society", History of Technology, Technische Universität Darmstadt, Germany.

Peter Timmerman, M.A., Philosopher, Ph.D. Candidate, Department of Philosophy, University of Twente in Enschede, Netherlands.

Aristotle Tympas, Lecturer on History of Technology in Modernity, Department of Philosophy and History of Science, National and Kapodistrian University of Athens, Greece.

Alexander Unger, M.A., Ph.D. Candidate in Educational Science, Associate of Post-Graduate College "Technology and Society", Assistant Professor at the Department of Media Research in Education, Otto-von-Guericke University, Magdeburg, Germany.

Claudia Wassmann, Ph.D., MD, Historian, Germany.

Marc Ziegler, M.A., Philosopher, Ph.D. Candidate, Associate of Post-Graduate College "Technology and Society", Technische Universität Darmstadt, Germany.

Science Studies

Sabine Maasen
Wissenssoziologie
(2., komplett überarbeitete Auflage)
Juli 2007, ca. 120 Seiten,
kart., ca. 12,80 €,
ISBN: 978-3-89942-421-8

Martin Carrier,
Johannes Roggenhofer (Hg.)
Wandel oder Niedergang?
Die Rolle der Intellektuellen in der Wissensgesellschaft
Juni 2007, ca. 175 Seiten,
kart., ca. 17,80 €,
ISBN: 978-3-89942-584-0

Carsten von Wissel
Hochschule als Organisationsproblem
Neue Modi universitärer Selbstbeschreibung in Deutschland
April 2007, ca. 320 Seiten,
kart., ca. 29,80 €,
ISBN: 978-3-89942-650-2

Thomas Gondermann
Evolution und Rasse
Theoretischer und institutioneller Wandel in der viktorianischen Anthropologie
April 2007, ca. 290 Seiten,
kart., ca. 29,80 €,
ISBN: 978-3-89942-663-2

Reinhard Heil,
Andreas Kaminski,
Marcus Stippak,
Alexander Unger,
Marc Ziegler (Hg.)
Tensions and Convergences
Technological and Aesthetic Transformations of Society
März 2007, 366 Seiten,
kart., 33,80 €,
ISBN: 978-3-89942-518-5

Jörg Potthast
Die Bodenhaftung der Netzwerkgesellschaft
Eine Ethnografie von Pannen an Großflughäfen
März 2007, 230 Seiten,
kart., 25,80 €,
ISBN: 978-3-89942-649-6

Tatjana Zimenkova
Die Praxis der Soziologie: Ausbildung, Wissenschaft, Beratung
Eine professionstheoretische Untersuchung
März 2007, ca. 300 Seiten,
kart., ca. 27,80 €,
ISBN: 978-3-89942-519-2

Christine Hanke
Zwischen Auflösung und Fixierung
Zur Konstitution von ›Rasse‹ und ›Geschlecht‹ in der physischen Anthropologie um 1900
März 2007, 298 Seiten,
kart., 29,80 €,
ISBN: 978-3-89942-626-7

Sebastian Linke
Darwins Erben in den Medien
Eine wissenschafts- und mediensoziologische Fallstudie zur Renaissance der Soziobiologie
Januar 2007, 262 Seiten,
kart., 26,80 €,
ISBN: 978-3-89942-542-0

Leseproben und weitere Informationen finden Sie unter:
www.transcript-verlag.de

Science Studies

Martin Voss,
Birgit Peuker (Hg.)
Verschwindet die Natur?
Die Akteur-Netzwerk-Theorie in der umweltsoziologischen Diskussion

2006, 264 Seiten,
kart., 25,80 €,
ISBN: 978-3-89942-528-4

Wolf-Andreas Liebert,
Marc-Denis Weitze (Hg.)
Kontroversen als Schlüssel zur Wissenschaft?
Wissenskulturen in sprachlicher Interaktion

2006, 214 Seiten,
kart., 24,80 €,
ISBN: 978-3-89942-448-5

Andréa Belliger,
David J. Krieger (Hg.)
ANThology
Ein einführendes Handbuch zur Akteur-Netzwerk-Theorie

2006, 584 Seiten,
kart., 29,80 €,
ISBN: 978-3-89942-479-9

Heide Volkening
Am Rand der Autobiographie
Ghostwriting – Signatur – Geschlecht

2006, 262 Seiten,
kart., 27,80 €,
ISBN: 978-3-89942-375-4

Niels C. Taubert
Produktive Anarchie?
Netzwerke freier Softwareentwicklung

2006, 250 Seiten,
kart., 27,80 €,
ISBN: 978-3-89942-418-8

Alexander Peine
Innovation und Paradigma
Epistemische Stile in Innovationsprozessen

2006, 274 Seiten,
kart., 25,80 €,
ISBN: 978-3-89942-458-4

Natàlia Cantó Milà
A Sociological Theory of Value
Georg Simmel's Sociological Relationism

2005, 242 Seiten,
kart., 28,80 €,
ISBN: 978-3-89942-373-0

Markus Buschhaus
Über den Körper im Bilde sein
Eine Medienarchäologie anatomischen Wissens

2005, 356 Seiten,
kart., zahlr. Abb., 28,80 €,
ISBN: 978-3-89942-370-9

Peter Weingart
Wissenschaftssoziologie

2003, 172 Seiten,
kart., 13,80 €,
ISBN: 978-3-933127-37-2